Genetics and the Social Behavior of the Dog

Genetics and the Social Behavior of the Dog

By JOHN PAUL SCOTT
and JOHN L. FULLER

THE UNIVERSITY OF CHICAGO PRESS
CHICAGO AND LONDON

This volume has also been published as *Dog Behavior: The Genetic Basis*

The University of Chicago Press, Chicago 60637
The University of Chicago Press, Ltd., London

10 09 08 07 06 05 04   8 9 10 11

ISBN: 0-226-74338-1 (Paperbound)
LCN: 64-23429

⊗ The paper used in this publication meets the minimum requirements of the
American National Standard for Information Sciences—Permanence of Paper
for Printed Library Materials, ANSI Z39.48–1992.

PREFACE

One of the major basic problems of psychology and biology is the effect of heredity upon behavior; yet in the 1930's and early 1940's research in this area had almost completely gone out of style. At that time it was possible to count on the fingers of one hand the persons who had done significant research in the field. Tryon had done an extensive experiment on heredity and maze running in rats, followed up by important studies from his students, Hall and Searle. Dawson had begun to work with behavior in dogs, but his project was terminated by the outbreak of war.

In the human area, interest in the Thirties was largely confined to a somewhat sterile controversy over the extent to which IQ scores could be modified by special training, although Thurstone and Strandskov had become interested in applying the technique of factorial analysis to problems of behavior genetics. Among most scientists there had been a revulsion against the exaggerated claims of the early eugenicists and the racist doctrines of Nazi Germany, and it was not infrequently stated that heredity had no effect upon human behavior.

It was against this general background that Dr. Alan Gregg, then Director for the Medical Sciences at the Rockefeller Foundation, went to lunch with Dr. C. C. Little, the director and founder of the Jackson Laboratory, and expressed the opinion that psychiatrists and sociologists were paying too little attention to the factor of heredity as it affected behavior. Dr. Little's imagination caught fire, and the outcome was a tentative arrangement for a long-range project, later known as "Genetics and the Social Behavior of Mammals."

The results of that project are now summarized in this book, which is primarily an account of an experiment lasting some thirteen years and involving intensive and extensive measurements on the behavior of hundreds of purebred and hybrid dogs, studied from birth to one year. As the experiment progressed, we were more and more impressed with the developmental approach to the problem of the genetics of behavior, and much of our work is presented from that viewpoint. In addition, we have analyzed the genetic results by most of the conventional statistical methods and have developed certain new ones as well. Finally, we have shown how these results relate to the general problems of human and animal behavior.

As well as presenting our own work, we have collected a great deal of background information on the origin and nature of genetic diversity in dogs. The book as a whole is therefore a source of general information on behavior of dogs and should be of interest to dog owners and breeders as well as theoretical scientists.

This is, naturally, not the last word on the subject. Our work has raised as many new questions as it has answered old ones, and we present our findings to date in the hope that they may provide a background for much needed future work. One thing which has impressed us is that the dog, for all its eight thousand years or so of association with human beings, is still in many respects a scientifically unknown animal. Despite the hundreds of papers that have been published about this animal, there are vast areas of information which are still unexplored. With the more generous support now available for research, it should eventually be possible to know instead of merely speculate about the nature of man's best friend.

While this experiment has been going on, much progress has been made elsewhere and in many scientific disciplines. Many workers have contributed to the revival of interest in the problem of genetics and behavior, in particular the ethologists under the leadership of Tinbergen and Lorenz, whose studies of instinct have reformulated the question of genetic differences between species. In other fields, symposia on behavior genetics have been held at the two 1963 International Congresses of genetics and psychology. A recent presidential address of the American Psychological Association was devoted to the subject of genetics and schizophrenia. A committee on behavior genetics has been set up by the Social Science Research Council. At a symposium of the American Psychiatric Society (published afterward as a volume entitled *Roots of Behavior*) one-fourth of the papers were directly concerned with behavior genetics, and eight of the thirty-one authors had some direct association with the Jackson Laboratory.

If Alan Gregg were alive today, we think that he would be pleased with the results.

J. P. Scott

J. L. Fuller

ACKNOWLEDGMENTS

The pronoun "we" is used deliberately in this book, not only because it has two authors, but in recognition of the contributions made by our numerous co-laborers on the project.

We particularly wish to recognize the efforts of our research assistants, on whom the accuracy and integrity of the data so largely depended. Many of them have since gone on to advanced training and have become independent investigators. We are especially indebted to Margaret Charles Higgins and Edna DuBuis for their long periods of service, and to C. L. Brace, Orville Elliot, Albert Pawlowski, and Joseph Royce for their contributions to data analysis. Others who made contributions over the 13-year span of the project were Duane Blume, Marian Burns, John Craig, Jr., Donald Dickerson, Clarice Easler, Alan E. Fisher, Priscilla Hinchcliffe, David King, Nancy M. King, Mona Meltzer, Robert Pettie, Maxine Schnitzer, Joyce Sprague, Mary E. Smith, and Elizabeth Williams.

We were fortunate in having over the whole period much the same group of general assistants, on whom a uniform system of care for the animals depended. We especially thank Frank Clark, Gordon Gilbert, Sheldon Ingalls, Pearl McFarland, and Dan Reynolds.

Many scientific investigators made notable contributions. C. C. Little, then the Director of the Jackson Laboratory, acquired the foundation stocks and worked out the color genetics of the breeds; C. S. Hall and W. T. James gave advice and helped organize the project; B. E. Ginsburg and Sherman Ross contributed much as long-time Visiting Investigators. Other colleagues and associates too numerous to mention here helped with encouragement and counsel.

Major financial support was given by the Rockefeller Foundation over a 13-year period during which the basic data were gathered. Continuing support enabling the processing of the data and writing this book was provided by the Ford Foundation and by the National Institutes of Health (Grants MH 01775 and MH 04481).

Certain figures and tables have been previously published in the following books and journals: *The Concept of Development* (University of Minnesota Press, 1957), Journal of Comparative and Physiological Psychology, Animal Behaviour, Journal of Heredity, American Zoologist, Monographs of the Society for Research in Child Development, Science, Journal of Genetic Psychology, Psychosomatic Medicine, and Annals of the New York Academy of Sciences. Permission to reproduce this material is hereby ackknowledged.

CONTENTS

LIST OF ILLUSTRATIONS / xiii

LIST OF TABLES / xvi

Part I   DEVELOPMENT OF BASIC BEHAVIOR PATTERNS

    1. A SCHOOL FOR DOGS / 3
    2. DOGS, WOLVES, AND MEN / 29
    3. THE SOCIAL BEHAVIOR OF DOGS AND WOLVES / 57
    4. THE DEVELOPMENT OF BEHAVIOR / 84
    5. THE CRITICAL PERIOD / 117
    6. THE DEVELOPMENT OF SOCIAL RELATIONSHIPS / 151

Part II   DEVELOPMENT AND EXPRESSION OF BREED DIFFERENCES

    7. ANALYSIS OF GENETIC DIFFERENCES / 185
    8. EMOTIONAL REACTIVITY / 194
    9. EXPERIMENTS ON TRAINABILITY / 205
    10. THE DEVELOPMENT AND DIFFERENTIATION OF PROBLEM-SOLVING BEHAVIOR / 224

Part III   INHERITANCE OF DIFFERENTIAL CAPACITIES AMONG HYBRIDS

    11. THE INHERITANCE OF BEHAVIOR PATTERNS: SINGLE FACTOR EXPLANATIONS / 261
    12. BEHAVIOR IN HYBRIDS: COMPLEX BEHAVIOR / 295
    13. DEVELOPMENT OF PHYSICAL DIFFERENCES AND THEIR RELATION TO BEHAVIOR / 326
    14. THE EFFECTS OF HEREDITY UPON THE BEHAVIOR OF DOGS / 356

Part IV   GENERAL IMPLICATIONS

15. IMPLICATIONS FOR THE ART OF DOG BREEDING / 383

16. THE EVOLUTION OF DOGS AND MEN / 397

17. TOWARD A SCIENCE OF SOCIAL GENETICS / 413

BIBLIOGRAPHY / 435

AUTHOR INDEX / 449

SUBJECT INDEX / 452

LIST OF ILLUSTRATIONS

PLATES

following page 176

The five pure breeds
Outside runs and nursery room interior
The Behavior Laboratory and staff
Breeding stock (cocker spaniels) and F_1 hybrids
Breeding stock (basenjis) and F_1 hybrids
Litter from backcross to cocker spaniel
Litter from backcross to basenji
BCS F_2 litter
CSB F_2 litter
Aboriginal dogs of Africa, Iraq, and Australia
Early English land spaniels and hounds
Forced training
Reward training
Problem-solving: detour and manipulation tests
Maze test and T-maze and delayed-response test
Trailing and spatial-orientation tests

FIGURES

1.1 Basic Mendelian cross / 9
1.2 Floor plan of nursery room / 25
1.3 Floor plan of all-weather dog kennel / 27
2.1 Geographical distribution of wild members of genus *Canis* / 32
2.2 Centers of domestication and adaptive radiation / 36
2.3 Skull width across cheek bones as related to skull length / 40
2.4 Greatest width of upper jaw relative to skull length / 41
2.5 Size of molar teeth relative to skull length / 41
2.6 Relative breadth of upper jaw in five modern breeds / 44
4.1 Rate of distress vocalization during first 4 weeks / 92
4.2 Development of walking and playful fighting / 95
4.3 Development of motor ability during transition period / 96
4.4 Development of learning capacities during neonatal and transition periods / 98
4.5 Change in response to conditioning between transition and socialization periods / 99
4.6 Development of distress vocalization of beagle puppies under 2 conditions of isolation / 103
5.1 Development of the heart rate / 121
5.2 Effect of socialization at different ages upon avoidance of an active handler / 125
5.3 Timing mechanisms limiting the process of socialization / 126
5.4 Performance in the leash-control test / 127

5.5   Eating during the leash-control test / 128
5.6   Avoidance and vocalization in response to handling / 134
5.7   Avoidance and fearful behavior in response to handling / 135
5.8   Agonistic behavior (playful fighting) in response to handling / 136
5.9   Attraction and following in response to handling / 138
5.10 Et-epimeletic behavior and social investigation in response to handling / 138
6.1   Percentage of occurrence of complete dominance in purebred litters / 157
6.2   Occurrence of fights or one-sided attacks during dominance tests / 158
6.3   Occurrence of complete dominance in male-female pairs / 160–62
6.4   Decline of nursing behavior by mothers during the neonatal period / 170
6.5   Average time spent with handler during the following test / 175
7.1   Distribution of running times / 188
7.2   Distribution of motivation-speed scores / 189
8.1   Total reactivity ratings for the five pure breeds / 198
9.1   Proportion of animals rated as quiet during weighing / 206
9.2   Types of demerits given during leash-control test / 209
9.3   Level of training on successive days of obedience test / 213
9.4   Mean time scores on goal-orientation test / 217
9.5   Proportion of animals succeeding on each trial of the motor-skill test / 220
10.1 Problems in the barrier test / 227
10.2 Results of the manipulation test on successive days / 231
10.3 Floor plan of the maze test, left-hand pattern / 233
10.4 Learning curves: errors in maze test / 235
10.5 Learning curves: times of solution of maze test / 235
10.6 Relative performance of different breeds in the maze / 238
10.7 T-maze used for cue-response, discrimination, and delayed-response tests / 239
10.8 Speed scores in cue-response test / 240
10.9 Apparatus used in the spatial-orientation test / 248
10.10 Breed scores on the spatial-orientation test / 252
10.11 Individual error scores on the spatial-orientation test / 252
10.12 Inter-trial and intraclass correlations in the spatial-orientation test / 255
11.1 Distribution of scores for playful fighting at 13–15 weeks / 270
11.2 The occurrence of playful fighting transformed to a scale of threshhold of stimulation / 272
11.3 Occurrence of barking during dominance tests at different ages / 275
11.4 Distribution of barking to excess in pure breeds and hybrids at 11 weeks / 277
11.5 Month in which breeding began in related cocker spaniels / 280
11.6 Intervals between first two estrus periods / 281
11.7 Proportion of cockers and basenjis rated as completely quiet during weighing / 288
11.8 Development and differentiation of postural responses / 289–90
12.1 Mean scores of hybrids on six measures of leash training / 300
12.2 Mean motivation speed scores of hybrids / 307
12.3 Attainment by hybrids of criterion in the cue-response test / 308

12.4   Attainment by hybrids of criterion in the delayed-response test / 310
12.5   Mean scores of hybrids on the obedience test / 311
12.6   Mean total reactivity scores of hybrids at 17, 34, and 51 weeks / 312
12.7   Mean scores of hybrids on two reactivity test measures / 314
12.8   Proportion of each hybrid group rated as "high" in tail wagging at 17, 34, and 51 weeks / 315
12.9   Mean correct choices scores of hybrids on the spatial-orientation test / 318
12.10 Mean persistence scores of hybrids on the spatial-orientation test / 319
12.11 Mean adjusted speed scores of hybrids on the spatial-orientation test / 320
13.1   Absolute weight gain per week in male puppies of the largest and smallest breeds / 327
13.2   Combined growth curve for basenji males / 328
13.3   Male and female growth curves in basenjis and cocker spaniels / 330
13.4   Growth curves of the five pure breeds / 331
13.5   Inheritance of hair length / 335
13.6   Distribution of differences in stanines between long and short-coated animals in relation to 34 behavioral variables / 337
13.7   Relative physical proportions of cocker spaniels and basenjis / 340
17.1   Functional differentiation of form during growth / 421
17.2   Functional differentiation of behavior in relation to the physical environment / 421
17.3   Functional differentiation of a social relationship / 422

LIST OF TABLES

1.1 Master list of experimental animals / 10
1.2 Schedule of feeding, cleaning, and disease prevention / 14
1.3 Schedule of observations, training, and testing / 24
2.1 Average skull measurements of different breeds of dogs / 43
2.2 Ratios of skull measurements to length of skull / 44
2.3 Coefficients of inbreeding of foundation stock / 49
2.4 Hybrids reported between species of the genus *Canis* / 53
3.1 Behavioral systems and known behavioral patterns in the family Canidae / 63–65
4.1 Time of complete opening of the eye / 90
4.2 Time of first function of the ear / 94
4.3 Time of eruption of upper canine teeth / 96
4.4 Time of development of allelomimetic behavior / 106
4.5 Occurrence of sexual behavior in 160 purebred puppies (0–16 weeks of age) / 107
4.6 Natural periods of development in dog and man / 114
5.1 Estimated variation in developmental events associated with the beginning of the periods of transition and socialization / 120
5.2 Relative adjustment of puppies at 1 year of age / 123
5.3 Rank order of puppies on tests given after 14 weeks of age / 128
5.4 Development of tail wagging / 139
6.1 Social relationships of dog and man / 153
6.2 Effect of size on dominance-subordination relationships / 164
6.3 Percentage of complete dominance in basenjis, cockers, and their hybrids at 15 weeks / 166
6.4 Occurrence of mothers nursing puppies under 2 weeks of age / 171
6.5 Occurrence of retrieving 1-week-old puppies by purebred and hybrid mothers / 172
7.1 Conversion table—original score to stanine / 189
7.2 Simplified interpretation of analysis of variance in breed comparison experiment / 192
8.1 Specific response ratings / 197
8.2 "Resting" heart rate / 199
8.3 Change in heart rate during quieting / 199
8.4 Mean stanine scores of five pure breeds on ten behavioral and four heart rate measures at three ages / 200
8.5 Percentage of subjects exceeding arbitrary thresholds on eight measures in reactivity test / 202
9.1 Mean number of demerits in leash training / 208
9.2 Intercorrelations of types of demerits during leash training / 210
9.3 Types of demerits in early, middle, and late phases of leash-control test / 211
9.4 Mean stanine scores on obedience for pure breeds / 214

9.5   Rank five pure breeds under forced training / 215

9.6   Correlations between tests of forced training / 216

9.7   Percentage of animals successfully trained in retrieving test / 219

9.8   Over-all performance on motor-skill test / 221

10.1   Performance of the different breeds on the detour test—first trials only / 229

10.2   Percentage of total variance of maze scores related to breed, mating, litter, and background variables / 236

10.3   Attainment of criterion in cue-response test / 241

10.4   Delayed-response performance of five pure breeds / 243

10.5   Animals showing perfect performance on the final trial of the trailing test / 246

10.6   Relative ranks of breeds on performance on most difficult trial of trailing test / 246

10.7   Distribution of fear reactions in trailing test / 247

10.8   Comparisons between pure breeds on spatial-orientation scores / 251

10.9   Analysis of variance of spatial-orientation measures / 253

10.10   Ranks of the different breeds in various problem-solving tests / 258

11.1   Modifications of the Castle-Wright formula for the analysis of quantitative inheritance / 264

11.2   Theoretical ratios for the differences between backcrosses, F_1's, and F_2's compared to the difference between parental strains / 266

11.3   Scores for avoidance and fearful vocalization at 5 weeks of age in response to a human handler / 267

11.4   Tendency to bite and jerk the leash / 269

11.5   Per cent of animals barking per opportunity, compared with expected percentage / 276

11.6   Per cent of animals barking 19 or fewer times at 11 weeks / 276

11.7   Tests involving persistent differences between maternal lines / 285

11.8   Distribution of adult weight measurements in basenjis, cockers, and their hybrids / 286

11.9   Distribution of average heart rates, 11–16 weeks / 287

12.1   Leash-training demerits of hybrids / 299

12.2   Within-litter variance of the fighting score of leash training / 301

12.3   Analysis of variance—leash-fighting scores of hybrids by matings and litters / 302

12.4   Leash-control fighting score—offspring of F_1 sires by different dams / 302

12.5   Leash-control fighting score—offspring of purebred dams by different sires / 303

12.6   Analysis of variance—fighting score in leash training / 304

12.7   Leash-control vocalization score—offspring of sires by different dams / 306

12.8   Leash-control vocalization score—offspring of dams by different sires / 306

12.9   Discrimination or cue response—analysis of variance of hybrid performance / 309

12.10   Analysis of variance—total reactivity scores of hybrids / 316

12.11   Total reactivity ratings: means of matings of purebred dams by purebred sires and F_1 sires / 317

12.12   Matings of F_1 sires by purebred and F_1 dams / 317

13.1   Percentage of adult weight reached at 16 weeks of age / 329
13.2   Breed rank in body measurements relative to weight / 342
13.3   Significant correlations between physical measurements and behavioral variables in pure breeds and hybrids / 345
13.4   Factor loadings of physical variables in the combined F_2 population / 346
13.5   Factor loadings of factor 1 (general size) in combined pure breeds / 348
13.6   Factor loadings of factor 1 (general size) in Shetland sheep dogs and fox terriers / 349
14.1   Composition of combined populations / 357
14.2   Number of variables in which the combined variance of segregating populations exceeds that of non-segregating populations / 358
14.3   Proportion of variance attributable to population differences / 360–61
14.4   Variance components of each selected backcross and F_2 population / 365
14.5   Mean variance components in variables showing differences between the backcrosses with a $P < .01$ / 366
14.6   Factor 2 (activity-success) in the combined F_2 generation / 372
14.7   Factor 3 (heart-rate) in the combined F_2 generation / 372
14.8   Factor 2 (reactivity) in the combined F_1 hybrids / 374
14.9   Correlations of poor performance on the trailing test with performance in other tests in two hybrid populations / 374
14.10 Common variance in trailing test as indicated by correlation with performance in T-maze / 377
16.1   Mortality rate per 100 births in pure breeds and hybrids / 406
17.1   Subdivisions of genetics / 413

PART I

THE DEVELOPMENT
OF BASIC
BEHAVIOR
PATTERNS

CHAPTER 1

A SCHOOL
FOR DOGS

For many years we have had psychologists and zoologists working side by side in our behavior laboratories. When we first brought them together, we often got stereotyped reactions to any new discovery in the field of behavior. A zoologist would say, "What a wonderful example of instinct," and a psychologist would counter, "My, what a remarkable example of learning!" This happened over and over again in the early stages of our work, and it illustrates the confusion which once existed, and may still persist, regarding the role of heredity in relation to behavior.

In the more remote past, scientists such as John B. Watson took the view that heredity had little importance and that the behavior of a human infant could be molded in any direction by learning. Others, following Sigmund Freud, held that human behavior was deeply rooted in instincts whose suppression could result in serious mental disorders. The vast majority of psychologists, psychiatrists, and other practical workers in the field of human behavior had no definite information about either the rapidly developing science of genetics or the ways in which genetic variation could affect behavior directly. Nor was the clinical observation of human behavior much help. True, many children reared in poor family backgrounds later developed inadequate and maladjusted behavior, but some of them were apparently unaffected, and still others worked harder than most people to compensate for their early experience. Was their ability to rise above environment due to heredity or to some unknown difference in early experience?

The question "What does heredity do to behavior?" is therefore a basic one, both theoretically and practically. There are many dif-

3

ferent ways of trying to answer it. One can either make comparisons between as many as possible of the million or so species of animals in the world or concentrate on one, as we did. Scientists working with birds have discovered many stereotyped behavior patterns which are largely organized and governed by heredity. We ourselves chose a mammal, partly because mammals are more closely related to human beings than are birds, and partly because the job had not yet been done. Among mammals, we chose the dog because it shows one of the basic hereditary characteristics of human behavior: a high degree of individual variability. We reasoned that if there were any animal most likely to show important effects of heredity on behavior, it was the dog, and that these effects could be assumed to be at least somewhat similar to those which appear in human beings.

To put this in more enthusiastic terms, the dog is a veritable genetic gold mine. Besides the enormous differences between breeds, all sorts of individual differences appear at the stroke of the geneticist's pickaxe, in this case the technique of mating two closely related animals. Anyone who wishes to understand a human behavior trait or hereditary disease can usually find the corresponding condition in dogs with very little effort. Dogs are timid or confident, peaceful or aggressive, and may be born with undershot jaws, club feet, or hemophilia (Dawson, 1937; Burns, 1952; Fuller, 1960).

Although it is true that a dog is a large, long-lived, and relatively expensive animal to keep, it is a good deal smaller, shorter-lived, and less expensive than some other mammals. Dogs breed more rapidly than primates or even the common herd animals, except for pigs. Dogs can pass through one complete generation within a year. A mature female continues to reproduce for five to seven years, bearing her young about twice a year in litters averaging four or five animals. Nevertheless, it requires very careful planning to get the maximum amount of information from an experiment with dogs, and one initial mistake can ruin years of effort.

Therefore, before our work had gone beyond the early planning stage, we asked a number of leading geneticists and psychologists in the United States to attend a research planning conference at Bar Harbor. Many of these men and women were doing related research, either on animals or on people. The chairman was Dr. Robert M. Yerkes, one of the pioneer workers on heredity and behavior, the founder of the Yerkes Laboratory of Primate Biology, and at that time the recognized dean of animal behaviorists in the United States. Equally distinguished workers made up the rest of the group. Dr. Frank Beach served as secretary, organizing the proceedings into a coherent whole. For three days we met with them on Mount Desert Island and discussed the direction which the research should take. Much of our final plan resulted from this conference (Scott and Beach, 1947).

At the start we had certain guiding principles and signposts derived from what had been discovered about other aspects of heredity. We knew that behavior itself is not inherited. In fact, the only things which are biologically inherited from the parents are the egg and sperm. We also knew that the basic factors of heredity are genes, which are nothing more than large organic molecules arranged in larger bodies called chromosomes. Even these are only parts of the nucleus of a living cell. The genes themselves can act only by controlling the synthesis of proteins, some of which, the enzymes, modify other chemical or physiological reactions, usually in the cells themselves. All this is a long way from the activity of a whole organism, or behavior, and the concrete fact of whether a dog bites you or wags its tail.

One of the primary questions we had to answer was whether heredity could produce an important effect upon behavior in a higher animal or whether it simply set the stage for behavioral activity which was then guided and molded by other causes. The question makes a great deal of sense in relation to the concept of levels of organization (see Table 17.1). Heredity stands at the most basic level, and therefore we can assume it to be very important. But assumptions are only the beginning of scientific inquiry, and the question still remains, What exactly can a factor operating on a low level of organization do to a phenomenon operating on a higher level? For behavior is primarily a phenomenon existing on the organismic or psychological level (Wright, 1953).

At the same time we realized that we could not neglect factors operating on still other levels. A major trend in animal behavior research in the last thirty years has been to demonstrate the importance of the effects of social organization upon behavior. The activity of an individual is understandable only in terms of his relationships to the members of a larger group. Our work was therefore designed in relation to levels of organization and can be so described.

In addition, we had the successes and mistakes of our predecessors to guide us. Professor Tryon (1930–41, 1963) of the Psychology Department of the University of California had previously done extensive and meticulous research on heredity and behavior in rats.

He devised a mechanical maze operating in such a way that the experimenter could not influence the results consciously or unconsciously and selected different strains of rats from those which learned the maze most quickly or slowly. These differences persisted for generation after generation, indicating that heredity could produce an important effect upon behavior. It appeared that selection had affected intelligence alone, but when a student of his, Searle (1949), tested these rats on different sorts of mazes, he came to the conclusion that one of the principal reasons why the two strains of rats differed in performance was that the slow running or "dull" rats were afraid of the mechanical maze. Instead of selecting purely for intellectual ability, Tryon had also selected for a difference in emotional reaction.

C. S. Hall (1941) had shown by his selection experiments on the defecation rates of rats taken from their cages and placed in an open space that heredity could alter another kind of emotionality. His work, and that of Searle, suggested that emotional capacities were highly variable and could produce important effects upon performance. From these previous experiments we knew that hereditary emotional differences might have important effects on behavior which at first glance did not appear to be emotional. We also realized that with such a long-lived and expensive animal as the dog, we would seldom be able to duplicate our experiments. In other words, there would be very little going back to take a second look. We would have to measure as many important characteristics as we could while the experiment was in progress.

Besides the work on other animals, there had been an earlier major attempt to work with heredity in dogs. The late Professor C. R. Stockard of the Cornell University Medical School had begun with the idea (a promising one at the time) that many of the differences between breeds of dogs were caused by hereditary disorders in the ductless glands (1941). He did, indeed, find breed differences in the endocrine glands, but anatomical peculiarities in the same breeds, such as dwarfism and shortleggedness, turned out not to be caused by glands but by hereditary factors working in other ways. Professor Stockard died before he could change the project to allow for this. Since his original experiment was not designed to test a Mendelian hypothesis, the results were inconclusive, and we decided to set up our experiment in the broadest possible way, realizing that we could not predict the important results in advance and that if we chose too narrow an approach we might end with nothing. Both from this viewpoint and the simple statistics of life insurance tables, we knew it was imperative that this project be a cooperative venture, not a one-man show, and in this book we shall try to recognize the innumerable contributions made by co-operators and co-workers in the years covered by the experiment.

Our over-all experimental design was to systematically vary the genetic constitution of the dogs while keeping all other factors as constant as possible. The dependent variable was behavior, but since behavior occurs only in reaction to stimulation, we had not only the problem of designing a uniform system of test situations which would measure the dog's reactions to specified changes in his social and physical environment, but also that of keeping all other environmental factors as constant as possible. Equally important was the task of keeping the dog's internal environment constant by seeing that each animal was adequately nourished and free of disease.

GENETIC METHODS

Choice of breeds.—Genetic variation was the primary variable in our experiment: the supposed cause which might or might not affect behavior. We therefore divided the experiment into two phases. In the first we made a survey of the different dog breeds in order to select those which showed the biggest behavioral differences. Thanks to the energy of Dr. C. C. Little, we had a wide variety from which to choose, ranging from the Great Pyrenees, a white dog roughly resembling a Saint Bernard, to the tiny Chihuahuas. We soon eliminated the abnormal sizes at both ends of the scale, realizing that the large breeds were too expensive for long experiments and that the toy or dwarf breeds were unsuitable because of their low fertility. Furthermore, if we chose dogs of approximately the same size, we could use the same apparatus for all.

Certain breeds were discarded early. Both dachshunds and Scottish terriers have a reputation for being stubborn and self-willed, but those we studied showed no trace of this as puppies, being uniformly friendly and demonstrative animals, not unlike beagles and cockers in this respect. We would obviously have to wait too long for the behavioral peculiarities of these breeds to develop.

We further decided to concentrate on breeds with normal physique. Although it is true that the shortleggedness of the dachshund, for example, has an effect upon its behavior, this does not require an elaborate scientific experiment for proof. We preferred to work with the more subtle and perhaps more important ways in which heredity affects behavior.

We finally selected five breeds for intensive study: basenjis, beagles, American cocker spaniels, Shetland sheep dogs (shelties), and wire-haired fox terriers. As it turned out, these were representatives of the major groups of dogs as recognized by dog breeders and omitted only the toy and non-sporting breeds.

In the first phase in the experiment, we developed our tests and raised all five breeds in the same environment, measuring their similarities and differences. We also studied the development of their behavior by making daily observations from birth up to 16 weeks of age. Wherever possible, we cross-fostered puppies between the breeds in an effort to see whether the maternal environment as well as biological heredity was affecting behavior. We also experimented with an even more drastic shift in the social environment by taking certain puppies out of the kennels and rearing them in homes. This not only gave us a chance to study the effect of maternal environment, but to see whether or not our kennel-reared dogs were reasonably normal compared to dogs reared outside. Unfortunately, we were able to do this only on a very few animals, partly because of the limited number of friends and fellow workers on whom we could impose, and partly because of the danger of disease transmission. Anticipating a little, we can say that Gyp, the basenji; George, the beagle; and Silver, the Shetland sheep dog, did not do strikingly better or worse on most of the tests, although they developed very different relationships with dogs and people.

A Mendelian cross.—The second phase of the experiment was to make a Mendelian cross between two of the breeds (Fig. 1.1). Preliminary results indicated that the American cocker spaniel and the African basenji showed a great many differences in behavior. On the one hand we had a gentle and pampered house pet descended from bird dogs, and on the other a breed which had recently come from the rough-and-tumble conditions of African village life. Furthermore, there was good evidence that the basenjis, having been first brought over from Africa in 1937, had never been crossed with the European breeds, whereas the cocker (as well as the other three) had originally come from the British Isles, where there had been abundant opportunity for crossing in the past and considerable historical evidence that this had actually taken place. These two breeds had been made genetically different by isolation as well as by selection.

We did not attempt to accentuate the differences between the breeds by further selection, as this would have been a ten-year program in itself. Instead, we decided to rely on the selection which the practical dog breeders had already done over a period of centuries. This selection, however, has not produced complete uniformity. In most cases the breed standards permit considerable variation, even in such obvious characteristics as color, and we had already found that there was a great deal of variation between both individuals and strains within the pure breeds. In order to limit this kind of variation, we decided to make all our crosses from the off-spring of a single pair of animals in each breed. This would limit the genetic variability in the parents to that which had existed in a single brother-sister pair rather than that which was characteristic

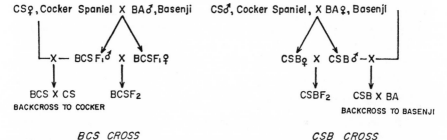

FIG. 1.1.—Diagram of Mendelian cross with symbols used for different genetic populations. The basic plan was to repeat each of the two crosses with four matings and obtain two litters from each mating. Because of deaths among the cocker spaniel females, replication was completed only three times in the BCS cross (see Table 1.1).

of the entire breed. We were able to carry out this plan successfully with the basenjis, but in the case of the cockers we were not able to obtain enough males from a single mating at the right time and had to modify our plan slightly by using as our foundation stock animals from two closely related matings in which the same male was mated to his sister and mother.

In order to control for the possible effects of the maternal environment and also to demonstrate sex-linked inheritance if it should exist, we made reciprocal crosses. The ideal plan was to mate four basenji males with four cocker females and vice versa. This would give us two F_1 populations, one from cocker mothers and one from basenji mothers. Then we took a first generation male from each mating and crossed him back to his mother. In this way we could compare two sets of offspring of the same mother. In the backcross group we expected that the genes would be assorting and recombining in different combinations according to the laws of Mendelian inheritance, whereas in the F_1 generation all animals would receive the same inheritance and be much less variable. Then we took the same F_1 males and mated them to their sisters in order to produce an F_2 generation in which the maximum genetic variability would occur.

We anticipated that we would be able to study the effects of heredity chiefly from the variability of behavior, and we therefore planned to raise at least twenty-five to thirty animals in each population, this being the smallest number in which a reasonably accurate sample of a population distributed in a normal curve can be obtained. These are not large numbers for genetic data; if we had been working with fruit flies, for example, we would have preferred to obtain at least a hundred in each group. Therefore, we made an effort to get an unusually large number of F_2's, the most crucial population for all types of genetic analysis. Furthermore, we planned to obtain at least two litters of puppies from each mating, so that we could compare variation in siblings which were litter mates and siblings which were not.

TABLE 1.1

MASTER LIST OF EXPERIMENTAL ANIMALS* AND KEY TO BREEDS

Breed or Hybrid, and Symbol	Inclusive Serial Nos.	Matings	Litters	Females	Males	Total
Basenji (BA or B)...............	737–3299	6	9	27	24	51
Beagle (BEA)...................	426–3529	7	18	31	39	70
Cocker spaniel (CS)............	546–3083	9	15	37	33	70
Shetland sheep dog or Sheltie (SH)	549–3447	5	9	16	18	34
Wire-haired fox terrier (WH).....	576–3440	5	11	24	20	44
F_1 BA♂ × CS♀ (BCS)..........	1834–3231	4	5	10	14	24
F_1 CS♂ × BA♀ (CSB)..........	1993–3748	5	5	16	13	29
Backcross to CS (BCS × CS).....	2249–3488	3	7	18	12	30
Backcross to BA (CSB × B)......	2415–3736	4	9	22	22	44
F_2 from BCS (BCS × BCS)......	2681–3675	3	6	18	14	32
F_2 from CSB (CSB × CSB)......	2580–3619	5	8	20	22	42
All......................	426–3748	56	102	239	231	470

* Includes all animals upon which some experimental work was done and which lived at least to 16 weeks.

The data could be analyzed in a number of ways. Analysis of variance would measure the changes in variation produced by genetic segregation, by differences in the environment produced by different mothers, and by the accident of being born at different times to the same mother. Correlational analysis would reveal the extent to which close relatives resembled each other, as well as the resemblance between different kinds of behavioral tests. In addition, we could analyze the effect of physical traits on behavior by calculating the correlations between body measurements and psychological measurements. We could also determine whether or not the external color factors were associated with behavioral differences. For example, the owners of cockers have often alleged that the reds and blacks have different temperaments.

Looking back on the experiment, we now see how we might have organized it in a better fashion. We should have allowed for a larger number of spare matings, as accidents and deaths will occur even in the best regulated laboratories. In addition, we had planned the experiment with the anticipation that most of the traits would show multiple-factor inheritance. Surprisingly, we found that many of the breed differences could be explained in terms of simple one- and two-factor inheritance. If we had the experiment to do again, we would emphasize the backcrosses, making a second backcross to the pure breeds in each case, this being a critical test for single-factor inheritance. We also found that our F_1 females became much superior mothers compared to the purebred animals, producing large quantities of milk and giving excellent care to their offspring, so that the F_2's had a better start in life. These genetic and environmental advantages of the F_1's and F_2's are reflected in decreased mortality (Table 16.1). It would have been interesting to have made backcrosses to the pure breeds from the F_1 females as well as from the males in order to determine the effect of superior early environment on the same genetic types.

Still another unanticipated result was the large amount of variation between litters. Litter mates tend to be alike, and, in an animal which rears its young in litters, the best unit of the population is not the individual but the litter. It would have been desirable to have had more litters than we did. However, all this would probably have taken twenty-five or thirty years to complete instead of thirteen.

THE INTERNAL ENVIRONMENT: CONTROL OF NUTRITION AND DISEASE

Feeding.—Since genetics was our chief experimental variable, we attempted to keep physiological factors as constant as possible. One of these is the state of nutrition, and, as we have seen, this is partly dependent upon the mother and hence upon her heredity. There was no way in which we could force the mothers to produce a constant amount of milk, but in all other ways we could provide a uniform diet.

When we began the experiment in 1945, meats of any kind were decidedly variable in supply. However, one prepared animal food with a guaranteed constant formula was readily available. This food was Purina Laboratory Chow, and we decided to make it the basic item in the diet, keeping it constantly available in a metal feeder. It was not a perfect food for dogs, as it had been developed as an all-purpose laboratory food for rats, mice, guinea pigs, and monkeys as well as dogs. It contained more roughage and less fat than was ideally suitable for dogs. It proved, however, to be an intrinsically sound diet, as our experimental animals developed into healthy and vigorous adults with every indication that their growth was equal to normal breed standards or better (see Chap. 13).

The young puppies and nursing mothers had to have supplementary food, and this was kept as simple as possible. One natural food which is extraordinarily constant in composition is milk, and this was the principal supplement. We used canned milk at first, but later switched to powdered whole milk because of its lower cost. To this we added a vitamin supplement, Abbott's Haliver Malt, in order to provide a source of vitamins A and D, which are those most likely to be lost from the aging of dry prepared foods. We also had available at all times for the puppies and mothers dry Purina Kibbled Meal, most of which was a fine powder which the young puppies began to eat as soon as they were able. This food did not have the standardized formula of the Laboratory Chow but was designed primarily as a dog food and was superior to it in that respect.

The dogs consequently had before them a constant supply of dry food which they could eat at any time. This reduced the labor of caring for the animals but also had another purpose. We wished to raise the puppies in natural social groups and kept them together in their litters rather than rearing them as isolated individuals. If allowed to compete for food, as frequently happens with dishes of meat or wet mash, unequal nutrition would soon result. As it turned out, puppies in a litter did not compete for the milk, as dogs never appear to fight for a liquid, and they rarely showed any disposition to quarrel over the dry food. In any case, no dog could keep guard over the dry food every minute of the day in a large pen.

In many psychological experiments we used a food reward, a small spoonful of canned herring having a strong fishy flavor similar to sardines. The dogs seemed to relish these tidbits, and the amounts taken were not large enough to disturb their nutrition.

Sanitation.—Our second concern was to keep the animals healthy and free from disease at all times. The first problem was sanitation. Many kennels maintain their dogs on bare floors and wash these daily. However, this means that the floors are constantly wet, dogs and attendants frequently step in feces, and there is a constant strong odor. Since our experimental crew had to work with the dogs in the pens, we had to consider the comfort of both. We decided to cover the floors of the pens with wood shavings which absorb moisture from urine and feces. This produced sweet-smelling rooms in which both dogs and attendants could easily keep themselves clean. Feces were picked up once a day after the animals were fed, leaving the room relatively clean for the rest of the day. This, of course, did not substitute for thorough cleaning, which necessitated moving the dogs from the pens, removing the shavings, and steam-cleaning the floors and walls. We wished to avoid disturbing puppies at critical times of development and at the same time to give them all uniform treatment. The pens were therefore thoroughly cleaned once every four weeks after the puppies were born. This system was a compromise between complete sanitation and psychological disturbance, but worked reasonably well.

Disease control.—The most common dangerous disease for young puppies is caused by intestinal roundworms which can make a young puppy seriously ill or even kill him. It is almost impossible to abolish these parasites completely, as they can be passed along to the puppies by the mother even before birth. We therefore set up a routine system of worming (see Table 1.2) for all puppies, whether or not they were infected, so that every puppy would have the same type of experience and the same chances for good health. Worming was correlated with the cleaning schedule, so that puppies were wormed at 4 and 8 weeks, before being put back into a clean pen, and again at 16 weeks, when they were transferred to a new pen.

Cleaning and health activities were thus organized around a 4-week unit, and this schedule resulted in the development of a 4-week module for planning the social experience of the puppies. This was later amplified and extended in the psychological testing procedure.

The second major threat to health was canine distemper. This is a highly infectious, airborne disease which attacks the respiratory and nervous system and produces 50- or 60-per cent fatalities, while exposing even the survivors to a lingering illness and prolonged convalescence. Its prevention is so important that in this one respect we deliberately departed from our ideal of uniform conditions

TABLE 1.2

SCHEDULE OF FEEDING, CLEANING, AND DISEASE PREVENTION FOR PUPPIES

Age (weeks)	Food	Cleaning	Disease Prevention*
0	Mother's Milk only until 3 weeks	Daily cleaning (continued through 52 weeks)	
1			
2			
3	Cow's Milk & Haliver Malt, Kibbled Meal, Laboratory Chow added		
4		Steam cleaning	Worming, Spratt's Capsules
5			
6			
7			
8		Steam cleaning	Worming, N-butyl chloride Anti-distemper serum
9			
10	Mother removed		Anti-distemper serum
11			
12		Steam cleaning	Killed distemper virus
13			
14			Live distemper virus
15			
16		Steam cleaning	Worming, N-butyl chloride
17–52	Laboratory Chow only	Daily cleaning only (moved to outside run)	

* Later modified. See text.

throughout the entire course of the experiment. At the time we started, the best system of prevention consisted of a series of three injections, two weeks apart. The first two injections consisted of a killed virus which produced a partial immunity. In the final one, the live virus produced a permanent immunity. Although no dogs died after this treatment, many of them became ill from the injection of the live virus, and this interfered with their behavioral tests. We therefore abandoned the use of the live virus. In the meantime the Lederle Company had developed an avianized live virus, one injection of which could produce permanent immunity without any signs of illness, and we switched to this system. Both methods were put to an actual test by the accidental introduction of a dog with latent distemper into the colony. Every animal which was not completely protected came down with the disease, including most of those dogs which had received only the injections of the killed virus and a few of those which had been protected by single injections of avianized virus. All research came to a standstill for six months while the disease was brought under control. As a result of this experience, a new system was set up. Beginning at 8 weeks of age, the time when most puppies begin to lose the immunity acquired by nursing from immune mothers (Baker *et al.*, 1959), the puppies were given temporary immunity through serum injections and at the same time given the avianized vaccine. Two weeks later the vaccine was re-injected, in case the first did not take. The whole colony of adults was reinjected every two years or oftener in order to maintain immunity, with completely effective results throughout the rest of the experiment. This supersafe system could not have been carried out without the co-operation of the Lederle Company, which generously contributed the necessary veterinary supplies.

Another highly fatal disease in dogs is infectious hepatitis. It is spread through urine rather than air and hence does not diffuse as rapidly as distemper. However, it provides a real threat to a dog colony because an individual which recovers may be a carrier and keep spreading the disease for years afterward. Injections of the killed virus responsible for this disease were later added to the distemper injections, providing immunity for approximately six months. This had to be maintained by repeated injections. At the present time a permanent type of vaccine combined with the distemper vaccine is available and is used in our colony.

To make sure that reasonably uniform physiological conditions were being maintained for the puppies, we weighed them once a week, beginning at birth and continuing through 16 weeks. In a young animal, continued growth is the best indication of good health, and failure to gain is often the first indication that all is not well. In addition, we counted the heart rates with a stethoscope. This measure turned out to be an important indication of probable survival in newborn puppies. Those with a slow heart rate usually did not survive more than a day or so. The heart rate also became an important indirect measure of behavioral development, being strongly correlated with major changes in behavior. The weekly weighings, besides functioning as a check on health, gave us an opportunity to make various behavioral observations upon development and to make sure that the puppies were truly normal in their sensory and motor capacities.

THE OBSERVATION OF DEVELOPMENT

From the very first it was obvious that our puppies were changing from day to day and week to week, and that while the heredity of a single puppy remained constant, it was acting upon a very different animal at birth than a few weeks later. We therefore began regular daily observations of our puppies, starting at birth and continuing up to 16 weeks of age, hoping to observe the very earliest manifestations of hereditary differences. We also hoped that we might in this way see the action of heredity pure and undefiled, before it became contaminated with the effects of experience.

The method was simple. Each day a trained observer watched the puppies for 10 minutes through the windows of the nursery rooms and wrote down everything they did. On the basis of our first observations, we compiled a check list on which the earliest appearance of various behavior patterns could be recorded. During the first two weeks after the birth of a litter, the observer went into the pen and sat quietly by the nest box in order to see the puppies more closely. After this time, when the eyes were open and the puppies were likely to respond to the observer, the observations were made from outside the room. Observations were always made in the morning, close to the time when the mothers were being fed and the pens cleaned, as this was the time when the puppies were likely to be most active. At other times of the day, the entire litter might be asleep during the whole 10-minute period. After the puppies had become 10 weeks old and were much more active and showed fewer changes from day to day, the observation period was reduced to 5 minutes, during which time as much or more activity could be recorded as before.

The results were quite unexpected but scientifically exciting. During the very early stages of development there was so little behavior observed that there was little opportunity for genetic differences to be expressed. When the complex patterns of behavior did appear, they did not show pure and uncontaminated effects of heredity. Instead, they were extraordinarily variable within an individual and surprisingly similar between individuals. In short, the evidence supported the conclusion that genetic differences in behavior do not appear all at once early in development, to be modified by later experience, but are themselves developed under the influence of environmental factors and may appear in full flower only relatively late in life.

More than this, we soon realized that we were dealing with a remarkable series of developmental changes; that the puppy comes into the world not as a simplified version of an adult but as an animal highly adapted to an infantile existence, and that he later undergoes a transformation in behavior which is almost as spectacular as the metamorphosis of the tadpole into a frog.

Thus the concept of development became a very important one in interpreting our results, both in relation to the development of social relationships and in respect to the increase in capacity for psychological performance. The concept of developmental change toward increasing complexity of organization ties together the action of the many factors which affect behavior on all levels of organization. For example, genetics does not act instantaneously but must work through physiology over long periods of time in order to affect behavior. Likewise, a puppy is born without experience of the outside world, and his behavior is organized and reorganized in relation to the social and physical environments throughout his lifetime. All these factors continually act together and upon each other to produce the capacity to react effectively to each new environmental situation throughout life.

THE SOCIAL ENVIRONMENT: TESTS OF SOCIAL BEHAVIOR

Controlling the social environment.—Our observations of development had shown us that the 4-week module did not coincide with important changes in social development. A major change in behavior takes place at about 3 weeks of age, when the sense organs first become completely functional and the puppy begins to form its primary social relationships and to eat solid food. Another basic social change is weaning, which may occur naturally as early as 7 weeks. We decided to keep the mothers with the puppies until 10 weeks so that the separation would not coincide with the cleaning and feeding schedule.

We also had to decide when to move the puppies to the large outdoor pens. We chose 16 weeks, primarily because this was a time when all the puppies were sufficiently developed physically to be able to stand the change to outdoor conditions at any time of year. From some other viewpoints, this was not an ideal time. Many of the dogs were just getting their second teeth and may have had some difficulty in handling solid food. Later observations on puppies raised in large one-acre pens indicated that they stayed very close to the kennel or "den" until about 12 weeks of age, after which they began to wander more widely. The experience of C. J. Pfaffenberger (1963) with guide dogs indicated that puppies kept in kennels beyond 14 weeks already begin to show the usual deleterious effects of prolonged kennel rearing: timidity and lack of confidence. The ideal time to move the puppies outdoors would probably have been between 12 and 14 weeks. However, since all animals were treated alike in this respect, the later date did not affect the behavioral comparisons.

A few exceptions were made to this plan. When females came into estrus before a year of age, they were temporarily removed from their litters and housed together in a separate pen. In the fox terrier litters, group attacks on certain individuals made it necessary to remove the victims. Sometimes the litter was divided, and in other cases a single animal was transferred to a group of more peaceful animals of the same age but a different breed.

Tests of social relationships.—During the first few weeks the puppies stayed with their mother and litter mates in their nursery rooms. Once a day the caretaker came in to feed and water the mother and to clean the room. Once a week a pair of experimenters came in for the weekly weighing. Otherwise the puppies had no contact with human beings. When they reached 5 weeks of age, we first began to develop and measure their social relationships. The first procedure was a handling test, in which we did all the sorts of things which people usually do to young puppies and noted their reactions on a check list. This test was given every two weeks from 5 through 15 weeks, and once again at a year.

A single pair of experimenters worked with the puppies up to 16 weeks of age. These were always a man and a girl, so that the puppies would have experience with both sexes and an opportunity to show differential reactions to them. A second pair of experimenters was added after 16 weeks, but many tests were still given by the first pair, so that each litter had contact with the same individuals throughout their tests.

When the puppies reached 5 weeks, we began testing for the development of social relationships between the puppies themselves. Here we chose only one aspect of such relationships, that of dominance and subordination. We began preparing for it at 2 weeks of age by placing a bone before all the puppies in the pen. This was done once a week thereafter. At 5 weeks of age, the puppies were paired in every possible combination within the litter and allowed to compete for the bone over a period of 10 minutes. This was repeated at 11 and 15 weeks, and again at one year. A puppy was not allowed to compete more than twice on the same day. This meant that the test took the greater part of a week, involving a great deal of handling and contact with the human experimenters in the process. It also meant that the larger litters obtained more contact than those in a small litter. Thus, in each dominance test, a puppy in a litter of 6 would be handled 5 times, whereas one in a litter of 4 would be handled only 3 times.

On the physiological and social levels, our chief concern was to keep conditions constant so that any differences might be caused principally by genetics. When we came to psychological processes such as the effect of learning and experience upon behavior, we carried out the same principle for the first few weeks of life, interfering with the puppies as little as possible and confining our data to the results of daily and weekly observations.

However, in such a standardized and simplified environment, there was very little opportunity for learning of any sort, except learning to do nothing. This posed a problem, as many of the major differences between the dog breeds show up as capacities to learn . various specialized tasks. We therefore deliberately began to subject the dogs to certain kinds of problems and training. In doing this we set up a secondary series of experiments. These did not interfere with the major genetic experiment because every dog was treated alike. In the secondary experiments, each dog was its own control, and the results consisted of changes in behavior in reaction to the new experience. Differences between individuals could, of course, be interpreted as the result of heredity.

The social tests were deliberately spaced at long intervals so that the effects of learning resulting from the test itself would be minimized. In contrast, each psychological test was done in a short and condensed period of time in order to get the maximum effect of learning and experience. While the effects of learning are long lasting, their greatest effect is immediate.

At first we thought of these tests as intelligence tests, but as time went on and we had more experience, we began to call them performance tests, since the animals seemed to solve their problems in many ways other than through pure thought or intellect.

First barrier or detour test.—We gave the first performance test at 6 weeks of age. It illustrates many of the general principles which we found useful in designing a good animal test. In the first place, the test had to fit the capacities of the species. There was no sense in requiring dogs to do things which demanded the use of hands. Since dogs are primarily hunting animals, we designed the test around a problem of finding an object containing food. Secondly, young puppies could not be required to go beyond their strength and general development. Until 3 weeks of age they are so immature that little can be done with them, and, indeed, they appear to be highly protected from psychological experience. We therefore decided to do the first performance test at 6 weeks of age. Before this time the young puppies had always lived inside their pens and had almost no experience with barriers of any kind except the walls, which they could neither surmount nor go around. We therefore attempted to discover whether or not there was any genetically produced ability to deal with barriers without experience. We took the puppies out of their pens and placed them in a strange room. For the first two days a girl experimenter simply took a puppy into the room, put it on the floor, and let it find a dish of food placed between her feet. The puppy was given two minutes to eat and explore the area. This was done once on the first day and three times on the second, by which time the puppies usually went rapidly to the food and ate it. On the third day, there was a barrier in the room. Experimenter and food were on one side, the puppy was placed on the other, and its task was to find its way around the obstacle. Since the puppy could see the food through the barrier, the difficulty lay in the fact that it had to go away from the food in order to reach it. Once it had succeeded, the test was repeated twice more, in order to reinforce memory of the event. On the next day a more difficult barrier was introduced, and on the third day the most difficult of all (Fig. 10.1).

The first two days were thus devoted to learning the location of the goal and to intensifying motivation by repeated reinforcement. With such young animals the performance could not be repeated more than two or three times in close succession without the puppy losing interest or becoming fatigued. Indeed, for many of them three times appeared to be too many.

Motivation.—In these early experiments a combination of rewards—food, the proximity of the experimenter, and being returned to the litter mates—were all used as part of the motivation. Because the experimenter might accidentally give cues to the puppies, she was instructed to sit quietly and do nothing but take notes. In experiments with older animals, the experimenter was if possible kept completely outside the room so that the puppy had to act independently.

The food reward was thus made the basic source of motivation, since it could be standardized a good deal better than any reward of approval or social contact. We used a teaspoon of canned fish for the reward and found it worked very successfully. There was no need to starve the puppies because they eagerly ate this addition to their regular diet and did not become satiated on the small amounts fed them. Making the puppies hungry would have been undesirable in any case because of possible bad effects on growth. The same canned fish was used in all subsequent experiments involving a food reward.

Relationship between performance tests.—As a general principle, each performance test was designed to provide pretraining for those to come. At 8 weeks there was a goal-orientation test in which the puppies learned to run to a particular spot in the room for food, and later to a different place. This test in turn was foundation training for a manipulation test in which food was placed in the same location but had to be obtained either by pulling the food dish out from under a cover or by pulling off the cover.

At 13 weeks of age the puppies got a second barrier test, this time in a more complex form. Thus the period up to 16 weeks was largely devoted to observation of development and tests of social behavior, with a few elementary psychological tests thrown in. The period from 16 weeks to 35 weeks, on the other hand, was devoted largely to performance tests, since the animals were now physiologically capable of a large number of co-ordinated movements. The puppies were also large enough by this time so that their physiological reactions could be easily measured.

At 16 weeks each litter was placed in a large outdoor run where it remained as a unit until the puppies reached the age of one year. They were at first somewhat disturbed in their new surroundings, and we found that they made the transition much more peacefully if we put them out for a few hours on one day, brought them back into their former pen overnight, and the next day permanently moved them outside.

Thereafter tests were done in one of three places. One was the home pen itself, in which the puppies soon became highly confident and relaxed. The other two areas were inside the Behavior Laboratory in places unfamiliar to the puppies, so that we had the problem of preparing them for working in an unfamiliar environment.

Emotional reactivity tests.—We first devised a test of emotional reactions to a totally new situation (Fuller, 1948). The puppies were carried into a controlled-environment laboratory and subjected to a variety of mildly frightening situations (loud noises, isolation, weak electric shock, etc.) while their external and internal emotional reactions were being recorded. This test was done at 17, 30, and 51 weeks of age.

Training tests.—The first of this group was leash training (Fuller, 1955). The puppy was taught to follow an experimenter into the building, given a food reward, and afterwards taken on a leash over the same route. Thereafter he was led all over the building without further rewards, in order to familiarize him with the new test situations. The method was essentially that of *forced training*, combined in the early phases with food rewards. Another experience of forced training was given in the obedience test, scheduled at 30 weeks of age, in which puppies were placed on a box and taught to "stay" until a command was given.

Tests for special abilities.—Still another group of tests attempted to measure certain special abilities peculiar to the individual breeds. The first of these was an artificial tracking test. Fish juice was smeared on metal plates attached to thin strips of board which could then be laid out in various patterns leading to a Syracuse dish holding a small bit of fish. Similar untreated plates were laid out as false trails. We wished to see whether beagles, which are generally used for tracking by scent, would give a superior performance in this situation.

A second test of this kind attempted to measure the special climbing ability of basenjis, which we had already observed in their successful attempts to climb upon their houses and to escape from their runs. In this test the dogs had to climb steeper and steeper ramps to reach the top of a pile of boxes on which food was placed. The final test required the animals to walk across a narrow plank stretched between the boxes and the top of the dog house in their home pen. As it turned out, the outcome of the test was strongly affected by motivational factors as well as motor skill.

A third test of this sort was the retrieving test, in which we attempted to measure the well-known ability of cocker spaniels to learn to retrieve objects more rapidly than some other breeds. Our original attempts to measure retrieving were done with young cocker pups between 8 and 10 weeks of age, and these animals did very well when tested again at 32 weeks. However, when we subsequently abandoned the early training, performance at the later age was very poor for all breeds, and the test therefore gave inconclusive results. There is an indication here of a critical period for learning this particular sort of skill.

Problem-solving tests.—Finally, there was a group of tests which attempted to measure more abstract kinds of intelligence. From Weeks 22 to 26, all the puppies were given a series of tests in a simple T-maze. After learning to run through the maze to receive food, the dogs were trained to run toward the arm of the T in which a panel was moved rhythmically. A moving object was chosen as the cue because the dogs appeared to notice moving objects more readily than stationary ones. Some animals showed extremely rapid learning in reaction to this cue, but others were afraid of it and, indeed, of the whole T-maze, and their learning was consequently retarded. In this, as in all other performance tests, emotional and motivational factors were always important. This training was used as the basis for a delayed response, or visual memory test.

Another test of intellectual capacity was developed from the motor-skill test and was called spatial orientation. The animals had to climb ramps and cross bridges similar to those used in the earlier test. The apparatus remained in the pens continuously throughout the test, so that the subjects could become completely familiar with it, and it often became a favorite place for the dogs to stand or sit. During the test the apparatus was arranged with a central food box to which there was only one open pathway. A human being could have visually inspected the barriers and immediately gone to the correct ramp. We hoped by this test to measure objectively the extent to which the various breeds are oriented by vision and capable of solving problems through the use of this sense. Our hopes were not completely realized. Striking individual and breed differences were obtained, but the results did not support the hypothesis of a simple difference in sensory dominance. Many of the animals appeared to solve the problem by routinely trying each possible solution in turn without visual inspection.

After the termination of the spatial-orientation test, at about 8 months of age, the dogs were given a vacation until they were approximately a year old. During this time they were observed regularly but given no formal tests. Beginning at 51 weeks, all the animals were given a final battery of tests, most of which repeated earlier ones, with the idea of finding out how stable the previously observed differences had become. The reactivity and handling tests were repeated, social-dominance ranks were again measured within each litter, physical measurements were taken for a third time, and a new test of frustration tolerance was given, in which dogs were confined to small cages and subjected to various sorts of stimulation.

Life history.—The testing schedule and, consequently, the life history of each animal in the experiment is summarized in Table 1.3. As there were some 30 major testing situations, each of which included multiple measurements resulting in from 5 to 40 scores, the data on each animal ran into the hundreds. Testing hundreds of animals, we eventually accumulated thousands of measurements.

TABLE 1.3

SCHEDULE OF OBSERVATIONS, TRAINING, AND TESTING

Test	Social	Emotional	Forced Training	Reward Training	Problem Solving	Physical or Physiological	Age (weeks)
Daily observation	x						0–16
Weekly observation, weighing			x			x	0–16
Maternal behavior	x						1, 7
Response to handling	x						5, 7, 9, 11, 13, 15, 52
Dominance Group	x						2, 3, 4, 6, 7, 8, 9, 10
Paired	x						5, 11, 15, 52
First barrier (detour)					x		6
Response to veterinary care		x					8, 10, 12, 14
Goal orientation or habit formation				x			9
Manipulation					x		10
Second barrier (maze)					x		14–15
(Transferred to outside runs)							17
Bi-weekly inspection, catching time		x					17–52
Somatotype, weighing						x	17, 34, 51
Reactivity		x					17, 34, 51
Following				x			18
Leash control and stair climbing		x					19–20
Motivation, T-maze				x			22
Discrimination, T-maze					x		23
Delayed response, T-maze					x		24–26
Trailing					x		27
Motor skill				x			29
Obedience		x					30
Retrieving				x			32
Spatial orientation					x		33–36
Physiological assessment						x	51
Response to confinement		x					51

In the following chapters we have selected for exhaustive treatment only those measures which seemed most meaningful in view of our general objective: to discover the influence of hereditary differences upon behavior. More detailed descriptions of some of the tests and measures are given later in the appropriate chapters.

By and large, this system of testing was very satisfactory. The puppies easily transferred their basic motivational training from one test to another, and there was no indication that they were solving their problems by watching the experimenter instead of using their brains. On the other hand, the puppies were continually being exposed to new situations and new problems, and they never had a chance to go back and practice familiar tasks. If the program had been set up so that the puppies could have gone back and repeated some of the things which they had already learned, they might have shown more confidence when they came up against new problems. Our subjective impression was that our puppies were not highly confident in psychological tests, contrasting with their behavior in social relationships with each other and with the experimenters, in which long familiarity made them highly confident.

THE ECOLOGICAL LEVEL: SPACE AND COMPLEXITY

Our chief concern here was to keep the physical environment as uniform as possible for all animals and at the same time maintain surroundings reasonably normal for dogs. One of our chief limitations was space. We could not give each dog the square mile or so of territory which he might have had if raised on a farm. During puppyhood we kept each litter in a large room measuring 10 by 18 feet, which is not far from the size of the space in which very young puppies confine their wanderings. At 16 weeks of age they were put into large outside runs, 20 feet wide by 75 feet long. In these, a

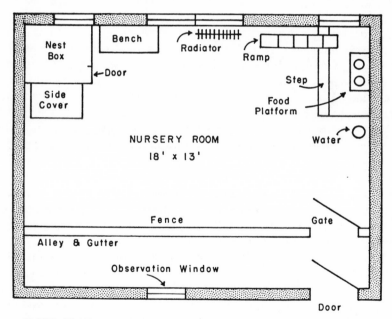

FLOOR PLAN

FIG. 1.2.—Floor plan of nursery room. The entire living space for the mother and puppies is 10 × 18 feet and so arranged that all pups are visible from the observation windows. One side of the nest box can be removed and laid down on the floor so that its interior is also visible.

litter of six dogs could live in close contact but not seriously interfere with each other's freedom of movement. Subsequent experience with smaller pens, 20 by 40 feet, indicates that in such pens interference does take place within groups of this size.

As to objects within the pens we provided only a minimum number, so that it would be easier to keep conditions uniform. In the nursery pens there was the nest box, the water pail, food dishes, and a small bench which was frequently used by the puppies as a hiding place. A piece of automobile tire casing was left in the pens for the puppies to chew on, and a similar piece hung from the ceiling where they could leap up and grab it. As soon as the puppies were old enough, their food was placed on a platform which could be mounted either by steps or along a ramp consisting of a cleated board about 6 inches wide. This gave the puppies experience with vertical as well as horizontal space.

In the outside pens there were very few objects except the nest boxes, water dishes, and a few stones mixed in with the gravel floor of the pen. The nest boxes themselves were constructed with insulated walls and roof for protection against the weather and were reached by an indirect passage (see Fig. 1.3) so that the dogs lived in what was essentially an artificial den kept warm by their bodies and provided with ample ventilation without drafts.

There were many physical factors which could not be completely controlled. These were entered in the records so that their possible effects could be later analyzed. One such variable was the season in which the puppies were born, this determining the season in which they were later tested and thus producing an artificial correlation between test results and temperature and other weather conditions. We always noted the temperature at which any test was given and maintained a permanent record of the outside temperature on a recording thermometer. When tests were given outdoors, we recorded the weather conditions, such as rain, snow, etc. Tests were never given during actual storms.

The environment in the outside runs was, of course, much more variable than that in the nursery rooms. This circumstance was not entirely bad, because it made the older dogs somewhat insensitive to random environmental stimulation from noises outside their pens, whereas an animal kept continuously in a more limited environment would be likely to over-react at the slightest new stimulation.

Hebb (1947) has shown in the McGill University Psychological Laboratories that rats raised as pets in a home are superior in their performance in psychological tests to rats which are raised in ordinary laboratory cages. He interprets this as the result of enriching their environment. Judged by this standard, the physical environment of our dogs was relatively barren. We attempted to enrich it in two ways: by bringing objects into the rearing pens during the experiments, and by taking the dogs out and introduc-

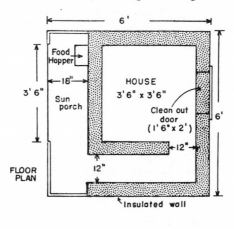

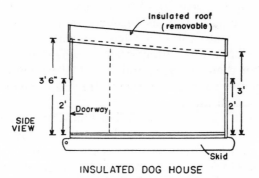

Fig. 1.3.—Floor plan of all-weather dog kennel. The indirect entrance is draft-free but provides ventilation. Insulated walls and roof keep temperatures at comfortable levels in both cold and warm weather.

ing them to other environments. Our animals certainly did not have as rich an experience as a dog running free on a farm, although their environment might compare quite favorably with a dog reared in a city apartment and never taken out except on a leash. Under our conditions the dogs probably never developed all of their capacities to a maximum degree. One of our major concerns was to protect them from environmental accidents which might either cripple or give an undue advantage to one individual. From an ecological viewpoint, we had to develop a sort of micro-climate and micro-environment which partially excluded the outside world. For example, our dogs in the outside runs were surrounded by fences 4 feet high which prevented their observing things at a distance and also prevented ceaseless running and barking at dogs in the neighboring pens. They could tell when dogs or people came close, but these contacts had little serious effect on their lives. The important events took place within their pens or in the testing rooms.

SUMMARY

When we stood back and looked at the completed design of our experiment, we discovered that we had evolved a virtual "school for dogs." It took us nearly 5 years to get everything running smoothly, and the school was to operate at its peak efficiency for almost 8 years thereafter, while we tested some five hundred purebred and hybrid puppies.

In the same amount of time, an elementary school teacher with a class of thirty pupils a year could have met and influenced the same number of individuals. We, as teachers in this unique establishment, learned a great deal which has application to the science of education, as will appear in later pages. However, we were primarily examiners rather than teachers, and our chief concern was to make sure that our pupils gave us answers to the basic scientific questions which we had asked them.

One of the first questions was, What kind of an animal is a dog? In evaluating the answers we would find that a dog is not a four-legged and childish human being dressed up in a fur coat. Our dogs could therefore give us answers to other questions only as dogs, closely related to human beings through social contacts but basically carnivores in their heredity.

CHAPTER 2

DOGS, WOLVES,
AND MEN

Dogs are extraordinarily variable animals in all visible respects. The range in size is almost incredible; a Chihuahua may weigh 4 pounds and a full-grown Saint Bernard 160, or forty times as much. Legs vary from the squat extremities of dachshunds to the long, graceful limbs of greyhounds and salukis. At opposite extremes we see the undershot jaws and foreshortened heads of bulldogs and pugs, and the long, narrow heads of the borzois. Tails vary from a tight curl to a sickle shape. Manifold variations in the color, length, and texture of hair exist and there is even a permanently bald breed, the Mexican hairless, contrasting with the poodle with its continuously growing hair.

Linnaeus, the great Swedish naturalist who originated our system of classification of animals and plants, placed the dog in one species and called it *Canis familiaris,* the familiar dog as contrasted with *Canis lupus,* the wolf dog. Later naturalists wondered whether such a highly variable species could really have had one common origin (Darwin, 1859; Packard, 1885; Hilzheimer, 1908).

This raised the questions, What is a dog? and Where did it come from? as well as the related problems of when and how this occurred. Since these are questions of prehistory, there can never be any complete and final answers to them, and the best we can do is to assemble the available evidence and draw the most probable conclusions. As we shall see, most of the evidence supports the idea that dogs have a unified common ancestry, being domesticated from wolves some eight to ten thousand years ago.

EVIDENCE FROM TAXONOMY
AND GEOGRAPHICAL DISTRIBUTION

Taxonomy is the science of classification. Beginning with Linnaeus, biologists began to classify animals according to genus and species on the basis of fundamental similarities. The idea of what constitutes a species has gradually developed until today we define a species as a population of animals which breed together or, to give a genetic definition, a population having access to a common store of genes. This means that groups of land animals (like most mammals) must show geographical continuity in their distribution in order to be considered species. Such populations can sometimes be divided into local races, but these are still part of the same species if there is continuous interbreeding between them. A local race is called a subspecies if it shows distinct differences from other such populations.

The definition of a species as a population has produced a gradual but dramatic revolution in the science of classification. The early taxonomists worked by first minutely describing a particular individual specimen, or "type," and then including in a species all individuals which closely resembled it. Since there is a great deal of variation within a population, it was easy for two different naturalists to pick up different specimens from the same population and describe them as separate species. Consequently, the older scientific literature on the subject is very confusing, and even today the classification scheme has not been completely revised. The following description is based on the latest available classification of dogs and their relatives.

To begin with the larger taxonomic divisions, dogs belong to the order Carnivora, which is divided into two large groups, the water-living forms like the seals and otters, and the land-living ones, which include seven different families: the bears, raccoons, weasels, civets, hyenas, cats, and dogs.

Members of the dog family typically run on their toes, in contrast to the bears and raccoons which walk on their heels. They are best adapted for swift running in open country, and the larger ones usually capture their prey by running it down. The family Canidae includes the foxes and the so-called "wild dogs"—the African hunting dog, the South American bush dog, and the dhole of India—as well as the true dogs and wolves.

The taxonomists originally placed several common animals in the genus *Canis: C. familiaris,* the domestic dog; *C. dingo,* the Australian wild dog, or dingo; *C. lupus,* the gray wolf; *C. latrans,* the coyote; and *C. aureus,* the jackal. All of these are still included in the genus *Canis,* except the jackal, which some experts place in a separate genus, *Thos.* Each of these species occupies a somewhat different ecological niche. Wolves are hunters of the large ungulates in the Northern Hemisphere, pursuing caribou in arctic regions and deer or moose in more southern climates. They live on tundras, in forests, or on plains, wherever their prey is found (Allen and Mech, 1963; Young and Goldman, 1944). Their close relatives, the coyotes, are smaller animals of the plains and deserts, chiefly hunting rabbits and small rodents which they find there, as well as being scavengers and carrion eaters (Young and Jackson, 1951). In the Old World, jackals are the counterparts of coyotes, being scavengers of the southern and equatorial deserts. Domestic dogs, of course, are found all over the world in close association with man, living in a variety of climates and eating a variety of food. The dingo is a hunter of marsupials on the Australian plains and deserts, and more recently has taken to preying on domestic sheep.

The fossil evidence shows that the family Canidae originally came from the Northern Hemisphere, and the distribution of its descendants is consistent with this. Wolves are found only in North America and Eurasia. Young and Goldman (1944) recognized two species of wolves in North America, *C. lupus,* the gray wolf, and *C. niger,* the red wolf formerly found in the Mississippi Valley and southern coastal parts of the United States. Each can be divided into various subspecies. The gray wolf, *C. lupus,* is also found in northern Europe and Asia, with a subspecies, *pallipes,* in India. This is a somewhat smaller animal, and some scientists consider it a separate species. There are no native wolves in Africa, South America, Australia, or Antarctica.

The coyote, *C. latrans,* is found only in North America, and originally ranged from Alaska nearly to Panama (Young and Jackson, 1951). The jackals have sometimes been divided into many species, but at present there are only three definitely recognized species (Hildebrand, 1954). Two of these live only in Africa. The common yellow jackal, *C. aureus,* ranges from northern Africa through Asia Minor to southeastern Europe, and in an easterly direction to India and beyond. The jackals are thus a more southern group than any other members of the genus.

The domestic dog has the widest distribution of any, being now found on all inhabited continents and always in close association with man. Before modern times, dogs were found on all continents except Australia and Antarctica. There were several kinds of native dogs in Africa and South America. Even in Australia there was the dingo, which is so similar to the domestic dog that it is supposed to have been introduced by the aborigines and later gone wild. This is made more probable by the fact that the dingo is one of the few

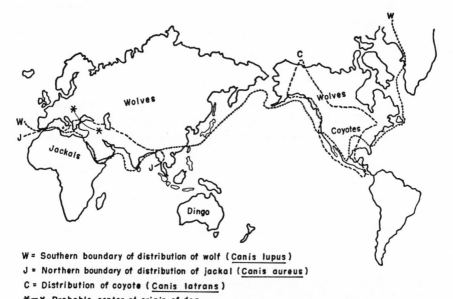

W = Southern boundary of distribution of wolf (Canis lupus)
J = Northern boundary of distribution of jackal (Canis aureus)
C = Distribution of coyote (Canis latrans)
- Probable center of origin of dog

© American Map Co., New York.

FIG. 2.1.—Geographical distribution of wild members of the genus *Canis*. Wolves are northern animals of Eurasia and North America, whereas jackals belong in Africa and southern Asia and coyotes are confined to North America. The earliest known remains of dogs (in Denmark) fall within the range of the wolf. (Modified from Werth, 1944)

placental mammals originally found in Australia. The others include man, several kinds of rats and mice, and a variety of bats.

Assuming that the dog must have been domesticated from some existing species belonging to the genus *Canis,* and that this species must have been domesticated at a place within its natural range, the geographical evidence eliminates South America as a possible domestication center, since no wild species of the genus *Canis* has ever lived there. We can also eliminate Australia, as the dingo is far more likely to be a domestic dog gone wild than it is to be an ancestral form (Werth, 1944). We cannot eliminate Africa, but if dogs were first developed on that continent, they can only have been domesticated from jackals, since jackal species are the only ones found in that area. In Eurasia, dogs could have been domesticated from either wolves or jackals, whereas in North America the only possible candidates are wolves and coyotes.

The evidence from geographical distribution of living animals thus narrows down the possible ancestors of the dog. There remains the possibility that the dog could have been domesticated from a wild species now extinct or from some of the so-called "wild dogs" now living.

EVIDENCE FROM FOSSILS

The "wild dogs" can easily be eliminated. These distant relatives of the dog include *Lycaon*, the African hunting dog; *Icticyon*, the South American bush dog; and *Cuon*, the dhole of India (the "red dog" of Kipling stories). According to Matthew (1930), the foxes and wolves have a common ancestor in the Miocene, some seven and a half million years ago, but their relationship with the wild dogs is much more remote, tracing back to a common ancestor in the Oligocene, some twenty million years ago. This pretty well excludes the "wild dogs" as ancestors of the domestic ones. They are actually more distantly related to the genus *Canis* than are foxes and are quite rightly placed in different genera from the dog.

At various times scientists have speculated that the modern dog is descended from a small wild dog which is now extinct (Allen, 1920). To date, no such hypothetical ancestor has been discovered. Matthew finds that wolves, coyotes, jackals, and foxes were present in essentially their modern forms in Pleistocene times, over half a million years ago according to recent estimates (Kulp, 1961). These four kinds of animals have remained distinct from each other ever since, and there are no skeletal remains until recent times which can be identified as domestic dogs. These dog remains are not fossils but true bones, discovered by archeologists in association with human skeletons, ruins, and artifacts.

EVIDENCE FROM PREHISTORY

Being a domestic animal, the dog is always found in close association with man, and the best evidence for the existence of prehistoric dogs comes from bones found with prehistoric human remains. The science of archeology is now in a state of ferment, partly because of new techniques such as radiocarbon dating, which makes it possible to estimate the age of bones and other organic remains on an objective scale, and partly because of a new interest in the domestication of animals and plants.

Braidwood and Reed (1957) describe the prehistoric "agricultural revolution" which produced as profound changes in the lives of early men as did the industrial revolution in the existence of modern man. Early men lived by gathering wild plants and other food, and occasionally hunting in packs. With the domestication of plants and animals, they had access to a greatly increased food supply, could live in much greater numbers, and began to settle in villages and towns. The dog is different from most domestic animals in that it would have been useful both to hunters and to prehistoric farmers and herdsmen.

The agricultural revolution took place in many parts of the world, but it first began in the "fertile crescent," the foothills of the mountains surrounding Mesopotamia and Palestine, and this is the area which Braidwood and Reed have studied most intensively. Reed (1959, 1960) states that the earliest authentic remains of a dog have been found in Jericho and dates them at approximately 6500 B.C. or eighty-five hundred years ago. In the contemporary agricultural village of Jarmo there are several clay figurines which are identifiable as dogs because of their curly tails. It is possible that even earlier remains of dogs will be found in the future, but this is a matter of conjecture. In contrast, the first evidence of dogs in Egypt is dated at only 3500 B.C., some three thousand years later. The evidence is a bowl with a recognizable picture of some greyhounds, or salukis. Long-limbed skeletons of the same date have been found in Mesopotamia, where salukis were probably first developed. It looks as if the use of the dog spread slowly from Palestine into Egypt.

The dogs in Iraq today include the saluki, a tall greyhound-like animal used for hunting gazelles, and a large heavily built guard and herd dog used by the Kurdish shepherds (Hatt, 1959). It is difficult or impossible to discriminate the bones of these latter dogs from those of the local wolves, and this region is definitely a possible center for the first domestication of the dog (Reed, 1959, 1960).

The only other area of the world which has been studied as intensively is Western Europe. The oldest authentic dog remains come from Denmark, discovered there by Degerbøl (1927). The bones and human artifacts found with them belong to a cultural period known as the Maglemosian, the latter part of the Mesolithic period during the transition to the Neolithic. These remains have not been carbon dated, but other estimates place them from 8,000 to 10,000 B.C. Reed (1959) thinks that they may be either older or somewhat more recent than the bones found in Jericho. As the evidence now stands, the earliest remains of domestic dogs come from Denmark, with those from the Middle East falling in second place; but this conclusion may have to be revised in the light of future discoveries.

More recent Stone Age dogs have been found in Europe associated with the Swiss lake dwellers, and others have been dug up from the bottom of Lake Ladoga in Russia. Using the type method, the older archeologists described each of these remains as separate subspecies of dogs, giving them such names as *"Canis familiaris matris optimae,"* and Studer (1901) tried to trace modern breeds back to them. Dahr (1937) found that these Stone Age dogs, far from being specialized breeds, were all very much alike as compared to the widely varying modern breeds, and most authors now agree that their remains are similar to the skeletons of modern Eskimo dogs. Degerbøl stated that his earlier specimens were distinctly different from wolves in having smaller teeth and jaws. This would argue against the dog being domesticated in that area, but again we must conclude that the evidence is still quite incomplete.

The original dog inhabitants of North America were extensively studied by Allen (1920), who found that all of them showed resemblances to wolves rather than to coyotes. Haag (1948) later estimated that the earliest dog remains in the Western Hemisphere can be dated about 1500 B.C., but this date may be too recent. At any rate, it is much later than even the Egyptian dogs, and it is logical to suppose that dogs were introduced into North America by trading or by migrating peoples coming across the Bering Strait into Alaska. Other areas of the world, including Southeast Asia, are all relatively unknown from the archeological point of view. Perhaps new discoveries will change the picture, but at present the only strong contenders for the center of domestication of the dog are Denmark and Mesopotamia.

There are two possible theories regarding the origin of the dog. One is that the dog was domesticated once and spread rapidly all over the world from this center. An alternate theory is that the dog was domesticated at several different times and places. If the first theory is correct and Denmark was the center of origin, the dog could only have been domesticated from the wolf, as that is the only wild member of the genus *Canis* which existed in that region. If Mesopotamia was the center, dogs could have been domesticated

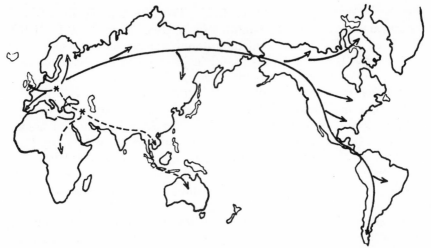

© *American Map Co., New York.*

FIG. 2.2.—Centers of domestication and hypothetical adaptive radiation of dogs from these centers. Solid lines represent the northern dogs, and broken lines the southern dogs, including the basenji and dingo.

from either the wolf or the jackal. However, if dogs were domesticated at various times and places, the field of possible ancestors is still wide open.

EVIDENCE FROM COMPARATIVE ANATOMY AND PHYSIOLOGY

One of the difficulties of studying prehistoric dogs is that their bones are sometimes indistinguishable from those of wolves or other members of the family Canidae. Biologists frequently have the same difficulty in identifying recently killed specimens. Is the large dog-like animal which was shot while killing deer a small timber wolf or a large German shepherd? In many cases the expert can only give an opinion rather than a definite identification.

Anyone who looks at wolves in zoological parks is immediately impressed with their doglike appearance. There are only a few noticeable differences. They have heavier coats than most dogs and long, bushy tails which are only slightly curved. The general shape of their heads is doglike, and their ears are erect. They are larger than most dogs and may weigh anywhere from 60 to 175 pounds (Young and Goldman, 1944). Their large, heavy heads and long, heavily boned legs contrast with those of most dogs. In short, they are powerful, rugged animals, not highly specialized for any one activity. They look like big dogs, but without any of the head or tail deformities which are characteristic of some dog breeds.

Coyotes, on the other hand, are much smaller animals. The average coyote weighs about 25 pounds, with the range extending between 18 and 30, except for some exceptionally large animals which may weigh as much as 74 pounds (Young and Jackson, 1951). Their general body proportions are the same as wolves (Hildebrand, 1954), but they have proportionately longer necks and narrower muzzles, giving them the general appearance of a large fox. They are specialized for fast running and for hunting small animals like rabbits and ground squirrels on the open plains.

The third group of wild relatives of the dogs are the jackals of the African and Asian deserts. They are very similar to the coyotes both in appearance and way of living but are even smaller in size, the average adult weighing only about 20 pounds.

We can conclude that the principal anatomical difference between these three wild species is a matter of size but that this is not an absolute difference, since coyotes overlap both wolves and jackals. Dogs, of course, overlap all these animals in size, the dwarf breeds being smaller than the smallest jackals, and the largest dogs being as large as the largest wolf. There is therefore no distinct "dog type" which can be immediately distinguished from the wild species. Of course, there are certain individual dogs and breeds which are distinctly different. No one would confuse a Chihuahua with a jackal, and a bulldog is immediately recognizable for what it is. Such extreme mutant types among dogs are easily identified, but the vast majority of the breeds are reasonably normal in basic anatomy, and if we are to discover anything regarding their relationships with one another, we must do it on a population basis.

A population can be described in terms of the average and the amount of variation from the average, usually expressed in statistical terms as the mean and standard deviation. When we describe a species, we must make many such measurements on many individuals in order to give an accurate statistical picture of the population. Few such studies have been done on dogs because of the immense amount of work involved, particularly if one has to work with hand calculators rather than computers. Most attention has been concentrated on the skull, partly because this is the portion of the skeleton which is most likely to be preserved, and partly because it shows the most variation.

Dahr (1937) measured the skulls of Stone Age dogs of northern

Europe and compared their dimensions with those of modern breeds. One of the highly variable characteristics of current dog breeds is the shape of the upper jaw. (Figs. 2.3, 2.4, 2.6). Bulldogs have short, broad jaws, and those of greyhounds are long and narrow compared with those of other breeds. When Dahr computed the ratio of snout length to breadth of the upper jaw at the narrowest point and graphed this ratio against skull length, he found that the Stone Age dogs all fell in a close group in the middle of the modern dog breeds. When he measured the height of the lower jaw from top to bottom in relation to the length of the row of molar teeth, he again found that the Stone Age dogs formed a compact group in the middle of the modern dog breeds. He concluded that these early European dogs all belonged in the same population and that the modern breed populations have diverged in all directions from them.

Another variable characteristic in dog skulls is the shape of the lower jaw. The wolves studied by Dahr had longer and heavier lower jaws than most modern breeds, although their average jaw thickness was exceeded by the very large breeds like the Saint Bernard. The length of the molar tooth row is a good indicator of jaw length. Not even the Saint Bernards have as long a row of molar teeth as do the wolves, so that the wolves stand at the top end of the scale with regard to jaw size. It is obvious that if wolves were ancestors of modern dogs, there must have been a very early selection for animals with smaller jaws and teeth. Dahr was inclined to think that the original dogs were a middle-sized race of wolves which are now extinct. Judging from the tendency for wolves to be smaller in more southern regions, this is a reasonable hypothesis, and, as mentioned earlier in this chapter, there are still such wolves living in Mesopotamia, indistinguishable from certain local dogs in their skeletal characteristics. Werth (1944) argues that the probable ancestor is the Indian wolf (*C. lupus pallipes*) which lives in that region, and Lawrence (1956) has independently reached the same conclusion.

Dahr based many of his conclusions on measurements made by Wagner (1930). The latter traveled around the museums in Oslo, Copenhagen, and Berlin, measuring the dogs' skulls found there. Some of these skulls date back as early as 1863 and may not reflect the modern breeds accurately. He was able to collect ten or more skulls from twelve different dog breeds, with smaller numbers from a large number of others. Because of the preference for males among dog owners, most of these must have been of one sex. He also measured seventeen wolf skulls which had been collected in Scandinavia, Greenland, and Siberia and therefore must have come from the large northern races of wolves.

Wagner made some forty different measurements on each skull. Of these, the best measure of over-all size is the basal length, the distance between the opening in the rear of the skull for the entrance of the spinal cord to the most anterior point on the upper jaw at the base of the upper incisor teeth. Other measurements can be graphed against this, as shown in Figures 2.3–2.6.

One of the first things we notice about living wolves is that they seem to have broad heads and heavy jaws. If we graph the greatest width across the cheekbones, which includes the jaw muscle, we find that wolves' skulls are wider than most dog breeds except for the Saint Bernards (Fig. 2.3). The bulldog breeds also have much the same proportions. Such wide cheekbones reflect the skeletal effect of having heavy jaw muscles.

When we graph the greatest breadth of the upper jaw (taken opposite the large carnassial teeth), we find that wolves are essentially no different in their proportions from normal dogs, and that Saint Bernards have jaws which are fully as wide (Fig. 2.4). The bulldog breeds are off in a class by themselves with a broader than normal upper jaw, but two other breeds are even more extreme, the Great Dane and German shepherd. These are the most broad-jawed of any breeds and far exceed the wolves. It is interesting that these two breeds have been developed in the same part of the world, and they possibly have some common ancestry.

Quite a different picture results when we graph the size of the molar teeth in relation to the size of the skull (Fig. 2.5). On the average, wolves not only have bigger teeth than any dog breed studied, but they are larger in proportion. In the majority of dog breeds, tooth size is very nicely correlated with the size of the skull, even in those breeds with deformed skulls such as bulldogs. The few exceptions are those breeds which have somewhat smaller teeth than the average. This means that Wagner's wolf population differed from all dog populations in only one respect, the average size of the teeth. As we have seen, individuals in other populations overlap. Little Red Riding Hood's classic remark to the wolf, "What big teeth you have, Grandma," seems to be partially justified by the facts. It follows that one of the earliest selection procedures followed by the early domesticators of wolves may have been to pick out animals with small teeth.

These measurements also bring out the fact that there are only

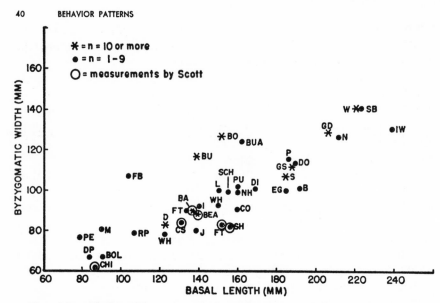

Fig. 2.3.—Skull width across cheek bones as related to skull length. Note that breeds with bulldog heads (having short, wide skulls) fall off to one side. Wolves have wide skulls, and coyotes and jackals narrow ones, but both are exceeded by various dog breeds. (Most data from Wagner.)

KEY TO BREEDS (FIGS. 2.3-2.5)

B = Borzoi	IW = Irish wolfhound
BA = Basenji	
BEA = Beagle	J = Jackal
BO = Boxer	
BOL = Bolognese	L = Lapland dog
BU = Bulldog	
BUA = Bulldog, old type	M = Mops (pug)
CO = Coyote	N = Newfoundland
CS = Cocker spaniel	NH = Norwegian hare hound
D = Dachshund	P = Pointer
DI = Dingo	PE = Pekinese
DO = Doberman	PU = Poodle
DP = Dwarf pinscher	
EG = English greyhound	RP = Rattle pinscher
FB = French bulldog	S = Setter
FT = Fox terrier	SB = Saint Bernard
	SCH = Schnauzer
GD = Great Dane	SH = Sheltie (Shetland sheep dog)
GS = German shepherd	
	W = Wolf
I = Icelandic dog	WH = Whippet (large and small)

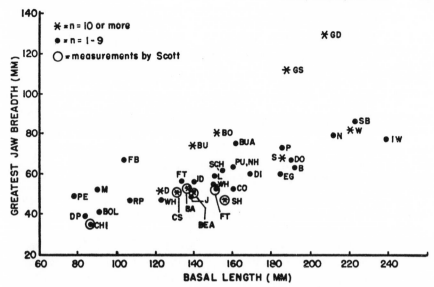

FIG. 2.4.—Greatest width of upper jaw relative to skull length. Note that Great Danes and German shepherds have unusually wide jaws. Coyotes and jackals have narrow jaws, but are not as extreme as some breeds. (Most data from Wagner.)

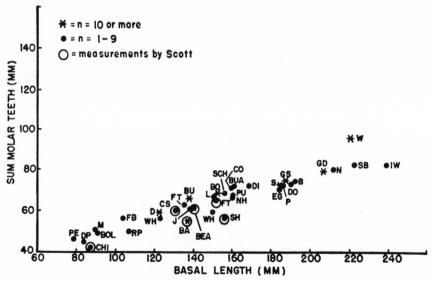

FIG. 2.5.—Size of molar teeth relative to skull length. Tooth size and skull size are highly correlated, but in large northern wolves the teeth are larger than in any dogs of similar size. (Most data from Wagner.)

a few real departures from normal proportions of the skull in the dog breeds. One of these is the bulldog type of head, and it is interesting that this deformity appears even in dwarf breeds like Pekingese and pugs. Another is the broad jaws of the Great Danes and German shepherds. The heads of the greyhound group are in better proportion than one would expect, and their biggest variation from the norm is in the width between the cheekbones. The narrow head of the borzois and Eastern greyhounds, or salukis, is the anatomical result of having narrow cheekbones and less space for jaw muscles rather than having the brain case squeezed together, as is often thought.

Another interesting point is that the dwarf breeds have brain cases which are almost as large as those of the big breeds. Reduction in body size has affected the brain very little, but the jaws and skull thickness have been reduced a great deal. Selection for small size has chiefly reduced those parts of the skull which are attached to the outside of the brain case. This again argues that the dog breeds all come from one species, for if the dwarf animals came from a small species, one would expect that the parts of their skulls would be in good proportion to each other.

Stockard (1941) studied the skulls of various modern breeds of dogs and came to the conclusion that the breeds could be divided into two types, those having long skulls and those having short ones. The proportions of the saluki, Great Dane, Saint Bernard, German shepherd, basset hound, and dachshund vary somewhat from each other but are very different from such animals as the English bulldog, French bulldog, Boston terrier, Pekingese, and Brussels griffon. The latter group of animals are the product of a gene or genes which produces abnormal development of the head.

A different gene or group of genes may shorten and deform the legs, as happens in the dachshund. It is probable that the mutations for deformed heads and disproportionately short legs occurred separately and were used in different combinations for the development of various breeds. Thus we have the boxer with a bulldog head and normal legs, as well as the short-legged English bulldog. Such animals, however, do not throw much light on the ancestry and relationships of the dog, except to indicate that breeds with bulldog heads may be related. Even this is not definite proof, as the bulldog mutation may have occurred more than once at different times and places.

These studies show that there are no distinct skull "types" except for the obvious distortions found in bulldog heads and the narrow skulls of the greyhound breeds. Wolves overlap with dogs in every skull measurement with the possible exception of the size of the teeth, and even here large dogs and wolves overlap as individuals. Given an unknown skull, a biologist can identify it only in terms of probability with respect to the measurements of a known population. Some skulls, like those of Chihuahuas and bulldogs, can be identified as dogs with very nearly 100 per cent probability, although even here there is always the very unlikely possibility of a similar mutation in a wild species. One way to increase accuracy is to determine probability on several different measurements. So far, this method has only been used to study the relationships between populations. Jolicoeur (1959) studied wolf populations in this way and found that the amount of overlap was inversely proportional to geographical distance, and he concluded that many of the subspecies of North American wolves which have been identified by the "type" method should be considered part of the same population. Such an analysis is yet to be applied to relationships between dog breeds, but we have every indication that there is a large amount of overlap between breeds of the same general size (Fig. 2.6).

Tables 2.1 and 2.2 summarize the results of measuring the skulls of sample animals from the five different breeds used in our experi-

TABLE 2.1

Average Skull Measurements of Different Breeds of Dogs (to nearest mm.)*

Measurement	Basenji (10♂, 9♀)	Beagle (5♂, 5♀)	Cocker (3♂, 5♀)	Sheltie (5♂, 4♀)	Fox Terrier (2♂, 2♀)
Condylobasal length............	147	148	138	164	161
Basal length..................	137	140	131	156	152
Brain cavity length............	76	78	74	83	80
Bizygomatic breadth..........	90	88	84	82	83
Greatest jaw breadth..........	53	51	51	47	53
Smallest jaw breadth..........	31	30	32	25	31
Snout length.................	65	66	59	79	83
Length of molar tooth row.....	54	57	54	62	65
Size of molar teeth............	55	60	60	56	65

* Note that shelties and fox terriers have longer skulls than the rest, and that the shelties show the smallest measures of any in skull breadth.

ments. These measurements chiefly bring out the facts that the Shetland sheep dogs (shelties) and wire-haired fox terriers have long skulls compared with the other breeds, and that shelties have narrow jaws in proportion to skull length. There is a great deal of individual variability within breeds and no clearcut separation between them except for the narrow-jawed condition of shelties (Fig. 2.6).

TABLE 2.2

RATIOS OF SKULL MEASUREMENTS TO LENGTH OF SKULL (basal length)

Measurement	Basenji (19)	Beagle (10)	Cocker (8)	Sheltie (9)	Fox Terrier (4)
Brain cavity length.	.56	.56	.57	.53	.53
Bizygomatic breadth.	.66	.63	.64	.53	.55
Greatest jaw breadth.	.39	.37	.39	.30	.35
Smallest jaw breadth.	.22	.22	.24	.16	.21
Snout length.	.47	.47	.45	.51	.55
Length of molar tooth row.	.40	.41	.41	.40	.43
Size of molar teeth.	.40	.43	.46	.36	.42

Another method of determining relationships between populations is to study the reaction of blood serum from one animal to that of another (Boyden, 1942). The technique is to take serum from one individual and inject it into another, which in due course develops antiserum. If the serum of a fox is injected into a chicken, the animal will develop antiserum against fox blood. The experimenter bleeds the chicken and extracts the antiserum. He now takes more serum from the fox, mixes it with the antiserum, and measures the amount of precipitate. This gives the amount of reaction against the original fox serum, or antigen. If he mixes dog serum with the antiserum, there is a smaller amount of precipitate. Comparison of the two figures provides an estimate of how alike the two species

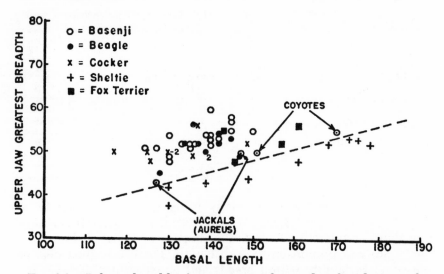

FIG. 2.6.—Relative breadth of upper jaw in five modern breeds. Note that a straight line will cut off all Shetland sheep dogs from the rest but that there is a great deal of overlap between the other four breeds.

may be. Using this method, Pauly and Wolfe (1957) compared three canine species and obtained the following figures: dog with wolf, 88 per cent; dog with fox, 50 per cent; fox with wolf, 52 per cent. Dogs are obviously more like wolves than foxes. Duerst (1942), however, experimenting with one wolf and one jackal, found that the two were similarly related to dogs.

The limitations of the method are that it must be done with extreme care, as small amounts of contamination can produce very great variation in results. Also, most results now available are based on antisera and sera obtained from only one member of each species. The population approach is still to be applied to this technique, and until this is done it provides only tentative evidence.

HISTORICAL EVIDENCE

By and large, historians have been little interested in dogs and refer to them only in passing. For example, Trevelyan, in his *English Social History* (1943), says that a complaint was made in Parliament in 1389 that laborers and servants kept greyhounds and other dogs and were wasting their time hunting. As a result a law was passed to prevent people with low incomes from keeping sporting nets or dogs. Such casual records indicate that dogs were popular and often used in hunting, particularly by the gentry.

In the same century Chaucer wrote his *Canterbury Tales*, which contain vivid descriptions of contemporary characters and their occupations. The Prioress kept "small hounds" as companions, and the Monk was a hunting man—"greyhounds he had, as swift as fowls in flight." The Wife of Bath metaphorically describes a woman's over-fondness for a man: "For as a spaynel she will on him lepe," indicating that exuberant affection was as characteristic of the spaniels of the fourteenth century as it is today.

The earliest English book on dogs was published in 1576, during the reign of Queen Elizabeth I. This was actually a translation of a paper in Latin written by Johannes Caius, a "Doctor of Phisicke in the Universitie of Cambridge," for whom the present Caius (pronounced "Keyes") College is named. The paper was originally written at the request of Conrad Gessner, the Swiss naturalist, who wanted information regarding the English breeds of dogs.

Caius was obviously influenced by the contemporary English social system, for he says that there are three kinds of dogs, a gentle kind serving the game, a homely kind for necessary uses, and a currish kind, "meete for many toys." Hunting was reserved for the nobility in dogs as well as in people. However, Caius describes many dogs which are the counterparts of modern breeds as well as many which have disappeared.

Classifying dogs according to their use, he first describes the hounds as hunting by scent and having large bagging lips and hanging ears. He places the bloodhound in a separate group because it was used to chase beasts that were wounded and to track down thieves who had stolen meat, particularly on the Scottish Border. Even at that time bloodhounds had an unusual reputation for being able to follow a trail.

In still another group were the "gase hounds," used for hunting by sight, and hence the name. Such breeds have apparently disappeared. Greyhounds were used somewhat differently, in coursing, or pursuit by sight, acting and appearing very much like their modern counterparts.

Terriers, deriving their name from the French "terre," were bred to creep into the ground and drive out small animals like foxes and badgers. The "Spaniells" originally came from Spain and were used both for hawking and hunting birds with nets. A group of closely related breeds were the setters. As Caius says, a setter was supposed to find birds, lie on the ground, and creep forward like a worm, lying down near them. The net was then prepared and the dog, on signal, would rise and frighten the birds, which then flew into the net. This description of setting birds for the net is somewhat different from that given by other writers, and perhaps they were used in several different ways.

He mentions the "Spaniells gentle" as neat, pretty dogs, probably referring to some of the toy breeds like the King Charles spaniel. A section is reserved for the "homely" dogs, including shepherd dogs and mastiffs or bandogs. This last name refers to the fact that a guard dog was usually kept tied up by a bond, or chain. Finally, there were mongrels, which Caius said "keep not their kind," including the "turnspits," used on a treadmill which turned the meat being roasted over a fire.

Among breeds which have now disappeared were the "lyemmer," which was midway between harriers and greyhounds, the "tumbler," used for hunting rabbits, and the "theevish" dog, which did not bark and was used by poachers for hunting in the dark. If this report is correct, barklessness has occurred in breeds other than the modern basenji.

We can see from these records that some of the principal breeds of modern dogs were already established four hundred years ago, and that the English were already importing breeds from other countries. How the breeds were related to each other and how much crossing was done between breeds we have no way of knowing except by inference.

Another book on dogs was published in 1686, over a century later. Richard Blome's *The Gentleman's Recreation* was a sort of encyclopedia, the first part covering the arts and sciences from grammar and rhetoric through horography, or "sun dialling," to fortification and heraldry, and the second including articles on various sorts of hunting and other special occupations of a country gentleman. As was the custom in those days, the expense of printing was borne by wealthy benefactors, and Blome honored them by picturing their coats of arms. The book was authorized by Charles II in 1682, but dedicated to James II, who was one of the sponsors. Among the 228 others, there appears the name of George, Lord Jeffries, he of the "Bloody Assizes" that followed the Duke of Monmouth's short-lived rebellion in 1685.

But Blome was not interested in politics. In his book he describes several varieties of beagles: the southern beagle; the "fleet, northern, or cat" beagle, somewhat smaller; and, finally, a very small type, the size of a lady's lap dog, sometimes used for hunting coneys (the old word for rabbits) and hares.

Like Caius, he describes terriers as small dogs used for following the fox or badger into the earth and says that a dog bred out of a beagle and a mongrel mastiff makes a good terrier. He also mentions "tumblers" and "lurchers," for hunting coneys; greyhounds, for coursing; and bloodhounds, used for "harbouring" a stag, i.e., locating the place where the stag spent the night, or "harboured."

His greatest enthusiasm was reserved for spaniels, which were employed for "springing" and retrieving of fowl.

Spaniels by Nature are very loveing, surpassing all other Creatures, for in *Heat* and *Cold*, *Wet* and *Dry*, *Day* and *Night*, they will not forsake their *Master*. There are many *Prodigious Relations*, made in several Grave and Credible *Authors*, of the strange Affections which *Dogs* have had, as well to their Dead and living *Masters*; but it is not my business of take notice of them here; but to apply myself to the Subject in Hand.

We can conclude that dog stories were as popular in those days as they are at present, and many of them just as difficult to believe. Blome goes on to write chapters on the use of water and land spaniels in the sport of fowling. His descriptions of training would do credit to any modern dog book.

Of all dogs there is none so fit and proper to be made a setting dog as a land spaniel, by reason of their natural inclination to ranging and beating about a field; but any dog, whether a water spaniel or a mongrel betwixt both, or a lurcher or tumbler or any dog that hath a perfect good scent, and naturally addicted to the hunting of fowl, may be brought to be a setter.

The land spaniel was first taught to "couch and lie close to the ground," then taught to lie still while a bird net was dragged across him, then taught to associate lying down with the scent of a partridge (which Blome says can be duplicated by a boiled "bullock's liver"). In the actual hunt, the spaniel first quartered the ground. When he indicated by his eagerness and wagging his stern that he had located the birds, he was ordered to lie down and remain until the net was thrown over both him and the birds. In hunting partridge, the technique described by Blome involved drawing a net over the covey before it took flight.

Blome calls the actual discovery of the birds "making a point" and remarks that "some dogs will stand up in their setting, which is a great fault." After shotguns were invented, birds were no longer taken with a net and new dogs were developed from the land spaniels which simply stopped and pointed when they located birds. Thus crouching became a fault in its turn, and all modern setters are really pointers. Only in the cocker spaniels does the original tendency to crouch remain, and these breeds readily crouch and crawl on their bellies when threatened. The modern springer spaniels were developed to "spring," or flush birds; and cockers, when they are used for hunting, are usually taught to flush the birds rather than to point. However, they still retain their old capacity for learning to crouch.

In contrast to spaniels, the basenji breed has a short written history. According to Victoria Tudor-Williams (1946), a prominent English breeder, basenjis were first successfully imported into England in 1937. The two animals used to form the foundation stock for the Jackson Laboratory were both descended from five basenjis imported into England in that year. Compared with most purebred dogs, their coefficient of inbreeding was surprisingly high (.23, as calculated from their pedigree which extends back to their five imported ancestors). Comparative figures for the foundation stocks of the other four breeds are shown in Table 2.3, based on available pedigree records covering at least four generations.

We are indebted to Dr. James P. Chapin (1958) of the American

TABLE 2.3

COEFFICIENTS OF INBREEDING OF FOUNDATION STOCK*

Basenji.............................	.23
Beagle.............................	.00
Cocker spaniel......................	.01
Sheltie.............................	.06
Fox terrier.........................	.00

* Cockers include only those animals used in the cross with the basenji.

Museum of Natural History, who observed this breed while on the Museum's Congo expeditions in 1909–15 and later years, for a description of basenjis in their native habitat. They originally had a wide distribution, from the French Congo and central Congo basin into the southern Sudan and Uganda, and perhaps further. They were owned and used in hunting by Pygmies and several other African tribes. Recently they have begun to cross with dogs brought in by European residents, and pure strains may be difficult to find in the future. There is no indication of crossing in the Jackson Laboratory stock, in which inbreeding has never revealed concealed recessives belonging to other breeds.

The name "basenji" in the Lingala trade dialect of the central Congo means "people of the bush." Evidently the early explorers asked the natives the name of the breed, and they replied that those were dogs belonging to the bush people. Chapin says that the Pygmies of the Ituri forest used basenjis for hunting in many ways. These tribes frequently hunted with nets. Men and dogs would go through the underbrush and drive small antelopes along the game trails to where a net had been set. Sometimes basenjis were used for flushing birds, which would then fly into the trees where they could be shot with arrows. They were also employed in tracking small game, and a dog which had a keen sense of smell was highly valued. Thus the basenjis are a general-purpose hunting dog and do not fit into any of the conventional divisions of the European breeds.

According to the breed standards, basenjis are barkless dogs. Those in Africa occasionally bark, but only when extremely excited. At night in the native villages they often make a tremendous noise, which Chapin describes as yodelling and wailing. Similar behavior occurred in our kennels.

We have not been able to get any authentic information regarding the breeding season in their normal habitat near the Equator, and perhaps there is none in this region of almost constant day length. In the north temperate zone, basenjis breed only once a year, close to the time of the autumnal equinox.

Living in the same regions as jackals, basenjis have had a much better opportunity for crossing with them than have most other breeds. The peculiar vocalization of basenjis lends some weight to this hypothesis. However, jackal noises are much more elaborate than those of the basenji and different in many ways. Van der Merwe (1953) says that the African black-backed jackal "yelps like a dog when startled," but gives an undoglike sound when cornered: "ke-ke-ke-kek." The danger signal when pups are present is a soft "wuf." In the mating season the female gives a "shrill but hearty laugh," and the male answers with a long howl. The jackal has a hunting cry or howl described as "ieaaaa-iea-iea-iea" or "nieaaaaa-niea-niea-niea." Whether jackal and basenji noises have some similarity remains an open question that will be answered only when detailed analysis of these sounds can be made with the sound spectograph.

In other respects, the behavior of basenjis and jackals is quite different. Jackals, like coyotes, are not strongly social animals, and the usual adult group is a mated pair. They are rarely found in packs. Basenjis, on the other hand, run in co-ordinated groups when raised in litters and behave more like pack animals.

Using historical information and such modern history as is included in the *American Kennel Club Dog Book,* we can say that modern breeds have originated in four principal ways. The first of these is the more or less accidental development of local varieties, such as the development of the Labrador retriever in modern times (Smith, 1945). A second way in which breeds have originated is by importing local varieties from distant parts. Thus the first chow was brought from China about 1780 (White, 1842), and the first basenjis were brought to England in 1937. Third, many modern breeds have been deliberately developed by crossbreeding and selection. Among these are the fox terrier, developed to help get foxes out of their dens when the sport of fox hunting became popular, and the golden retriever, whose basic stock reportedly was brought to England by a group of Russian circus performers. The Shetland sheep dogs, or shelties, were originally small nondescript dogs living in the Shetland Islands. These were crossed with the large Scotch collies to improve their appearance and later crossed with some smaller breeds in order to reduce the size again (Coleman, 1943). The sheltie breeders are now selecting for a small dog with the general appearance of a collie. Other newly developed breeds are the Airedale and Doberman pinscher. Finally there are the breeds with known ancient histories, such as the salukis of the Middle East, which the Crusaders probably brought back to England to be the ancestors of English greyhounds and their relatives. Spaniels and hounds also have historical records dating back many centuries and have been subdivided into several modern breeds.

Incidentally, the classification scheme of the American Kennel Club, as well as the somewhat different one of the Kennel Club of Great Britain, is not based on common ancestry, but on similar uses (Scott, 1963b). For example, the non-sporting breeds include such animals of widely different ancestry as the poodles, descended from hunting dogs of western Europe, and the Pekingese, which are toy dogs imported from China. Thus the modern dog fanciers follow Caius in classifying breeds by function rather than origin.

EVIDENCE FROM GENETICS

Beginning with Darwin, students of heredity have wondered how the domestic dog could be so variable, and Darwin himself thought that dogs must have been descended from at least two species in order to account for the variation of modern breeds (Darwin, 1859).

At that time there was very little accurate information about the nature of wolves and none concerning the possibility of variation provided by simple Mendelian genetics. Recent studies show that the wolf is a highly variable species in the wild. Murie (1944) could recognize individuals as being distinctly different even within the same pack. Jolicoeur (1959) finds great variation in the color and skull measurements of Canadian wolves from one locality to another. Young and Goldman (1944) point out how different wolves can be in physical measurements. There were endless local varieties before wolves were exterminated over much of North America, and the northern wolves were much larger than the southern. This means that wolves are naturally highly variable and can be truly called a polymorphic species. The original domesticators of the wolf thus had plenty of variability from which to choose, except in the proportions of the skeleton, which appear to be highly uniform (Hildebrand, 1954).

As we have seen, the wide variation in structure in dog breeds is produced by a relatively small number of mutations such as the bulldog head, short legs and lop ears. In addition to these, there are all the variations in coat color and hair length and texture.

Finally, there is a graded series in body size, from the dwarf through middle-sized dogs to the giant breeds. If we assume that there are five different mutations of body form and perhaps ten of coat color, we could obtain from them a total of 2^{15} combinations of traits. This amounts to 32,668. Even with only 10 mutations 1,024 combinations are possible, far more than the few hundred known varieties of dogs. It is not unreasonable to suppose that as many as 10 to 20 major mutations have occurred and persisted in dogs in the last ten thousand years, especially in view of the tendency of people to be interested in and attempt to preserve new and freakish animals. In short, mutation, selection, and Mendelian genetics will account for the vast amount of variation in dogs without the need to suppose a dual ancestry. As a matter of fact, the differences between species in the genus *Canis* are not the sort which differentiate dog breeds. There are no wild species with lop ears, curly tails, or short legs. A cross with a new wild species would bring in relatively little variation of this type.

A more precise way of obtaining evidence on basic genetic relationships would be to make a detailed study of chromosomes in different dog breeds and wild species. Chromosome structure and number affect the capacity to hybridize. If animals from two species mate, their offspring receives a set of chromosomes from both parents. The embryo may be able to develop even if the two sets do not match each other, but when its germ cells are formed, the chromosomes must pair exactly in the process of meiosis or the germ cells will not develop. Thus in a cross between the horse and the donkey, the hybrid mules are usually sterile. However, once in a long while chromosomes will be lost or accidentally arranged in a compatible order, so that some mules have become parents (Anderson, 1939). Sterility of the F_1 hybrid is thus an indirect indication of chromosome incompatibility and the fact that the parents really belong to two different genetic populations.

In the genus *Canis*, Iljin (1941) crossed dogs and wolves and carried their offspring to the second generation with no indication of sterility. This is the only controlled genetic experiment with hybrids, but according to naturalists' reports (Gray, 1954), dogs have been crossed with both coyotes and jackals to produce fertile F_1 hybrids. The result with jackals is inconsistent with Matthey's (1954) report of chromosome differences, and this report needs to be verified by direct experiment.

While dogs have been crossed with every other member of the genus *Canis*, wolves have not been crossed with either coyotes or jackals, probably because of differences in size. Seitz (1959) made a successful mating between a coyote and jackal but did not determine the fertility of the F₁ hybrid. At this point we cannot eliminate the possibility that at some point dogs have been crossed with a species other than wolves, and that some of the introduced genes have persisted.

TABLE 2.4

HYBRIDS REPORTED BETWEEN SPECIES OF THE GENUS *Canis*

	Dogs	Wolves	Coyotes
Wolves..................	F_2	...	...
Coyotes.................	F_2	...	...
Jackals (*C. aureus*).......	F_2	...	F_1

With the methods of studying chromosomes recently developed in studies of human genetics, it should be possible to make accurate comparative studies of the genus *Canis*. Such information should throw considerable light on the relationships between the dog and its wild relatives as well as those between the dog breeds. However, this must be done on populations rather than on scattered individuals in order to provide worthwhile evidence.

Little information on this point is now available. The dog itself has a very large number of chromosomes, 78 in all, but no one has examined the chromosomes of the wolf. The yellow jackal (*C. aureus*) of Asia and North Africa has only 74 chromosomes (Matthey, 1954), which indicates that it could not be closely related to the dog breeds that have been studied so far. Ahmed (1941) studied the chromosomes and their behavior in Sealyham terriers, spaniels (breed not reported), and a cross between spaniels and Manchester terriers. All the animals had the same number of chromosomes, 78, but Ahmed found indications of differences in appearance and behavior of the chromosomes in the different breeds. These were all breeds from the British Isles, and it is possible that less closely related dogs might show more extreme chromosomal differences.

Other species of Canidae are quite different from the dog (Matthey, 1954). The red fox has 38 chromosomes, and another species of fox (*Vulpes ruppelli*) has 40. The rare "raccoon dog" of Europe has 42, while the fennec, a small foxlike animal of Africa, has 64 chromosomes. Summing up the present evidence, we can only say that there is no necessity to postulate more than one wild ancestral species, and the most likely candidate is a local variety of the wolf, *Canis lupus*.

CONCLUSION

No one can accurately reconstruct the past, and particularly the prehistoric past. The most that we can do is to take the available evidence and draw a probable conclusion. From what we have seen in this chapter, all the evidence points toward the wolf as the most probable ancestor and closest relative of the domesticated dog.

We cannot say exactly where domestication first occurred, for the archeological study of prehistoric human remains and domestic animals is far from complete. However, the oldest authentic skeletons of dogs are found in Denmark, and it seems likely that the first domestic dogs were produced somewhere nearby in Central Europe, or possibly in the "fertile crescent" of Mesopotamia. Once domesticated, they spread slowly over the world. The first dog was a valuable invention for a primitive people, and we can suppose that neighboring tribes heard of these new animals and gradually acquired them.

Present knowledge does not give us an accurate date for when this occurred. The dogs of Denmark have been dated as early as 8000 B.C., but not by modern radiocarbon methods. Even this technique is still in its infancy, and many more facts are needed before we come to final conclusions. Assuming that the above date is correct, it was fifteen hundred years before dogs spread to Mesopotamia. In 6750 B.C., the primitive farming village of Jarmo in Iraq had domesticated goats. At the same site was found a dog-like figurine with a curly tail. In the ruins of the Jericho of 6500 B.C., there were actual dog bones. After another three thousand years, dogs had spread to neighboring Egypt.

From Egypt the dog spread southward and eastward into tropical regions, becoming adapted to existence in a hot, humid climate which their wolf ancestors have never seen. Werth (1944) argues that the basenji, dingo, and certain medium-sized dogs of Papua and the East Indies are descendants of these early southern dogs. They are outwardly similar in appearance, being usually yellow in color with short hair, prick ears, and curly tails. While such superficial similarity is not perfect proof, the hypothesis is a plausible one.

From Central Europe dogs spread northward and got into the hands of hunting tribes. There are no dog remains from the English hunting village of Star Carr dated at 7200 B.C., but dogs eventually were acquired by Eskimos and their eastern cousins who went over the Bering Strait into North America.

We can conclude on the basis of present facts that the domestication of the dog took place about 8000 B.C. This means that the dog has had a domestic history of some ten thousand years, and that some four to eight thousand generations of dogs have lived upon the earth with abundant opportunities for mutations and genetic variations to occur.

As to how domestication first took place, we can only guess. Probably it happened very simply. In Alaska and other northern areas where people still live close to wolves, wolf cubs are often captured even today. Some of them become acceptable as pets or sled dogs. Primitive peoples everywhere in the world frequently adopt young birds or mammals as pets. We can suppose that wolves hung around the primitive agricultural villages of Europe scavenging any waste food or bones that were thrown away, and that the human inhabitants might frequently come across wolf cubs in the spring. Men seem to have domesticated dogs about the same time that they began to live in permanent villages. Contrary to what one might expect, there is no evidence that earlier hunting tribes had dogs (Werth, 1944), and the usefulness of the dog as a hunting animal was probably discovered later.

We can imagine a wolf puppy growing up in a village, fed at first and later existing on scraps. As wolves and dogs still do today, it became adopted into human society and established a territory around its home. Its sensitivity to the approach of strange animals and people at night must have been immediately valuable. Later, when goats were also domesticated, its dog-like descendants could warn their owners of the approach of wild wolves which might attack the herd.

At first there must have been some confusion between tame wolves and wild ones, but very early in domestication a mutation for curly tail must have occurred. Wild and domestic wolves could then be easily told apart. Also, there must have been an early selection for animals which could be more easily controlled, and this meant the development of small or medium-sized wolves with smaller and less dangerous teeth.

Thus the basic traits of domestic dogs were established. As dogs spread from village to village throughout the world, there was an ideal opportunity for what the evolutionists call *adaptive radiation*, a phenomenon which always occurs when a group of animals moves into a vacant habitat (Fig. 2.2). Such conditions are ideal for the survival of variations because whole new populations can arise from a few variable individuals.

As a result, each village had a small population of dogs somewhat different from those of the others. Subdivisions into small populations would be ideal for further genetic change. All this, coupled with the human tendency to select and save animals with individual peculiarities, could account for the origin of different breeds. There are many early historical references to localities famous for different dogs, such as the Molossian hounds of classical antiquity.

At the same time, we cannot conclude that each breed was completely pure and separate from the time of its origin. On the contrary, the promiscuous habits of dogs, as well as the deliberate crosses made by their owners, tended to make all dogs in a given area somewhat related to each other. It is only within the last century that breeders' clubs have attempted to produce pure breeds of dogs. Even so, the history of the modern European breeds shows that many of them were deliberately produced by crossing several breeds together so that while each breed may have certain distinctive characteristics, European dogs are actually interrelated in many ways.

Size prevents large dogs from interbreeding with very small ones, but when we place size measurements on a graph, we see that there is continuous variation from the Chihuahua to the Great Dane, with the possibility of crossing anywhere along the line between animals of similar size.

All this means that dogs belong to only one species, as Linnaeus originally supposed, in spite of their enormous physical and behavioral variability. The only other animal which shows anything like the same degree of variation is man himself. Dogs are dogs and are basically related to wolves. This is true of their behavior as well as their physical structure, as will be seen in the next chapter.

CHAPTER 3

THE SOCIAL BEHAVIOR
OF DOGS AND WOLVES

INTRODUCTION

Anyone who owns a dog is impressed with the extraordinary degree to which such an animal becomes part of a human group—docile, affectionate, and protective. One may well wonder whether the family pet is really descended from a wolf and, even if so, whether selection has not changed his behavior so completely that it is now almost human.

We attempted to answer this question by observing what groups of dogs did when off by themselves with no human beings around them (Scott, 1950). We were fortunate in having plenty of space, and for equipment we had three one-acre fields surrounded by 7-foot wooden fences. The dogs inside only saw people who came close enough to be visible through the cracks, and we made sure this happened as seldom as possible. We built some observation platforms in the neighboring trees and from these vantage points watched what the dogs were doing for over a year. In one pen we put a litter of young fox terriers, in another a litter of Scottish terriers, and in a third a litter of beagles. As the months passed, puppies were born in two of the groups, and we were able to watch the behavior of the dogs and their offspring as they lived under conditions of considerable freedom.

One of the difficulties with studying dogs is that everyone knows dogs. Each of us tried to write down what the dogs did. Being well-trained scientists, we tried to be as accurate and conscientious as possible. Our first notes usually indicated that dog B_1 moved across the field, while dog B_2 moved down the field, with very little other information. We were so familiar with the details of dog behavior, such as tail wagging, sniffing the ground, etc., that we overlooked them entirely. As time went on, however, we began to notice these details and realized that these were precisely the kind of information we wanted. We did not care so much where a dog went as how he did it. And so in subsequent notes we began to describe such details as how a male might approach a female and sniff her tail while she growled at him. To give an actual example: "B₇ again plays with pup. Pup rolls over—cries. B₇ sniffs pup—pup runs—pup wags tail—cries loudly. B₇ playing with it—chasing it. Pup growls at B₇."

Thus we had the beginning of a collection of the kinds of behavior peculiar to dogs. Indeed, this is the first task in studying the behavior of any animal. It is done through observation and is primarily a collector's job, a good deal like collecting specimens of the various kinds of insects or other animals in an area. Instead of making a biological survey of a region, we were making a psychological (or ethological) survey of behavior.

THE UNIT OF BEHAVIOR STUDY IS THE BEHAVIOR PATTERN

As we leaf through the notes we often find such items as: "Dog T₁ scratched." Usually the observer did not have time to describe this in detail, but the phrase conjures up a picture of the dog lifting one hind leg and scratching its shoulder, neck, or any region that it can reach in this way. Anyone who has had a pet dog knows that you can produce this behavior by tickling the dog along its side, setting off a simple reflex which the dog cannot control in any voluntary way. This appears to be a simple and unified piece of behavior having the general function of grooming.

An observant dog owner also knows that this is not the only way in which a dog responds to an itching spot. If it is on his hind leg, he may attack it with his teeth; if it is on his back, he may roll over and over on the ground. Thus we have a list of three alternate ways in which dogs respond to the same general type of stimulus.

In another note we find that "someone walked down the road. B₂ ran to the fence and began to bark." Here is another simple piece of behavior—barking. The effect was to arouse all the other dogs in the group, which rushed over and began to bark also. It looks as if barking has the function of an alarm signal. Another time the Scottish terriers began chasing butterflies, leaping into the air and snapping at them as they hovered over the field. And so we continued to collect a long list of canine habits which make up a dog's repertory of behavior.

In the end, we had an inventory of patterns of behavior (Table 3.1). By "behavior pattern" we mean a unique and independent piece of behavior having a complete adaptive function. It may be a simple reflex like the scratching reflex, or it may be a piece of voluntary behavior such as investigating another dog with the nose.

Behavior is a general and elastic term which can include almost any sort of activity exhibited by an entire individual. A *behavior pattern* means something more definite—a natural unit of behavior which has a function by itself. One test for completeness of function is to attempt to subdivide the behavior. For example, the scratch reflex can be divided into lifting the hind paw and drawing it downward. Lifting the hind paw is a subdivision of behavior, and it may have any one of a dozen different functions as part of different patterns of behavior. As we attempt to reduce behavior to its simplest parts, we know that we are no longer dealing with a behavior pattern when specific function disappears.

Once we have isolated simple patterns of behavior, we can see that they can be combined: either in series, as when a dog goes through the patterns of behavior leading to mating, or (more rarely) simultaneously, by performing two actions at the same time. One test of the independence of a behavior pattern is to see whether or not it can be combined with others in different sequences. In higher animals like dogs there are characteristically a number of basic behavior patterns which can be combined and recombined in a number of ways. We seldom find long sequences of stereotyped behavior such as are common in the lower animals.

Behavioral systems.—Any collector is faced with the problem of classification. If he collects postage stamps, they have to be sorted by country and by year. If it is insect specimens, he sorts them according to their resemblances to one another, placing all similar specimens in the same pile and calling them a species. Behavior patterns can be classified in many ways, but the most basic and important method is by function—the adaptive effect of behavior. When we do this we find that almost all behavior patterns can be sorted into nine groups (Scott, 1950). These are essentially the same as what Krushinskii (1962) has called "general biological forms of behavior."

One basic function of behavior is to provide for the intake of nutritive materials into the body. This includes both solids and liquids. A dog characteristically laps water and liquid foods with his tongue, standing with the tail down. On the other hand, he deals with solid food by biting and chewing. One characteristic pattern is to lie flat, holding his food with his forepaws while he tears off small pieces with his teeth. The patterns of drinking and eating have a common function and are called ingestive behavior.

A very different sort of function is involved in agonistic behavior, i.e., adaptation to a situation of conflict with another animal. The dog's responses may include the patterns of barking, growling, biting, running away, or rolling on the back and yelping. Each of these behavior patterns has a specific function, but they all have a common general function and bear at least a loose relationship to each other. Taken together, they form a *behavioral system*, by which we mean a group of related behavior patterns having a common general function.

These are not the same as the organ systems of physiology, although they may be related to them. Ingestive behavior is closely related to the alimentary system within the body, but, as with all behavior, it involves the nervous system and muscular system as well. A system of behavior is not something which can be dissected out of the body, but rather an attribute of the body as a whole. Each one includes a group of related behavior patterns with underlying physiological reactions in various organ systems. In short, these are functional systems at a higher level of organization—that of behavior. Scientifically speaking, this is the domain of the twin sciences of psychology and ethology.

Instinct.—These concepts of the behavioral pattern and the behavioral system are replacing the older and less exact idea of instinct. People used to say that a dog had an instinct to scratch, or to herd sheep, or to fight. An instinct might thus refer to either a simple behavior pattern or an organized group of patterns or even the unnamed and unknown impulses which caused an animal to act. While it had considerable utility for early biologists, the concept is too inexact to be of much value in modern scientific work in the description of behavior. It employs the same word for either the part, the whole, or the cause of an organized piece of behavior.

In addition, instinct was often used as a final explanation of behavior. At one time, scientists classified behavior as either instinctive or learned and often felt that if behavior could be labeled instinctive they had explained it. In contrast to this older usage, the terms "pattern of behavior" and "behavioral system" do not imply explanations, but are simply names for what we observe. Explaining them is a matter of working out their causes, and we find that it is very seldom that any piece of behavior can be explained on the basis of one simple cause.

The relationship between behavioral systems.—Anyone who observes dogs running in the streets knows that eliminative and sexual behavior are related to each other in these animals. Males are highly stimulated by smelling the urine of a receptive female, and a male about to mate with a female may go through an elaborate pattern of eliminative behavior before doing so. Are we dealing here with one behavioral system or with two? The answer is that in other animal species eliminative behavior may be entirely unrelated to sex, as it is in birds, and hence definitely belongs to a separate system. Even in dogs, most eliminative behavior has little to do with sexual behavior. A behavioral system is a group of behavior patterns which is in most species unrelated to other systems; however, the possibility that a new relationship between behavior patterns can be formed through the evolution of behavior is always present.

Another example of associated behavioral systems in dogs is the relationship between agonistic behavior and eating, since dogs fight with their teeth. All behavior systems are to some extent related to each other by belonging to a common system, the individual organism which does the behaving.

MODIFICATION OF BEHAVIORAL SYSTEMS OF THE WOLF IN VARIOUS BREEDS OF DOGS

The natural social group of the wolf is the pack, which can be as small as a single pair or as large as twenty-five or thirty individuals. For the most part, the groups are small. The wolf pack observed by Murie (1944) had six members—four males and two females. In one year there was only one litter, but in the next both females had young. The pack hunted a range at least fifty miles across, usually going out together in the evening and returning the next morning. The cubs were raised in a den, and the adults guarded it and the immediately surrounding area as a territory.

In our large outdoor fields we established artificial packs of dogs, but since their range was limited by fences, they could not show all the behavior of wild wolves. Sometimes they would wander over the field as individuals, apparently investigating each corner and tuft of grass. At other times they might be excited by a sound from outside the pen and all rush toward it as a pack. No strange dogs could enter, but if a strange person came in, the group would retreat a short distance and run back and forth, barking at the intruder. In a later experiment, King (1954) introduced strange dogs into such a group and saw considerable hostility. The entire field was evidently home territory to the pack, its size being possibly slightly smaller than the home territory defended by Murie's wolf pack.

Except where they are allowed to run wild, domestic dogs rarely form stable packs, although dogs in a neighborhood sometimes join together and run in temporary groups. A house dog usually has its closest social relationships with its owners, so that they correspond to the wolf pack. The den area is the house, and the dogs defend the yard around it as a territory just as do wolves around the den. Most dogs are well fed and have no need to hunt, but will nevertheless make regular journeys away from the house, marking scent posts as they go. Their range is usually much smaller than that of wolves, often not more than a mile or two across, although some dogs move over much greater distances.

Thus dogs in general show the same basic living habits as their wild ancestors. We may now raise the question whether selection and domestication have importantly modified behavior in the various dog breeds. To obtain the answer we may compare the behavior of domestic dogs with that of wolves, chiefly as described by Murie (1944) and Crisler (1958) in Alaska, by Schenkel (1947) in the zoological garden of the city of Basel in Switzerland, and more recently by Ginsburg (1963) in the Brookfield Zoo in Chicago.

Table 3.1 compares dog and wolf behavior, but also includes the behavior patterns of foxes and coyotes when these have been described. Foxes show all the behavior patterns of dogs and wolves, at least in a slightly modified form (Tembrock, 1957), with the significant exception of attitudes of dominance and subordination. It will be recalled that foxes do not run in packs. Both foxes and coyotes show an agonistic pattern not found in dogs or wolves, in which the animal stands with arched back and lowered head, like a spitting cat except that the tail is down.

Sexual behavior.—The courtship and mating behavior of wolves has never been observed in detail in the wild, but observations on captive animals show essentially the same behavior patterns as in dogs. Little sexual behavior takes place except during the period when the female is receptive. The first physiological sign of estrus is a slight bleeding from the vagina, which may appear as early as December. The entire period from this point up to the time when the female ceases to be receptive may be as long as forty-five days,

TABLE 3.1

BEHAVIORAL SYSTEMS AND KNOWN BEHAVIORAL PATTERNS IN THE FAMILY CANIDAE

System and Pattern (I = Infantile only)	Dog	Wolf	Coyote	Fox
Investigative behavior:				
Walking or running with nose to ground, sniffing	F, N	M		T
Head in air, sniffing, may run from side to side	F			T
Sniffing anal and/or genital region	F, N	Y, S		T
Sniffing nose or face	F, N	M, S		T
Head raised, ears erect (listening and looking)	F, N	M		T
Nosing and sniffing urine or feces	F, N	Y, S		T
Crawling forward, moving head from side to side, sniffing (I)	N	M?		T
Epimeletic behavior:				
Shelter building:				
Turning around before lying down	F, N	M		T
Digging bed in dirt		M		T
Digging enlargement of den	F	M, Y		T
Grooming:				
Scratching self	F, N			T
Rubbing against object	F			T
Rubbing or rolling on ground	E		M	
Biting own fur	F, N			T
Shaking self	F			T
Licking own genital or anal region	F, N			T
Licking puppies, chiefly in genital and anal regions (eating excreta)	N	Y		T
Feeding:				
Allowing puppies to nurse	F, N			T
Vomiting food for puppies	N	Y		
Carrying food to puppies		M, Y		T
Food caching (burying food)	N	M, C	M	T
Miscellaneous:				
Carrying puppies to nest	N	Y		T
Pushing puppies with nose	N	Y		T
Whining (possibly warning)	N	Y		
Et-epimeletic behavior (attention getting or care soliciting):				
Whining	F, N	M		
Yelping	N	S		
Tail wagging (special kinds also seen with agonistic behavior)	F, N	M, Y, S		T
Licking face or hands of person, usually with tail wagging	F, N	S, C		T
Touching with paws	N	M		T
Allelomimetic behavior (often combined with et-epimeletic, investigative, and agonistic behavior):				
Walking or running together	F, N	M	M	
Sitting or lying down together	F, N	M		
Getting up together	F, N	M		
Sleeping together	F, N	M		
Howling in unison	N	M, Y, S, C	M*	
Howling, solitary (loneliness)	N	M, Y		T

F = Field; N = Nursery; K = Kennel; E = Elsewhere (where observed in dogs)
M = Murie; Y = Young; S = Schenkel; C = Crisler; T = Tembrock (where reported in other canids)
? = Behavior implied, not directly described.
* Characteristically different from wolves or dogs—higher pitched, interspersed with barking.

TABLE 3.1—Continued

System and Pattern (I = Infantile only)	Dog	Wolf	Coyote	Fox
Agonistic behavior (patterns associated with conflict):				
Fighting and predation:				
Chasing	F	M, Y	M	T
Biting	F	M, Y, C	M	T
Snapping teeth	K	Y, S		
Pawing (sometimes in standing position)	N			T
Snarling (showing teeth)	F, N	M, Y, S		
Growling	F, N	M, Y, C		T
Barking	F, N	M, Y		T
Wagging tip of tail	K	S		T
Tail switching		S		T
Playful fighting, similar to above, but less intense, includes panting	F, N			T
Pouncing or springing		M, S	M	T
Tossing small game into air		M		T
Herding	K	M, Y, C		
Defense and escape reactions:				
Sitting	N	Y		T
Crouching	F, N	M		T
Running away	F, N	M		T
Yelping and showing teeth	N	S		
Tail between legs	F, N	Y, S		
Rolling on back, pawing and extending legs	F, N	M		T
Attitudes of dominance:				
Forepaws on back, growling, tail erect (may bite neck)	F, N	S		
Standing over dog on ground, growling	F			
Standing or walking stiff-legged with tail erect	F, N	S		
Head down, back arched, tail down			M	T
Mounting, tail down, neck biting, without pelvic thrusts				T
Attitudes of subordination:				
Allowing dominant animal to place feet on back, tail erect	N			
Tail down	F, N	S		
Tail between legs, crouching, ears depressed	N	M?, S		
Roll on back, legs extended, tail between legs	N	M, S		
Miscellaneous: hair raising	K	S		
Sexual behavior:				
Male:				
Running with ♀	F	Y		T (follow)
Forepaws extended, body thrown back on haunches, head to one side	F, N	C		
Licking ♀ genitalia	F, N	S		T
Mounting	F, N	Y, S		T
Clasping	F, N	Y?		T
Pelvic thrusts	F, N	Y?		T
Copulatory tie	F, N	Y		T
Female:				
Running with ♂	F	Y		
Forepaws extended, body thrown back on haunches, head to one side	F	M?, C		T

F = Field; N = Nursery; K = Kennel; E = Elsewhere (where observed in dogs)
M = Murie; Y = Young; S = Schenkel; C = Crisler; T = Tembrock (where reported in other canids)
? = Behavior implied, not directly described.
* Characteristically different from wolves or dogs—higher pitched, interspersed with barking.
Note differences from wolves and dogs in dominance-subordination behavior of foxes and coyotes. Information on coyotes is relatively incomplete.

TABLE 3.1—Continued

System and Pattern (I = Infantile only)	Dog	Wolf	Coyote	Fox
Mounting	N	S		T
Clasping	N	M		T
Pelvic thrusts	N			T
Standing for ♂	F, N			T
Tail moved to one side	F	S		T
Both:				
"Wrestling," forepaws around each other's necks	E			
Eliminative behavior (see closely related patterns of investigative behavior):				
Male:				
Micturition with all 4 legs extended (I)	N			T
Micturition with lifting of hind leg, usually in places used by other ♂'s	K	S		T
Female: micturition in squatting position	F, N			T
Both sexes:				
Wandering, nosing ground before defecation	F, N			
Defecation	F, N	Y, S		T
Scratching ground with all 4 feet following defecation (rare in ♀)	K, N	Y		
Defecation and urination in places previously used	F, N	Y		T
Ingestive behavior:				
Lapping, tail out and down	F, N	Y, S		T
Chewing and swallowing, same tail position	N	S	M	T
Gnawing, holding food with paws	K		M	T
Eating grass	F	M		T
Sucking, pushing with head, alternately pushing with forepaws, hind feet pushing, tail out and down (I)	N			T
Comfort-seeking behavior (shelter-seeking):				
Lying in a heap (I)	N	M		T
Lying close together	F	M		T
Curling up	F, N	M		T
Miscellaneous motor activities:				
Twitching while asleep (I)	N			T
Stretching	F, N	M		T
Yawning	F, N	M, C		T
Rolling over	F, N	M	M	T

F = Field; N = Nursery; K = Kennel; E = Elsewhere (where observed in dogs)
M = Murie; Y = Young; S = Schenkel; C = Crisler; T = Tembrock (where reported in other canids)
? = Behavior implied, not directly described.
* Characteristically different from wolves or dogs—higher pitched, interspersed with barking.

or roughly six weeks (Young and Goldman, 1944). A captive wolf described by Murie (1944) had a cycle three weeks long, being receptive only in the third week. Most cubs are born in May, which means that actual mating behavior usually takes place two months previously, in February or March.

The patterns of sexual behavior associated with copulation are relatively simple, beginning with mutual investigation of the genital and anal regions. The female soon stands still, holding her tail to one side, and her vagina may move slightly when touched. The male mounts the female from the rear and attempts to insert his penis while clasping the female with his forepaws. Rapid pelvic thrusts follow insertion, and stimulation of the base of the penis causes rapid enlargement of this region (the bulbus glandis) through engorgement with blood, so that the two animals become locked together. The male then turns around so that the two may stand tail to tail for some minutes while ejaculation goes on. All this behavior is essentially the same as in the dog.

Before the female becomes receptive there may be considerable courtship behavior. A typical pattern involves extending the fore-paws on the ground while keeping the rear legs semi-erect and throwing the head to one side with the tongue out. Both animals may exhibit this pattern of behavior, then one darts to one side, and the other follows. An associated pattern consists of throwing the forelegs around each other's necks in a sort of playful wrestling. According to Crisler (1958), these patterns of behavior are very similar in wolves and dogs, except that the wolves appear more serious and dignified.

The sexual behavior patterns of wolves and dogs are relatively simple, and they are not linked in sequences by any strong degree of genetic organization. Courtship behavior is not necessary for copulation, for a male meeting a receptive female will often mate with very little preliminary behavior. This contrasts with the be-havior of some species of birds in which elaborate courtship is necessary in order to synchronize the later parental care by the two sexes. However, the physiological changes which produce sexual receptivity in the female are such as to produce, under natural conditions, a long continued association between a pair. The odor-ous substance which stimulates the male is produced long before the female is receptive, so that the male stays in close attendance. The general course of mating begins with a period in which the female is attractive but as yet repulses the male. She is apparently also attracted to him in this stage, since she will sometimes initiate courtship behavior, but she rejects the male when he attempts to mount. She finally becomes receptive for a period of days or weeks, so that sexual behavior can be repeated over and over again. This long association and constant interaction would be expected to produce a strong attachment between the pair.

Whether wild wolves are monogamous is still a matter of con-jecture. Certainly dogs are not, under the highly artificial conditions in which they live and are usually bred. At any rate, there is nothing like the situation in a flock of bighorn sheep or herd of elk, in which females come into heat for only a day or so and a single male normally mates with a large number of females. In the wolf pack, all the females are receptive for long periods. There is considerable evidence that each male is quite possessive of one female and may mate only with her. Ginsburg (1963) finds that although wolves in captivity are polygamous, consortships may be formed by either member's ability to keep others away from its mate.

Wolf and dog patterns of sexual behavior are thus quite similar. The principal modification produced by selection is in the seasonal nature of the sexual cycle. In nearly all dog breeds, the females come into heat at approximately six-month periods during any season of the year. Furthermore, the females usually mature in the latter part of the first year, whereas wolves do not come into estrus until the second or even the third year. There has obviously been selection toward early maturity and increased fecundity in domestic dogs. This trait may also be associated with selection for smaller size, as rapid growth usually ceases with sexual maturity.

One breed of dog, the African basenji, still shows a seasonal cycle of sexual behavior. In these animals the majority of females reared in the latitudes of England and the United States come into heat in September and produce puppies in December. The Australian dingo, which is usually assumed to be a domestic dog that went wild, shows similar seasonal breeding in northern latitudes, but in its native Southern Hemisphere it breeds in the corresponding autumn season, which, of course, is March in Australia. The breeding periods of both the basenji and the dingo are obviously controlled by changes in seasonal conditions, and Fuller (1956b) has shown experimentally that artificially shortening the length of day in the spring will produce heat periods in the basenji as early as July.

It is possible that these animals are descended from very early dog breeds which originated before the seasonal habit was lost, and that the time of breeding was modified by selection to correspond to tropical conditions. The breeding cycle of basenjis at the equator is as yet unknown. Here the length of day is constant, but there are seasonal rises in temperature at the equinoxes.

The underlying physiological basis for the sexual cycle of female domestic dogs consists of balanced reactions between the hormones of the pituitary gland in the brain and the hormones of the ovary, or female sex gland. Hormones of the pituitary stimulate the ovaries to produce and expel eggs. At the scar left by each egg, a mass of cells (the corpus luteum) continues to grow and produces progesterone, a hormone which acts chiefly on the uterus to maintain pregnancy, but also suppresses the hormones of the pituitary. The corpus luteum continues to function for about four months, after which it gradually disappears, permitting the pituitary hormones to act again to start a new cycle. In wolves, basenjis, and dingos, the pituitary must be stimulated by changes in the light cycle before it will act, even when progesterone is absent.

To summarize, sexual behavior in dogs has been chiefly modified away from seasonal breeding and toward early sexual maturity, both changes resulting in increased fecundity. While domestic dogs rarely have an opportunity to exhibit all the possible patterns of sexual behavior under the usual conditions in which they are mated, the basic patterns of behavior are the same, with no obvious difference between various breeds.

Eliminative behavior.—As they travel over their hunting range, male wolves regularly visit certain "scent posts," which may be small stones or bushes as well as actual posts or trees. Here they lift the leg to urinate or squat to defecate, and scratch the ground thereafter. Similar behavior is familiar to the owner of any domestic dog. However, we were surprised to find that dogs kept in our large runs and fields with board fences showed eliminative behavior quite rarely compared to house dogs, which seem to spend most of their time going from bush to bush. In the male the typical pattern of micturition is to approach a scent post, smell this carefully, then lift one hind leg and spray a small quantity of urine on the object. Dogs confined in a pen or field that is not entered by other animals rarely do this, and males raised in such an environment may continue to show the puppy pattern of squatting urination well into maturity. The primary stimulus (or releaser) for the leg-raising pattern is apparently the odor of a strange dog's urine in combination with a visual landmark. Dogs in a strange locality which other dogs have not previously visited, will approach and urinate on any objects which are slightly elevated, such as stones, bushes, and tree trunks. Once he has urinated on an object, the dog is inhibited by the smell of his own urine from doing so again until another dog has used the site (von Uexküll and Sarris, 1931).

The pattern of behavior associated with defecation is different and much less frequent. The male squats to defecate near a scent post and may follow this by scratching the ground close by. This scratching never has the effect of covering the feces, as it does in cats, and if it has any function, it is probably to add another visual mark to the site.

Scent also seems important. All dogs, both male and female, have a pair of glands just inside the anus. It is possible that these impart a characteristic odor to the feces. This is in addition to the scent gland described by Seton (1925) on the dorsal side of the base of the tail. At any rate, dogs characteristically sniff fresh feces, and part of the investigatory behavior between dogs is a mutual sniffing of the tail region.

The eliminative behavior of females is quite different from that of males. In house dogs, a female will visit scent posts, but only near her own home, and usually in only one place. She characteristically urinates in a squatting posture, but some females also lift their legs. The female uses much the same defecation posture as the male, but rarely if ever scratches afterward.

In the period of sexual receptivity, the behavior of the female changes. She now wanders more widely than usual, visiting several scent posts and urinating on each. When males subsequently visit these posts they become highly excited, pursue the female, and remain in the vicinity. The householder who owns a female may expect to find half a dozen males regularly camped on his doorstep as soon as the female comes into heat. Apparently all the males within miles become aware of the fact within a short time.

We had expected to have trouble of this sort with our dog colony and were surprised to find that even though several females might be in heat inside our pens, we never had more than casual visits from male house dogs in the neighborhood, and that they made no attempt to stay nearby. This suggests that the males have to come into close contact with the urine of a female in order to be sexually stimulated, and that the scent is not carried on the air for any distance. Once in close contact, males can readily distinguish one type of urine from another. Beach and Gilmore (1949) placed two samples of female urine in an experimental room and allowed one male to enter at a time. Every male spent more time examining that of a female in estrus than the urine from another female.

The principal social function of eliminative behavior in dogs is therefore to bring a receptive female together with a male. It has never been experimentally studied in wolves, but elimination certainly does not function as territorial marking in any strict sense, since wolves make no attempt to defend anything but the area close to their dens. Nor does it seem to have any territorial significance in domestic dogs. As long as the resident male is away from home, strange males will freely enter the yard and mark it; the male returning to the scene shows no signs of looking for an intruder. Aside from breeding, the behavior seems to function simply to keep animals aware that others exist and indicate where they can be found. For wolves living in packs there is ordinarily no reason for one strange wolf to try to find another, and these scent posts are probably useful only to young animals not attached to a pack. Such signs would help them either to find a pack or form a new one with similar isolated individuals.

This is somewhat conjectural, because the behavior has never been fully observed in wild wolves nor has their mating behavior been adequately studied. What happens, for instance, when a female in a pack containing several males comes into heat? Observers often assert that a female mates with only one male, but our only direct evidence comes from observations on captive animals. In any case, under the conditions of a wild pack we would expect that the female would deposit her urine only close enough to attract the males in her own group.

Another peculiarity of eliminative behavior in dogs and wolves is that they never soil their sleeping places. This is, of course, highly adaptive for animals which live in dens, and the development of this trait in young animals will be discussed in a later chapter.

The patterns of eliminative behavior have been even less modified by selection than those of sexual behavior. Although some minor peculiarities may exist, there are no widespread differences between the breeds.

Epimeletic behavior.—This is the giving of care and attention. In dogs such behavior is principally directed toward young puppies, with some self-care. There is almost no epimeletic behavior between one adult and another.

The typical behavior of a female toward her young puppies after she has been away is to walk toward them, nose them, and lie down on her side with her feet toward them. Then she begins nosing and licking each puppy, which stimulates urination and defecation. The results are cleaned up by the mother's tongue, thus keeping the nest clean. In the meantime, aroused by the tactile stimulation, the puppies work their way toward the mother and attempt to nurse. In this, the mother is completely passive, not attempting to place the pups in any particular position. However, if a puppy gets away from the others and begins to whine or yelp, the mother will usually go to it, pick it up carefully with her jaws and carry it back to the rest. The puppy is always carried with its whole body in her mouth, feet dangling down, rather than by the skin as cats carry their young.

This behavior is strongly influenced by the maternal hormone prolactin, which also stimulates the flow of milk. One young basenji mother had her first puppies and dropped them in various places all over the nursery room. She was highly excited but paid no further attention to them even when we brought them all together in the nest box and tried to get her to lie down beside them. Finally we gave her an injection of prolactin. Within an hour she had settled down with the puppies, and peace reigned.

As the puppies grow older these patterns of behavior begin to fade and others take their place. When the puppies are about three weeks of age, the mother begins vomiting food for them. Many mothers will eat this themselves if the puppies do not finish it all. The mother allows the puppies to nurse less often, and they sometimes do this while she is standing. The mothers still clean the puppies if they soil themselves, although by this time the puppies urinate and defecate by themselves outside the nest box.

In the domestic dog, males rarely show any interest in puppies, but in wolves males as well as females vomit food for the cubs. Lois Crisler (1958), who raised and observed two sets of captured wolf cubs in Alaska, reported that her yearling male and female both fed the younger cubs, even though sexually immature themselves. We can conclude that the patterns of care-giving behavior exhibited toward older puppies are common to both sexes and are not produced by hormones.

The principal differences between dogs and wolves in these patterns of behavior are that they are less well developed in dogs. Human owners have largely taken over the care of older puppies, which means that the pattern of vomiting can be weakened without serious consequences. It seems to have been much reduced in males, although this may simply be because male domestic dogs are rarely given the opportunity to develop relationships with young puppies. However, in our limited observations of puppies reared in large fields by both parents, we never saw the male caring for the young.

Another set of patterns of epimeletic behavior is concerned with self-grooming. Dogs will lick themselves in the anal and genital regions and will also lick wounds. They scratch areas which are irritated or attempt to bite them with their teeth. However, there is no elaborate cleaning and grooming such as one sees in cats and mice. Grooming another adult animal is very rare, although one dog will lick an open wound on another's body. These patterns of behavior are essentially the same in dogs and wolves, with no obvious species or breed differences.

Et-epimeletic behavior.—The epimeletic behavior of the mothers is associated with the et-epimeletic behavior of puppies—calling for care and attention. For the most part, this consists of distress calls by young puppies, that is, whines and yelps of different degrees of loudness. Similar noises are made by young wolf cubs, although possibly not so readily.

As the puppies grow older, they will run to the returning mother, wagging their tails rapidly and leaping up to paw and lick her face and breast. The mother frequently vomits food for them on these occasions, and the pattern of behavior of the puppies probably has the function of food begging. This is the same sort of behavior which puppies exhibit toward their human masters, and there may be some tendency to prolong it into adult life in certain breeds which are uncommonly "playful" as adults. This, however, is just an impression and still lacks any objective proof.

Ingestive behavior.—The patterns of behavior associated with taking in solid food and liquids are quite similar in dogs and wolves. Liquids are ingested by lapping. The dog or wolf stands with his tail down and scoops up water or liquid food with his tongue, making considerable noise in the process. Semi-solid foods are managed in a somewhat similar way, the dog seizing part of the food in his teeth, releasing it and lowering his head suddenly to shift the food into the back of his mouth, and gulping it quickly. He deals with bones or tough pieces of meat by lying down, holding the food in his paws, and either tearing off strips with his front teeth or gnawing on the object with his heavy back teeth.

Another pattern is carrying food in the jaws, the animal trotting along with head held high and rolling his eyes in either direction. Wolves, with their stronger jaws and neck muscles, are able to accomplish prodigious feats in this way. A wolf will pick up a piece of bone and meat weighing twenty pounds or more and carry it with little effort. Another common habit of wolves is to bury food around the den. In a cold climate, this preserves the meat from birds and other carrion eaters, but in warm weather the food spoils rapidly. This trait persists in most house dogs, which bury food around the yards of their owners. In our laboratory, some dogs persistently used their noses to bury their food dishes with shavings from the floor.

Being fairly simple, these behavior patterns can be used with many different kinds of food, and may be combined and modified in various ways. Physiologically the dog is primarily adapted for a meat diet and a hunting existence. According to McCay (1946), dogs under ordinary conditions can go for at least a week without food or water and suffer no serious harm. When food is available, they eat rapidly, making little effort to chew their food. They have a large gullet which permits them not only to eat big chunks but to vomit it back easily for the benefit of puppies. The ingestive behavior of dogs and wolves is so organized that the animal eats a great deal when food is available, but is able to go for long periods without it. Consequently, the idea of a "hunger drive" measured by the amount of hours since eating does not apply to the dog. The dog is, in a sense, always hungry, but he is not driven to eat. One of our investigators gave a dog an electric shock when it came near its food dish and then waited to see how long it would take before the dog came back to eat. The dog never came back. After waiting several days for the dog to eat, the experimenter stopped the experiments for fear of harming the animal. This agrees with trappers' reports of wolf behavior. If a wolf gets into a trap and escapes, it is hopeless to use this type of trap again, as the wolf always avoids it.

We found that one of the best ways to use food motivation in an experimental situation was to give a puppy a taste of some special food each day (we used sardines). By the third day, the puppy was highly motivated. Elliot and King (1960) found that, when they fed puppies at regular mealtimes, the animals ate eagerly, even when so grossly overfed that they could not have been physiologically hungry. The "habit of eating," or food reinforcement, is thus a very effective motivating force, even in the absence of physiological hunger. This contrasts with the situation in other animals like rats or sheep, which normally eat continuously for long periods. In these species, a high state of motivation can be produced simply by keeping them from food for a few hours.

There are few breed differences in these basic patterns of behavior, and most of them are connected with food choice. Before scientific information concerning nutrition was available, dog owners fed their charges almost anything—porridge, crusts of bread, or perhaps nothing but bones. There must have been selection for animals which were not fussy about food, and many of today's hunting and working breeds will eat and thrive on food which others will reject. For example, we found no difficulty in getting young puppies to start eating ordinary dog food, except basenjis, although the basenjis readily ate raw meat.

Shelter-seeking behavior.—This is not highly developed in wolves. Wolf dens are usually formed by enlargement of natural caves or of holes started by burrowing animals. The wolf does a little digging to make the area comfortable but makes no effort to build a nest. In cold weather, wolves sleep in a curled position with their noses buried in the long hair of their tails, and they will also sleep near each other for warmth. However, their mechanisms for maintaining heat are much better than in dogs, and wolves have been seen asleep on the snow with legs outstretched (Allen and Mech, 1963; Ginsburg, 1963).

In our large outdoor runs at the Jackson Laboratory we constructed artificial dens (see Fig. 1.3). Litters raised together would sleep together in these enclosures and stay inside when the weather was bad. A well-known pattern of behavior connected with sleeping is the tendency of dogs to turn around several times before lying down. This probably has the adaptive function of feeling with the paws for a smooth area on which to rest.

These fundamental patterns of behavior are not greatly modified in the different dog breeds. Perhaps the most noticeable variation is in the modification of sleeping attitudes in certain breeds. Many dogs will simply lie on their sides with feet outstretched, but some will lie on their bellies with fore and hind feet extended, or sometimes even lie on their backs. These postures probably represent adjustments to anatomical peculiarities such as leg structure or distribution of body hair.

Allelomimetic behavior.—This behavior is defined as doing what the other animals in a group do, with some degree of mutual stimulation. Puppies first do this at about five weeks of age, when the litter begins to run in a group. This foreshadows running in a pack, one of the outstanding characteristics of dog and wolf behavior. To do so, the animals must maintain contact with each other, primarily through vision, but also through hearing and touch.

As well as running, dogs and wolves are found lying down together, getting up together, and even barking and howling in unison. Allelomimetic behavior is also important in predation when a pack makes a combined attack on a large animal.

We had expected to find more of this behavior in breeds which hunt in packs (as, for example, beagles and foxhounds) than in breeds which hunt in a more solitary fashion. However, it turned out that the principal modification of behavior in hounds is not a positive increase of allelomimetic behavior but simply a lessening of agonistic behavior, so that strange animals can run together in the same pack without fighting.

Allelomimetic behavior is thus a basic part of the social life of dogs and wolves. If an animal keeps in constant contact with others of its kind, behavior of this sort will inevitably result. The tendency to stay with others is especially strong in young animals. Adult wolves and dogs will occasionally go off on solitary hunting expeditions, as when a pack will split up to find game. Even then, they usually maintain vocal contact, so that if one animal is successful, the others soon find it. Allelomimetic behavior is useful in hunting, since a group is able to attack large animals more successfully than is an individual, but its primary function is to provide safety. As long as the members of a group keep in constant contact, they can react together in emergencies. When something threatens the den, the whole pack defends it.

Agonistic behavior.—Since wolves are primarily carnivorous animals, a large part of their behavior is concerned with predation. Getting food involves three systems of behavior: investigatory behavior in finding game, agonistic behavior in attacking it, and ingestive behavior in eating it. Under natural conditions, wolves are the chief predators of large hoofed animals: deer, moose, mountain sheep, or caribou. These food preferences are easily transferred to domestic stock, and wolves can be highly destructive to domestic cattle and sheep. Murie (1944) has described how wolves hunt mountain sheep in Alaska, and Lois Crisler (1956) their hunting of caribou. Adult animals in good condition can easily escape from wolves, which are relatively slow runners. The best opportunity for the wolf is to find a young or lame animal away from the rest. In attacking a large animal, a wolf avoids the head and makes quick dashes at the hind legs, springing back if unsuccessful. If an animal is cornered, several wolves may join together in the attack. Usually the prey cannot successfully avoid all of them.

The exact method of hunting depends upon the pack and its usual prey. Allen and Mech (1963; Mech, 1962 and 1963) watched a pack of wolves for several years as it hunted moose on Isle Royale in Lake Superior. This pack was unusually large, containing 15 or 16 members, and always attacked in a group. In an attack the wolves would dart in and back, avoiding the hooves of the moose, then swarming in for a mass assault if the animal attempted to run, clinging to its rump and flanks and eventually its nose. Only one out of every thirteen moose approached was killed, as the pack soon left any animal which fought back vigorously. Out of 68 kills, 20 were calves, and only one of the remaining adults was under six years of age.

Wolves sometimes appear to herd their prey. Since the wolves often separate, a hunted animal may unwittingly come close to one wolf while avoiding another. This pattern of pursuing herd animals is used in the domestic herding dogs, but the sheep or stock dog is not allowed to actually attack. The herd dog must be aggressive enough to chase sheep but timid enough to be inhibited from attacking them by a distant shout or gesture from the herder.

Wolves, of course, do not always have large game available and eat a variety of other foods including berries and carrion. They occasionally show a special pattern of behavior for hunting meadow mice, in which they leap on the mouse with all four feet, pinning it to the ground. This pattern is seen in many wild Canidae, such as coyotes and foxes, but is rarely seen in domestic dogs, which usually pounce without leaping off the ground, as do wolves on most occasions (Murie, 1944).

The patterns of agonistic behavior directed against other wolves are somewhat different from those used against prey, possibly because of the different size, shape, and behavior of wolves as compared to the prey animals. For example, two fighting wolves will show much in-and-out fighting, slamming the adversary with the hips, and then diving for his legs (Ginsburg, 1963); but two dogs may come together head to head, each attempting to get past the snapping jaws of the other and slashing at the nearest available part, usually the neck or shoulder. If one animal gets a good hold on the other's neck, he will usually hang on as the other animal cannot bite him in this position. Meanwhile each tries to force the other to the ground. Much learning and adaptation are involved in fighting, but the basic patterns of behavior are essentially the same in dogs and wolves.

There are several common situations which arouse agonistic behavior in wolves and dogs. One of these is the possession of food. An animal feeding on a bone will growl at any other which comes near, and will sometimes make a short rush and snap, which the intruder easily avoids. We have seen this pattern beginning in young puppies as early as two or three weeks of age, when a ridiculously small puppy will growl over a fresh bone.

Another set of behavior patterns is associated with the intrusion of strangers into the territory near the den. The first reaction is barking, which seems to be primarily a warning signal. The whole pack joins in, and the continuous noise may in itself have an aversive effect. If the intruder keeps advancing, the defending wolf pack will usually first investigate him and then attack. The intruder runs, tail between his legs, and the others rush after, biting at his flanks.

If two strange animals approach each other on neutral territory, each walks slowly and stiff-leggedly toward the other, tail held straight up and waving slightly from side to side. They touch noses and then may cautiously nose each other's tail and genital region. Such behavior may lead to mutual acceptance but more often results in an attack by one animal or the other.

Within a natural social group, agonistic behavior is reduced to a relationship of dominance and subordination. This may take several forms, depending on the degree of dominance. Some dogs simply growl at each other and move apart. More typically, the dominant dog places his feet on the back of the other, growling as he does so, while the subordinate one keeps his head and tail lowered. A still more subordinate animal may roll over on his back while the dominant one stands over him, head to head; the subordinate animal rapidly snaps his teeth and yelps. Sometimes the dominant animal makes a few threatening snaps at the subordinate one. Wolves in the Brookfield Zoo (Ginsburg, 1963) show a pattern of behavior not commonly seen in dogs, where the dominant animal pins the other to the ground with his jaws around the other's throat.

Another pattern is exhibited by a strange animal approaching in a subordinate way. The stranger turns his head away, his eyes closed and ears held back, and attempts to make close contact with the other animal by weaving around him and leaping in the air with the back curved. This behavior is often described as "courting" and indicates a friendly approach.

While the above patterns of behavior can be seen in almost any breed of dog, the frequency of their expression has been highly modified by selection. In the old English sport of bullbaiting, dogs were urged to attack a bull. The bulldog breed was selected for a tendency to attack the nose of the bull and hang on instead of using the slashing attack from the rear preferred by wolves and most dogs. Again, in attacking small animals the usual behavior is to dash in, snap, and withdraw, avoiding any risk of injury. The terrier breeds have been selected for their courage, i.e., the tendency to attack prey and keep on attacking regardless of any injury suffered. This behavior depends in part on the possession of unusually tough and insensitive skin on the neck and shoulders. The same tendency appears in their fights with other dogs, so that fights between terriers often go on to the death.

Other breeds have been selected in the opposite direction. The scent hounds are remarkably peaceable animals, rarely getting into serious fights even among strangers. This trait is useful in managing a large pack and enables them to be kept in groups in a kennel. The hounds also have long, bagging lips which could easily be bitten by the animal himself in the course of fighting. These may be a result of selection for greater powers of scent, the animal tasting as well as scenting the air, but they should also tend to inhibit fighting.

The bird-dog breeds are likewise unusually peaceable animals, having been selected for peaceful coexistence in kennel life. In addition, the setters in medieval times were selected for showing the pattern of crouching behavior useful in hunting birds with a net, rather than that of attack. Retrievers are still selected for a "soft mouth," an inhibited bite such that birds will not be damaged when carried back to the hunter. In the modern pointers, the hunting dog is still not allowed to attack the birds but must stand still when he finds them. In all bird hunting breeds, the dog has to be restrained while at a distance from its master. All these selected traits result in a great reduction of agonistic behavior.

As previously mentioned, the herding dogs have also been selected for their ability to be trained to restrain their attacks. At the same time, making a threatened attack is an essential part of getting the sheep to move. In many parts of the world, herd dogs also guard against large predators, including wolves, so that the older herding breeds were often large and aggressive animals. In this respect they were closely related to the guard dogs once used to protect houses and dwellings. Shepherd dogs still serve this function on many farms.

In ancient times, guard dogs and war dogs were selected for their ferocity as well as size. Mastiffs and Great Danes are modern descendants of such breeds. Their ancestors were used for attacks on thieves and marauders, but in recent times these giant breeds have been selected in the opposite direction, and most of them are unusually gentle.

The patterns of agonistic behavior have thus been subjected to great modifications in the different dog breeds. Compared with wolves they are highly specialized in their choice of patterns of agonistic behavior. Most of this specialization, however, has acted to intensify or diminish the patterns, rather than to alter the basic organization of the behavior. The result is that, although all dogs show similar patterns of fighting when sufficiently aroused, it is easy to incite some breeds to fight and extremely difficult to stimulate others.

Investigatory behavior.—Wolves and dogs are primarily hunting animals. They find their prey by searching for it rather than by waiting for it to come to them, and since they frequently spend most of their days and nights in hunting, they show the patterns of investigatory behavior more frequently than those of any other system.

Wolves are unspecialized animals. They hunt a large variety of game and eat almost anything available when food is scarce. In hunting they use all their senses; eyes, ears, and nose, whichever is appropriate. By contrast, various dog breeds have been selected for their capacity to learn special kinds of hunting. The scent hounds have been selected for their ability to follow a trail, although in actual field work the hounds will use their ears and eyes as well and do a great deal of random exploration. For example, a beagle hunting for a particular object will circle over a great deal of ground with its head held high before it finally strikes a strong air scent, then drop its nose to the ground and go over it inch by inch until it finds the object which it has smelled. Such dogs rarely follow a scent trail step by step but run rapidly along it, weaving back and forth across the trail with the head held moderately high. If they lose the scent, they circle until they pick it up again.

At the opposite extreme are the sight hounds. These long-legged dogs were first bred in the Middle East, where the Arabs still use salukis for hunting gazelles. They are primarily adapted for running after swift prey in open country, where scent is of little importance. A similar breed is the Russian wolfhound, or borzoi, which can actually outrun wolves, tiring them until they can be cornered and killed by a hunter.

The bird dogs use their senses much more equally. Since birds leave few tracks on the ground, the bird dog finds its prey by rapid quartering of the ground, often under direction, and usually locates it by scent when a few paces away. Retrievers have to mark the spot where birds fall, and this requires using the eyes.

In addition to hunting and environmental investigation, there are special patterns of social investigation in dogs and wolves. As Schenkel (1947) points out, there are two areas of particular interest, the head and the tail, although some observers report that wolves pay more attention to the head than do dogs (Crisler, 1958). One of the most prominent patterns is the mutual investigation of the anal and genital regions with the nose. Unlike the varied investigatory behavior connected with hunting, social investigation is very similar in all breeds.

In general, the investigatory behavior of the dog breeds is not strikingly different from that of wolves. The changes have been chiefly produced by emphasizing or diminishing certain patterns and particularly by strengthening or reducing the effect of certain kinds of stimulation. For example, the shepherd breeds seem to be highly stimulated by the smell of sheep or even deer, and they occasionally become sheep killers or deer hunters. Likewise, the bird dogs are highly stimulated by birds and sometimes become chicken killers. By contrast, many terriers have little interest in scent. We put a live mouse in a one-acre field and then let in a group of beagles. They found it in less than a minute. It took a group of fox terriers a quarter of an hour to accomplish the same task, and Scottish terriers were never successful at all. At one point, one of them actually stepped on the mouse without noticing it. How these differences in behavior were brought about is not known exactly. From their actions, fox terriers appear to be not so much deficient in the capacity for scent as simply uninterested. They are, however, highly stimulated by sounds.

Conclusion.—Behavioral patterns in the dog and wolf are essentially the same. Selection has particularly modified the agonistic and investigatory systems of behavior and to some extent the sexual system. These modifications are usually quantitative rather than qualitative, and most of them involve the diminution or exaggeration of an existing pattern without creating anything essentially new.

The nine behavioral systems of dogs and wolves are related to each other in characteristic ways. Sexual and eliminative behavior are associated, the latter assisting in the location of mates. Likewise, in these predatory animals, investigative, agonistic, and ingestive behavior are closely associated with each other. The last three can be performed in unison as well as individually, so that allelomimetic behavior is often also associated with them. Thus we have a picture of the basic behavioral organization of dog and wolf. Essentially, this is a systematic and objective way of describing what is commonly called "personality."

COMPARISON WITH HUMAN BEINGS

We can use the major systems of social behavior as an outline for comparing human and dog behavior patterns, for it is obvious that all nine behavioral systems are also well developed in human beings. When we do this, we immediately find that the detailed patterns of behavior are very different. After all, the dog is a four-footed animal with a well-developed tail and no hands. Particular behavior patterns, therefore, are bound to be different. No human being can wag his nonexistent tail, and no dog can pick up things with his paws. Human beings and dogs are basically different in anatomy, physiology, and behavior. At the same time, social behavior patterns are similar enough so that many of them are mutually recognizable and each species can give appropriate responses to the other's behavior in many situations.

In attempting to ascertain the possible behavior of prehistoric man, much has been made of the fact that many primitive human societies still existing in historic times get their food by a sort of pack hunting (Etkin, 1954). This is different from dog behavior in that it is done entirely by males, the females and infants making their contribution either by food gathering or by helping prepare meat once it has been obtained. In dogs and wolves both sexes take part. Dogs readily join in human pack hunting without any particular training, and even pet dogs will join in a fight between two children. This last is a case where the dog recognizes the human form of agonistic behavior. Likewise, it is easy for humans to recognize the function of canine growls and bites.

Other patterns of behavior are not so easy to understand. When a dog jumps up with extended paws and wagging tail, most adults recognize this as "friendly" behavior, but a small child may be frightened by it. Even an adult can be misled. When a dog advances slowly with tail held stiffly erect and wagging slowly from side to side, an inexperienced person may conclude that the dog is trying to be friendly. Only close observation of such behavior between dogs reveals that this latter pattern usually precedes a fight.

One general characteristic of behavior which makes dogs highly adaptable as domestic animals is the tendency to treat human beings as though they were fellow members of a pack, even if the "pack" is reduced to one other member. With this goes allelomimetic behavior and the tendency to join in group attacks. A second is the dog's tendency to use the human home as a den and defend it against strangers. The latter behavior is almost universal in humans, and may go back to primitive life when the only commonly available shelters were caves.

We can see that there are certain basic similarities between dog and human behavior patterns and systems, and we may now consider the problem of whether there are resemblances in the genetic systems which underlie these. In human prehistory, and indeed in much of historical time, the majority of human societies were tribal villages, each consisting of a few hundred individuals, and each virtually isolated from others by distance, language, and social organization regulating marriage. This kind of population is ideal for genetic change, and it may account for the rapid evolution of the human capacity for language. The beginnings of urban civilization, some 10,000 years ago, marked the beginning of the end of these small isolated groups, and modern human populations with their enormous numbers, great mobility, and increasing tendency toward crossbreeding, represent a situation designed to produce relative genetic stability. Dogs first became domesticated about the time urban civilization began, but for centuries afterward each town and village had its own group of dogs, somewhat cut off from others by the same sort of behavioral mechanisms as their human masters, but with populations even smaller and generations turning over much more rapidly.

This was an ideal situation for genetic change, with each village tending to produce its own variety of dog. In addition to this, there was the factor of human selection for what were considered attractive, useful, or fashionable traits of appearance and behavior. Some of the things which were selected were obviously non-adaptive to an animal living under natural conditions, such as bulldog heads and short, thin hair. Such selection made genetic change much more rapid in dogs than in human populations. Under modern conditions with scientific methods of selection and artificial isolation of breeds, such changes can proceed even more rapidly.

As we have seen previously, the diversity of physical form produced by these methods is very great, much greater than the diversity in behavior patterns. All dog breeds show the same fundamental general patterns of behavior. This means that behavioral organization is relatively resistant to genetic change.

The dog breeds have sometimes been compared to human races. They are, however, basically different in that the human races have never been subjected to controlled systematic selection in favor of particular kinds of individuals. There are a few physical characteristics which are adaptive to the climatic conditions in which the races were originally found. For example, a person with a dark skin and tightly curled hair can control his body temperature better under hot, humid conditions than can a person with light skin and straight hair (Baker, 1958), but many other physical differences seem to be accidental. Both the tallest and shortest of human beings are found in equatorial Africa.

One would expect that population differences in behavior patterns would be even less marked, and this indeed seems to be the case. Basic human behavior patterns are mutually recognizable between all human races, even those which are most diverse physically. In short, we would expect that behavioral differences between human races would be much less than those between dog breeds. In other words, this is not a useful type of comparison. The dog breeds do not correspond to human races.

On the other hand, in any human population, even from a supposedly pure race, there is an enormous amount of individual variability, both in form and behavior. Here the concept of polymorphism is a useful one. It was originally applied to cases such as certain butterflies in which two different color varieties existed in the same species. The human race is obviously polymorphic with regard to the two sexes, and between mature and immature individuals. In addition, there are all sorts of differences between individuals of the same sex and age. In a human social group, it is an obvious advantage to have individuals of different abilities and skills.

When we look at wolves, the wild ancestors of dogs, we see that they are also polymorphic in that they show a great deal of variation in size and color (Jolicoeur, 1959). We can also see that it would be theoretically advantageous for a wolf pack to be behaviorally polymorphic: to have one member which would be highly timid and react to the slightest suspicion of danger and thus keep the pack on the alert, and another which might be bold enough to go in and obtain food when danger was slight and the need for food great.

What selection has done is to take this individual variability of wolves and accentuate it in the dog breeds. We can think of each breed as representing one of many possible individual behavioral variations.

This is the most useful comparison we can make. A dog breed represents a large group of genetically similar but not replicated individuals. Human families are also groups of genetically similar individuals but are unlike dog breeds both in their small size and in the fact that outbreeding is enforced rather than prevented. In short, we can learn relatively little about the differences between human populations through a study of the dog breeds, but a great deal about the possibilities of *individual* variation of human behavior.

CHAPTER 4

THE DEVELOPMENT OF BEHAVIOR

Since the basic behavior patterns of dogs are so similar to those of wolves, we wondered how soon the differences would appear. Would young puppies be like little wolves, or would they show doglike characteristics from the very first? Again, how soon would breed characteristics appear? Would all young puppies be essentially alike, or would they begin developing in a different fashion from birth? These questions are related to the important practical problem of the prediction of later performance from early behavior. We expected that by watching animals as they grew, we would see the interaction of hereditary and environmental factors as they molded behavior, with the process of learning producing a more and more important effect on behavior as the animal grew older. We realized many of these expectations from our observations, but the results also turned up some fascinating facts which we had not foreseen.

In order to get our information on development, we began systematic daily observations of each litter, as described in Chapter 1. The result was some ninety-six pages of notes on every litter raised, and we analyzed these according to the age of the puppies. As we watched the animals from day to day, they hardly seemed to change. But when we began to assemble our notes and observations, we saw that there were certain times when the behavior of a puppy would change overnight and that development fell into distinct natural periods, each with its own characteristic behavior.

THE NEONATAL PERIOD

Social behavior.—When we take a puppy two or three days old away from its mother and place it on the floor a short distance away, it begins to crawl slowly, throwing its head from side to side and whining or yelping as it goes. This vocalization is, of course, care soliciting or et-epimeletic behavior, and usually attracts the attention of the mother. If she does not respond, the puppy keeps on crawling. It does not orient itself toward the mother and may go in a circle after moving a few inches. However, if its swinging head touches the mother or one of the other pups, it stops and crawls toward them.

We have here an infantile pattern of investigative behavior which seems to be based entirely on the sense of touch. Comparing this with adult behavior, we notice two things: each movement is extremely slow, and the behavior itself is quite inefficient, depending chiefly on a process of trial and error. It works well enough if the puppies are confined to a nest box or some den-like enclosure, but if they are kept in a large room, a puppy separated from the rest is likely to move in any direction and end up far away from the mother. Of course, maternal behavior will ordinarily take care of the situation, as most mothers will pick up young puppies and return them to the nest.

A second sort of situation occurs when the mother has left the puppies for a time. The puppies are heaped together in a ball in the center of the nest, resting quietly. The mother comes toward them, lies down beside them, and starts poking them with her nose, turning them over and licking the underparts of their bodies. This stimulates the puppies to urinate and defecate, and the mother keeps on licking them until they are completely clean. For these young animals, eliminative behavior is a reflex stimulated by anything similar to the mother's tongue. Puppies raised away from their mothers get into serious trouble unless they are properly stimulated with a warm wet towel. The net result of these patterns of behavior by mother and offspring is that the nest area is always entirely clean, showing no trace of urine or feces.

Meanwhile, the puppies have begun to crawl; and as they do so, they come into contact with the mother's breast and begin sucking movements. In the most complete form of this pattern of behavior, the puppy pushes on the mother's breast with alternate forepaws and occasionally pulls back with its head, bracing with its forefeet and pushing with its hind feet. This activity probably stimulates lactation, and it also disturbs any puppies which have not been awakened, so that they also push forward to eat. This is the infantile form of ingestive behavior.

Newborn puppies actually show very little activity other than these simple patterns of et-epimeletic, investigatory, eliminative, and ingestive behavior. They react to pain, cold, or hunger with the same limited repertory, yelping and moving at random. We see that all behavior at this age is adapted to infantile life and that the characteristic patterns of adult behavior are completely missing. In fact, if one had only behavior to go by, one might assume that the neonatal puppy belonged to an entirely different species from adult dogs.

Sensory capacities.—These observations stimulated us to study the basic capacities underlying the development of behavior. The first of these was obviously the sense organs. Most people know that puppies are born with the eyes closed, which means that they cannot see anything, in the usual sense of the word. Some puppies react to a very strong light, particularly those with light skin pigment and whose eyelids are therefore more transparent.

When we examine a newborn puppy closely, we find that its ears are also tightly shut, so that there is no external opening through which sound can enter. When we test it with sudden loud noises, there is no reaction to either high or low tones. The newborn puppy appears to be completely deaf, in spite of the fact that it makes considerable noise itself. It probably does not even hear its own yelps. In keeping with this, we have never observed mothers calling or vocalizing to their young puppies.

Since adult dogs make so much use of their noses, we might expect that the sense of smell would develop early. However, puppies cannot locate their own mothers by odor from any distance (James, 1952), although Troshikhin (1955) got responses to odorous substances from a few centimeters. The neurologist P. J. Harman examined the brains of some of our newborn puppies for myelination and concluded that the olfactory nerves and region of the brain connected to them were so undeveloped as to make it unlikely that the true sense of smell had any important function. We tested this with a commercial compound developed to repel dogs from furniture and bushes. The active ingredient is related to oil of citronella and is almost odorless to most human beings although there is a slightly nauseating aftertaste in the throat. It is possible therefore that this substance primarily affects the taste buds rather than the nose. Newborn puppies react to it by a characteristic withdrawal reflex, drawing back the head as far as possible.

The sense of taste is obviously present. Newborn puppies will lick a glass rod smeared with fish or meat juice or milk but reject a bitter substance such as quinine. They will suck on any smooth warm object such as the human finger but keep this up only if milk is forth-coming.

As to the other sensory capacities, the puppies react strongly to cold (Welker, 1959) and pain. Given a choice between a warm heating pad and a cold one next to it, they come to rest on the warmer one. They also react negatively to extreme heat but have poor temperature control, so that one must take care, in keeping young puppies warm by artificial means, not to raise their body temperature too high and kill them. Another well-developed sense is that of balance, or response to gravity. Turned on its back, the puppy immediately struggles to turn over, and if it crawls to the edge of a table so that part of its body is unsupported, it yelps in distress. As shown by responses to the mother, reactions to touch are well developed.

Compared with an adult, the neonatal puppy is greatly deficient in sensory capacities, being in fact most deficient in those senses which are most important to an adult dog, i.e., hearing, sight, and smell. From a sensory viewpoint, the young puppy is primarily a tactile animal, responding to touch, pain, and cold. Even in his chemical senses, he is largely limited to taste, which is effective only on direct contact. The puppy is in touch with only that part of his environment which actually touches him.

Motor capacities.—The motor capacities of the newborn pup are likewise limited. Its only method of locomotion is a slow crawl, the action of the front legs being better developed than that of the hind ones. There is no tail wagging, and sucking and licking are the only oral activities other than vocalization. The latter consists chiefly of distress vocalization, a series of rapid whines or yelps. Neonatal puppies frequently make these noises at a rate faster than one per second while being weighed (Fig. 4.1). The puppy is thus quite helpless at birth, although more advanced than newborn human babies in that it is at least able to crawl.

Capacities for organization of behavior.—Early in our observations of newborn puppies, we noticed that they did not seem to learn by experience. A puppy would crawl to the edge of the scale platform, fall off, and begin to yelp in distress. When placed in the middle of the platform, it would do the same thing over again. Likewise, the only change we saw in sucking behavior was that the puppies began to nurse more strongly and efficiently after a week or so, which could have been caused simply by development of the muscular system. Some recent experiments (Stanley *et al.*, 1963) indicate that the puppy is capable of some degree of slow learning with regard to sucking. A puppy which is given milk after sucking a rubber nipple will eventually begin to suck more often than a puppy which is not so rewarded, and one which is given quinine instead of milk will eventually refuse to suck the nipple at all.

Observing behavior is not the only way of inferring the function of the central nervous system. Another method is to make a microscopic examination of brain tissue. Nerve fibers of young animals frequently lack the myelin sheath characteristic of adult tissue. This sheath is an outer layer of fatlike material associated with speedy transmission of stimuli. The fact that a nerve fiber is unmyelinated does not mean that it cannot function but only that it functions more slowly. Myelinated fibers in adults transmit stimuli 50 to 100 times faster than the unmyelinated fibers of the sympathetic nerves.

Harman's (1958) investigation of the brain of newborn puppies shows that the only areas which are well myelinated are those connected with the trigeminal nerve—which goes to the mouth and includes the sensory nerves of taste as well as the motor fibers to the jaw muscles—and the non-acoustic portion of the auditory nerve, which is the part connected with the organs of balance. In the cerebral cortex, the convolutions are simple and the underlying fibers almost totally unmyelinated. The development of myelination therefore seems to be correlated with the development of function. We would expect that stimuli passing through the unmyelinated nerves of puppies would move slowly, and can conclude that the slow actions and delayed response times of newborn puppies result from this.

The undeveloped nature of the brain, sense organs, and motor organs all result in a greatly reduced capacity for learning compared with that of older puppies. The only sense organs through which stimuli could be associated are those of touch and taste. These are likewise the only ones which could be used for making discriminations. The number of responses is so limited that there are only a few activities which could possibly be affected by learning: sucking, crawling, yelping, and elimination. The response time is often so slow, occurring seconds after stimulation, that a puppy might have considerable difficulty in associating a response with a given stimulus. We must conclude that the capacity for learning in these newborn animals, if it exists at all, is quite limited and possibly of a different nature from that in the adult.

Summary.—The social behavior patterns of the newborn pup are limited to those connected with neonatal existence and are quite different from those found in adult animals. The entire neonatal period is primarily devoted to one function, that of obtaining nutrition by nursing. Supplementing the behavior of the puppies is that of the mothers, who rarely leave them even for a few minutes during this time. The neonatal puppy is not a self-sufficient organism. Its temperature fluctuates with the environment, and it needs the warmth of the other pups and the mother's body. It even needs stimulation in order to feed properly, as newborn pups which are kept warm will lie quietly for hours, only attempting to nurse when stimulated by the mother.

An observer of neonatal existence is strongly impressed by the effective way in which the young puppy is normally shielded from the effects of the external environment, both by maternal care and its own limited sensory, motor, and intellectual capacities. The puppy can be greatly disturbed physiologically by adverse conditions, but there are few ways in which it can be affected psychologically.

THE TRANSITION PERIOD

During the entire neonatal period the puppy grows in size and strength but still retains the same patterns of behavior as at birth. The first change in behavior occurs after the eyes open, when the puppy for the first time crawls backward as well as forward. Other changes follow in rapid succession, so that the puppy undergoes a behavioral metamorphosis which is almost as spectacular as the metamorphosis of form from a tadpole to a frog. Behavioral patterns adapted for neonatal life are decreased or abandoned, and the characteristic patterns of adult behavior begin to appear. The period is one of transition from neonatal to a more adult form of existence. This process begins with the complete opening of the eyes, occurring on the average at 13 days of age, with much breed and individual variation (Table 4.1).

Changes in social behavior patterns.—The neonatal puppy is characteristically difficult to feed by hand. It sucks only halfheartedly on a baby's nursing bottle, apparently because it is difficult to duplicate the exact primary stimuli or releasers which trigger the nursing pattern. It can be fed a few drops at a time with an eye dropper, the puppy making futile sucking movements. It does not lap milk, and the easiest way to feed a neonatal puppy is to insert a small stomach tube and inject milk directly into the stomach.

By contrast, the puppy at 2 weeks of age is highly adaptable. It will readily nurse from a bottle and can even lap up milk or soft food from a dish, albeit in a clumsy fashion, usually plastering itself

TABLE 4.1

Time of Complete Opening of the Eye

	Number Tested	Per Cent Completely Open at		
		1 week	2 weeks	3 weeks
Pure Breeds:				
Basenji................	43	0	65	100
Beagle................	49	0	94	100
Cocker spaniel..........	51	2	94	100
Sheltie................	25	0	31	100
Fox terrier.............	27	0	11	100
Total...............	195	0.5	67(59*)	100
Hybrids:				
BCS F$_1$..............	24	0	42	100
CSB F$_1$..............	29	0	83†	100
BCS × CS.............	30	0	90	100
CSB × BA.............	44	0	86	100
BCS F$_2$..............	32	0	63	100
CSB F$_2$..............	42	2	98	100
Total...............	201	0.5	79(77*)	100

* All breeds (or hybrids) weighted equally.
† Includes one retarded litter.
Opening of the eye occurs early in beagles and cocker spaniels, more slowly in basenjis, and still more slowly in shelties and fox terriers. In the hybrids, BCS F₁'s and F₂'s are relatively slow. When all hybrids are taken together, the average is intermediate between the two parent breeds.

with food and occasionally choking. By 3 weeks of age the puppy is able to stand and drink milk or eat in a fairly efficient fashion. However, its first teeth have only just begun to come through, and it is still incapable of doing any effective chewing. When her puppies are about 3 weeks of age, the mother begins to vomit warm, semiliquid food from her stomach, this being the natural supplemental food at this age.

We have extensive evidence, obtained by sudden weaning at different ages (Scott, Ross, *et al.*, 1959), on the ability of young puppies in the transition period to eat in ways other than nursing. Artificial bitches' milk and powdered dry dog food was available at all times to the puppies, and we helped them to find the dish when they were too young to locate it by themselves. Before and after weaning, we tested them twice a day by sticking a finger into their mouths and recording the amount and force of sucking. Until 19 or 20 days of age, the puppies would suck fingers readily, particularly when hungry. After this age, there was no difference between experimentals and controls left with the mother, and by 4 weeks of age all the puppies had quit finger-sucking entirely. They might gently chew, but the behavior of sucking on objects which did not produce milk had disappeared. All of this evidence means that the only pattern of ingestive behavior available to the newborn puppy is sucking. Alternate patterns of lapping and chewing begin to appear in the transition period, and the puppy is capable of abandoning sucking at any time thereafter, providing the proper sort of food is available.

As part of the same experiment, part of a litter was temporarily removed from the mother and placed in the other half of the same room, separated by a wire barrier through which the animals could see but not touch the others. We left them for several hours, and then came back to observe. During the first week of life, they might be found anywhere in their half of the room and often separated from each other. During the second week, we usually found them together in the same spot where we had left them. In the third week, after their eyes were open, we always found puppies close to the fence, indicating that they were now able to orient themselves in the room. The development of the eyes is thus accompanied by a change in the pattern of investigative behavior at the start of the transition period. However, the most striking change in investigative behavior occurs later, at about 3 weeks of age, when the puppies begin to respond to people or other animals at a distance. All this evidence indicates that the early pattern of investigative behavior based on touch is giving way to adult patterns of behavior employing other sense organs.

Et-epimeletic behavior does not disappear, but some of the situations which once evoked it are no longer effective. For example, the amount of vocalization while the puppies are being weighed drops to a low level by 2 weeks of age, and almost disappears after 3 weeks (Fig. 4.1). At the same time, they begin to wander around and yelp when moved to a strange place, even though warm and comfortable. Until this age, they would be quiet anywhere as long as they were not cold or hungry.

Eliminative behavior also begins to change toward the adult pattern (Ross, 1950). By 3 weeks the puppies are beginning to walk outside the nest to urinate and defecate. They no longer require stimulation by the mother, although they still have not begun to use specific spots. Most mothers continue to lick and clean their puppies, but elimination is no longer dependent on this stimulation.

Agonistic behavior also begins to appear. Some puppies will growl as early as 2 weeks of age if given a bone; and after 3 weeks, playful fighting with other puppies becomes increasingly common. The puppies paw and mouth each other clumsily, a pattern which will eventually become more and more like the fighting of adults. Before 2 weeks there is very little indication of escape behavior. The first indication is backward movement in reaction to visual stimulation, which we can think of as a visual startle reaction. Later, at 18 to 20 days on the average, the puppies first begin to show a startle response to sound.

One of the neonatal patterns, shelter-seeking, changes relatively little except that the puppies are more likely to be found separated

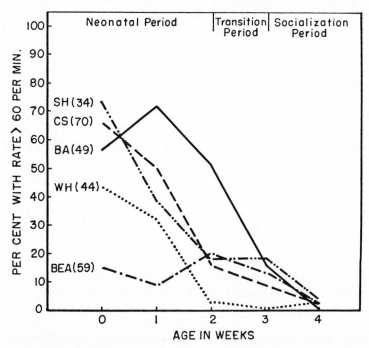

Fig. 4.1.—Decline of the rate of distress vocalization during the first 4 weeks of life. Note breed differences. These reactions were obtained while the pups were weighed and probably measure discomfort resulting from contact with the cold metal scale platform.

from each other when the room is warm. This probably reflects an increasing ability to maintain their own body temperatures.

At 3 weeks of age there are still three important types of adult behavior which are completely missing: sexual, allelomimetic, and epimeletic. The first two will appear in the next few weeks, but the third will be confined to self-grooming until the animals become adults.

This is the general picture of change in social behavior patterns. The puppy at 3 weeks is a far different animal from a week earlier.

Since such important changes occur, their timing becomes quite an important matter for understanding behavioral development, and one of the best ways to approach this problem is to examine changes in the underlying basic capacities of the animal.

Sense organs.—As soon as the eyes open we can demonstrate that they are functional by shining a bright light into them and observing the contraction of the pupils. However, in some puppies this is slow, and the pupil keeps fluctuating, indicating that the mechanism is not completely developed. The puppies will show nystagmus, which means that if we hold a puppy's head and body firmly and swing the whole animal in a horizontal arc, his eyes will flick back and forth as if he were fixing his eyes on an object and following it as it moves past. However, this may mean only that a reflex involving the semicircular canals of the ear is functioning and not that the puppy is responding to sight. A histological examination of the eyes shows that the retina at two weeks of age is quite undeveloped and, in fact, is not completely formed until approximately 4 weeks of age, indicating that complete visual function has not yet been achieved during the transition period (Blume, 1956; Parry, 1953).

Another indirect line of evidence is provided by the alpha rhythm of the brain waves. This electrical activity of the brain is associated with the development of visual function and indicates activity in that part of the cerebral cortex associated with vision. Charles and Fuller (1956) measured the electrical output of the brain in developing puppies, using surface electrodes. The newborn show almost no brain waves and no difference between the sleeping and waking states. Then at approximately 3 weeks of age, the electroencephalographic picture changes radically. The brain waves increase in amplitude, and an EEG taken in sleep becomes different from that recorded during the waking state. However, the EEG does not show its adult form until 7 or 8 weeks of age, when the nervous system is probably like that of an adult as far as vision is concerned.

Still another way of measuring the development of perception is the "visual cliff" test of Walk and Gibson (1961). The apparatus consists of a 6-inch wide board laid across a piece of plate glass. On one side of the board the glass is laid directly over a sheet of checkered cloth; on the other side the cloth falls away, producing the illusion of a drop-off from the board to the floor below. Disregarding the glass, one side appears deep and the other shallow. Mr. Frank Clark tested puppies on this apparatus from the time of eye opening until they showed an aversion to the "deep" side. Until they were about 30 days old, puppies stepped from the board indiscriminately onto either side. After this age, most puppies uniformly went to the shallow side in spite of previous safe landings on the deep one.

Summarizing this evidence, we can conclude that while the puppy responds to light as soon as the eyes open and sometimes before, it is probably not fully capable of observing form until about 4 weeks of age, and that complete visual capacity is not present until about 8 weeks.

The startle response to sound, which is the first indication that the sense of hearing has been developed, appears at 19.5 days on the average (Table 4.2). Whether the sense of hearing is completely developed at this point we do not know, but the puppies seem to respond to any loud noise, whether it is a relatively high-pitched sound of the Galton whistle or a low-pitched noise from any other source. The startle reaction to sound is a definite and easily recognized reaction, and since it is closely associated with a number of other changes, it can be taken as the best measure of the end of the transition period.

Changes in motor capacities.—During the transition period, the

TABLE 4.2

TIME OF FIRST FUNCTION OF THE EAR: ANIMALS GIVING A STARTLE RESPONSE TO SOUND

	NUMBER TESTED	PER CENT RESPONDING AT		
		2 weeks	3 weeks	4 weeks
Pure Breeds:				
Basenji................	43	0	72	100
Beagle................	49	0	84	100
Cocker spaniel.........	57	2	61	100
Sheltie................	24	0	62	100
Fox terrier............	27	4	92	100
Total..............	200	1	84(74*)	100
Hybrids:				
BCS F_1................	24	0	88	100
CSB F_1................	29	7	97†	100
BCS × CS.............	30	0	93	100
CSB × BA.............	44	5	95	100
BCS F_2................	32	6	78	97
CSB F_2................	42	7	88	100
Total..............	201	4	90(90*)	99.5

* All breeds (or hybrids) weighted equally.
† Includes one retarded litter.
The development of this capacity shows much less variability between breeds than the opening of the eye or eruption of the teeth. Thus the startle response can be taken as an excellent developmental marker, closely associated with the beginning of the period of socialization. Contrary to other indicators, fox terriers are more advanced in this respect than other breeds. The hybrids develop faster than either parent strain, but there is still a tendency for the BCS group of hybrids to fall behind, as in other measures. This difference between the results of the reciprocal crosses might be attributed to either a maternal effect or accidental selection of slow and rapidly developing parents.

puppy begins to get up on his feet and walk instead of crawl (Fig. 4.2). The most accurate objective evidence which we have on the onset of this ability is the record of the posture of puppies when weighed each week (Fig. 4.3). At birth and for the next two weekly weighings, almost 100 per cent of the puppies lie flat when placed on the scales. At 3 weeks, 50 per cent of them adopt some other posture; and by 4 weeks, very few of the puppies still lie flat. This does not mean, of course, that they cannot stand, as observation shows that all the puppies can walk by 3 weeks of age, but only that

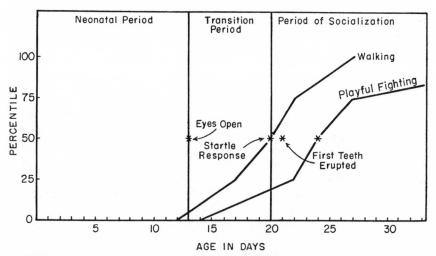

Fig. 4.2.—Development of walking and playful fighting in relation to the opening of the eyes, startle response to sound, and eruption of first teeth. The graphs represent cumulative figures of first occurrences in animals observed in 10-minute daily periods. The zero point is the day before the first animal was observed walking or showing playful fighting.

some of them still react by lying down when placed in a strange situation.

Another important change in motor capacities arrives with the eruption of the first teeth (Table 4.3). In puppies these are the upper canines, which can first be felt through the gums at approximately 20 days of age on the average. Along with this the puppies begin to bite and chew as well as suck. All movements are still slow and clumsy compared with those of adults, but the puppies can at least chew and move in the adult fashion. Another motor capacity which appears as one of the new patterns of social behavior in this period is tail wagging.

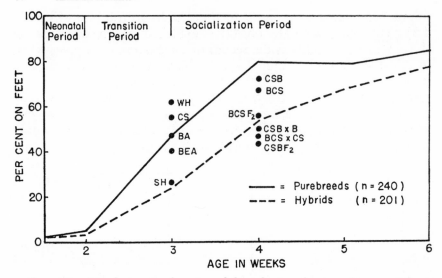

Fig. 4.3.—Development of motor ability during the transition period, as shown by the percentage of animals which stayed on their feet in some posture (standing, crouching, or sitting) when first placed on the scale to be weighed. Note that hybrids developed more slowly than pure breeds, possibly because larger and heavier.

TABLE 4.3

Time of Eruption of Upper Canine Teeth

	Number Tested	Per Cent Erupted at		
		2 weeks	3 weeks	4 weeks
Pure Breeds:				
Basenji................	51	0	79	100
Beagle................	54	0	74	100
Cocker spaniel.........	67	0	22	100
Sheltie................	30	0	30	100
Fox terrier............	31	0	14	89
Total................	233	0	47(44*)	99
Hybrids:				
BCS F_1................	24	0	63	100
CSB F_1................	29	0	59†	97†
BCS × CS............	30	0	63	100
CSB × BA............	44	0	75	100
BCS F_2................	32	0	53	100
CSB F_2................	42	0	86	100
Total................	201	0	68(67*)	99

* All breeds (or hybrids) weighted equally.
† Includes one retarded litter.
 The pure breeds vary widely in this characteristic, basenjis and beagles being fast, and the rest slow, particularly fox terriers. In the hybrids there is a tendency for all populations in the BCS cross (started from cocker mothers) to develop more slowly. If the retarded litter is omitted from the CSB F_1, the percentage at 3 weeks is 74, almost exactly the same as that in the backcross, suggesting that maternal environment (perhaps prenatal) has an effect on the eruption of teeth.

Learning capacities.—One of the best ways to study learning objectively is by the method of conditioning, originally developed on the salivary reflex of the dog by Pavlov. In this type of experiment the dog is stimulated in some way, usually by a sound or light, and presented immediately afterward with a piece of meat. Before long the dog's mouth begins to water as soon as the stimulus appears, and before he has a chance to see the meat. This process of forming an association between a neutral or initially meaningless stimulus and a response, is called conditioning. In order to do it effectively there must be a *primary stimulus,* which produces a response without previous experience, and another *secondary* or *neutral stimulus,* which has no such effect. The latter is difficult to find for a young puppy, since there are so few stimuli to which it can respond, and many of these produce primary responses themselves.

Changes in the ability to be conditioned can be measured in the following ways. One is to take the number of times that the neutral and primary stimuli have to be presented together before a response is obtained to the secondary stimulus alone. Adult dogs will frequently make such an association with one experience. A second way to measure this capacity is to measure the number of pairings necessary before the association with the neutral stimulus becomes stable. In stable conditioning the animal responds to the neutral stimulus in very nearly 100 per cent of the trials. Still a third method is to set some arbitrary standard of accomplishment and see how long it takes the animal to meet the criterion.

In our own experiments we first set the arbitrary standard of obtaining a stable response after ten trials (Fuller, Easler, and Banks, 1950). Instead of using the salivary reflex, which would be difficult to measure in young puppies, we employed the leg-withdrawal reaction to a mild electric shock and applied this technique to twenty-five puppies, including cocker spaniels, shelties, wire-haired fox terriers, springer spaniels, and a beagle × Kerry blue terrier hybrid. A variety of neutral stimuli were tested, including the sound of a buzzer, light, touch, the taste of Karo syrup, and the odor of a dog repellant. None of the puppies developed stable conditioned reflexes after ten trials until they were between 18 and 21 days old, although there were signs of unstable responses as early as 14 days. In each case the onset of stable conditioning was quite sudden. One day the puppy might give occasional responses and on the next it would give a stable reaction after ten trials.

The type of reaction which the puppies gave after 18 days is typical of older pups and adults, and we can conclude that there is an important change in the capacity for conditioning at this age. In a later experiment, we worked with a less extreme criterion for conditioning, with the idea of obtaining a better measure of the early development of the learning capacity (Cornwell and Fuller, 1960). In this case the puppy met the standard when it began to respond to the neutral stimulus in 50 per cent of the trials. We also used a puff of air as the neutral stimulus, this being something to which the puppies could react very early in development. A sample of five fox terrier puppies met the 50 per cent requirement at anywhere from 10 to 19 days of age, with an average of about 15 days (Fig. 4.4). This kind of criterion apparently gives a much more

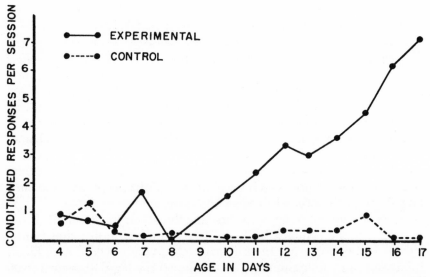

Fig. 4.4.—Development of learning capacities during the neonatal and transition periods. Each curve represents the average score of five fox terrier puppies given 10 trials per day. Note that there is no evidence of improvement until at least 10 days of age (after 50 previous training trials) and that the experimental animals were still not making perfect scores at 17 days. There was considerable individual variation, and numbers are too small to give a clear-cut developmental picture.

variable result than the more rigorous one, but it nevertheless demonstrates that conditioning of an unstable sort does appear at some time in the transition period.

These results esentially agree with work by Russian experimenters (Klyavina, *et al.,* 1958). Working with auditory stimulation at 15 days of age (at which time our results indicate that only a few puppies would respond to sound), they found that the leg-with-drawal response appeared at about the sixteenth pairing and became stable by the eightieth. This is consistent with our results as these puppies would not have passed a ten trial criterion. At one month, learning was as rapid as that observed by us beyond the age of 3 weeks.

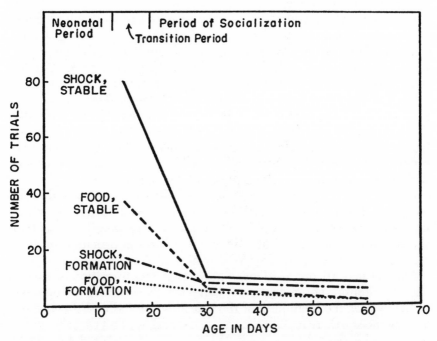

Fig. 4.5.—Change in response to conditioning between the transition and socialization periods. The capacity for forming a conditioned response to food develops before that for electric shock. (Data from Klyavina *et al.*, 1958.)

The Russians also found that a motor response to food could be conditioned more rapidly than the withdrawal to shock. At 15 days this response appeared on the average after the ninth pairing and became stable by the thirty-seventh. These puppies still would not have met a ten-trial criterion for a stable response, but the experiment indicates that the ability for conditioning ingestive behavior appears sooner than that affecting agonistic behavior. While these findings are not as clear-cut as we might desire, all results indicate that the puppy develops the ability to quickly make an association concerning a painful experience near the end of the transition period.

This raises the question of whether an emotional response to pain might be learned earlier in development. However, when we tried to condition the increase in heart rate in response to pain, we found that this ability actually appeared later in development (Fuller and Christake, 1959). If this is true, it suggests that the puppy in early development is highly protected from the psychological after-effects of unpleasant experiences.

Stanley (1963) has developed a new method of studying changes in capacities for conditioning, based on operant conditioning. In the operant technique, first used by Skinner on adult animals, the puppy must associate what he does with the resulting change in stimulation. This avoids the problem of finding neutral stimuli and is closely related to the normal process of nursing. The puppy first sucks and, as a result, is stimulated by milk. There is a theoretical difficulty arising from the possibility that milk itself may be a primary stimulus for sucking. If so, the nursing process can be explained as a circular reflex: sucking is stimulated by contact with a soft, warm surface; sucking produces milk; and milk stimulates sucking.

Stanley has therefore devised an electronic mother, or rather an electronic breast, which can be set to release milk after any given number of sucks. The puppies respond as if they were learning that the machine gives milk, but we have as yet no information on how soon this ability to learn develops or how it changes with age.

Summary.—All our evidence shows that the transition period is one of profound reorganization of behavior. By selecting the opening of the eyes as the beginning of the transition period and the appearance of the startle reaction to sound as the end, we find that the whole process takes less than a week, the eyes opening completely at 13 days and the startle reaction appearing at 19.5 days. These figures are the average of the five breeds studied. Among them, the process may be faster in some breeds than in others. Furthermore, a great many of these changes seem to be concentrated—again on the average—around 18 to 19 days of age, particularly those involving a change in behavior. The puppy thus becomes a very different organism within the space of a few days.

The most essential change in function is from the neonatal to the adult form of nutrition. The change is not absolute, as the puppies still continue to nurse; but they now begin to eat like adult animals as well. Another fundamental change is from the neonatal to the adult form of locomotion. The puppies can now walk and are capable of leaving the nest area. Another important change takes place in the relationship with the mother. Previously, the relationship was quite simple and involved ingestive behavior almost entirely. Now it becomes more complex as other behavior patterns appear. In addition, the puppy changes from an animal which is highly protected from its environment to one which is extremely sensitive. As we shall see, this change is associated with, and perhaps necessary to, the formation of primary social relationships. It also means that any one sort of environmental stimulation has a very different effect on the behavior of the puppy at different ages.

PERIOD OF SOCIALIZATION

This is a period of rapid development of social behavior patterns, in contrast to the transition period, which is chiefly one of changes in basic sensory and motor capacities. Most of the new patterns of behavior are directly connected with the mother and litter mates and form a part of the animal's rapidly developing social relationships.

Ingestive behavior.—In the neonatal period, the mother constantly attended the puppies. Now she begins to leave them for long periods, and when she returns, she may vomit food for the puppies as well as allowing them to nurse. This behavior typically begins between 3 and 4 weeks after the birth of the pups. Frequently she will not lie down with them, and they have to run after her and nurse as she stands erect. The puppies readily eat the vomited food and will also lap up water or milk when available.

Weaning is a gradual process, starting with vomiting by the mother and the taking of solid or semisolid food by the pups. When they are about 5 weeks old, the mother may begin to growl at her puppies when they try to nurse. The puppies walk up to her, start to nurse, and the mother quickly turns, growls, and snaps in their faces without actually touching them. The surprised puppies often roll over on their backs, yelping, and soon learn to stay away. Some mothers stop producing milk as early as 7 weeks after the puppies are born, but a few still produce it at 10 weeks, so that final weaning to solid food normally takes place sometime within this period.

The puppy is still not capable of eating like an adult. Its small baby teeth are sharp, but incapable of producing any effect on solid bones. Likewise, the puppy is still incapable of doing its own hunting.

Eliminative behavior.—Early in the socialization period, the puppy begins to leave the nest to urinate and defecate. At first this is done in any spot close by, but by 8.5 weeks, the puppy is beginning to use definite spots for defecation (Ross, 1950). It will wander around the pen with its nose to the ground, apparently following a trail, but more probably smelling traces of urine or feces so that an eliminative reflex is set off. The puppy at this age will go for many hours without soiling its sleeping place but will urinate and defecate frequently while awake and active. In most cases the chosen spot is as far as possible from the food area in the pen. The male puppy still does not show the leg-lifting reaction.

These basic patterns of developing eliminative behavior have considerable practical importance to the dog owner who is attempting to housebreak a puppy. The important facts are: first, that a puppy will not soil its own sleeping place and will not urinate or defecate if shut up in it overnight. This probably happens because part of the normal pattern of elimination is the preliminary moving around which is not possible in a restricted area. The second fact is the response of the puppy to odors of previous elimination in particular spots. This latter behavior is the preliminary to the "scent post" behavior of adult animals. Once these spots are established and not completely deodorized, urination and defecation can be readily controlled. The wise dog owner helps the puppy establish these spots, either outdoors or indoors on papers, and leads the puppy to them at frequent intervals. Until it is 12 weeks old, a puppy is likely to urinate or defecate every hour or two, as long as it is awake.

Et-epimeletic behavior.—The distress vocalization of puppies now becomes more differentiated, the puppies making a greater variety of sounds. They still yelp when hungry or hurt but are not so likely to become cold and vocalize for this reason. In addition, certain new reactions appear. One is yelping in reaction to restraint. We first noticed this in puppies which accidentally got caught behind a nest box and soon found that simply confining a puppy in a small box would produce the same reaction. Removal of the mother and litter mates would also cause the puppies to yelp, but one of the most effective situations was a strange place away from the home pen. Puppies placed in a cage in the hall outside their home pen made an ear-splitting racket.

Using this information, we (Elliot and Scott, 1961) made a developmental study whose results are shown in Figure 4.6. Puppies left alone in their home pens do only a moderate amount of yelping at 3 weeks of age, and this tends to decrease as they grow older. However, a puppy left alone in a strange place yelps loudly and continuously, producing the maximum number of vocalizations when

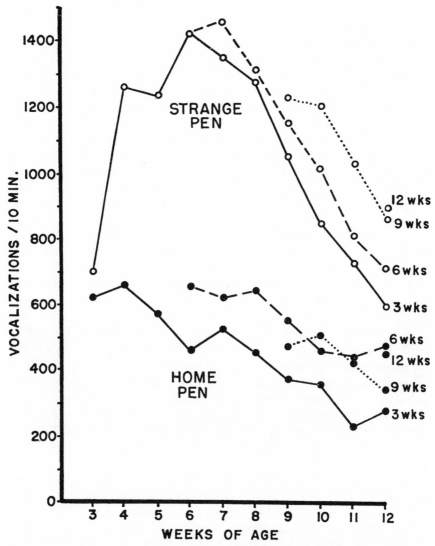

Fig. 4.6.—Development of distress vocalization of beagle puppies under two conditions of isolation. Note that the peak of vocalization in a strange pen comes at 6–7 weeks of age, which other evidence shows to be the peak of development of the capacity for rapid socialization.

it is 6 to 7 weeks old and gradually decreasing them thereafter. By 12 weeks, it makes very little noise in a strange place. This trend reflects a process of maturational change rather than becoming habituated to the situation, because a puppy given the experience for the first time at the later ages shows much the same yelping rate as those which have been isolated before.

We can interpret this behavior as a developing emotional reaction. For a highly dependent and helpless animal like a young puppy, the most dangerous possible situation is one in which it is completely alone in a strange place. Under natural conditions a lost wolf puppy would be vulnerable to any predator as well as in danger of eventually starving to death.

This reaction also indicates another function of the period of socialization—that of becoming attached to a particular locality. The puppy is at first little disturbed by a change in locality, becomes greatly disturbed around 6 or 7 weeks of age, and still later, at a time when puppies normally begin to explore their environment at a distance from their sleeping places, becomes less disturbed by it.

Investigatory behavior.—One of the outstanding changes in behavior at the beginning of the period of socialization is the tendency of puppies to respond to the sight or sound of persons or other animals at a distance. The 3-week-old puppy approaches slowly and cautiously toward a human observer seated quietly in its pen. It finally comes close and starts nosing his shoes and clothes. After this, it may start to wag its tail rapidly back and forth. The tail wagging itself appears to have no directly adaptive function, but is simply an expression of pleasurable emotion toward a social object. What effect it has on other dogs is difficult to tell, but it seems to have the same effect on human observers as the smile of a child; i.e., it is a reward for the person who has initiated a social contact. For the puppy, this whole pattern of behavior is the typical method of initiating a social relationship, and it may develop further into the playful behavior described below.

The puppy will investigate any new inanimate object in its pen in much the same way, except that no tail wagging is involved. However, the puppy is in no sense a hunting animal at this age. Even when raised in our large one-acre fields, puppies stayed within 10 to 20 feet of their nest box until approximately 12 weeks old, when they began to spread out and investigate the rest of the field.

Agonistic behavior.—The first evidences of agonistic behavior are the startle reactions to sound and sudden movement. From being an animal which does not respond to these stimuli at all, the puppy now responds to all loud sounds and sudden movements indiscriminately. Its new capacities for learning soon enable it to discriminate between those situations which are actually dangerous and those which have no significance; and as the puppy grows older, it no longer responds to irrelevant sounds and movements. The mother stimulates fear responses in weaning her pups, and they can often be elicited by a human handler walking rapidly toward the puppy, even though it has never been stepped upon.

The development of these fear responses depends a good deal upon the environment of the puppy. Under our normal methods of rearing, many puppies become quite fearful of human handlers at 5 weeks of age. This fear almost completely disappeared in reaction to daily handling throughout the next two weeks. However, when puppies were raised in large fields with few human contacts, the fear responses became progressively more extreme. In one experiment Freedman, King, and Elliot (1961) brought puppies in from such a field at different ages and placed them in a room with a passive observer for 10 minutes each day. At 3 weeks of age, the puppies came to the observer almost at once; but by 7 weeks, it took two days on the average before the puppies would make a positive social approach. The 14-week-old puppies were so fearful that they never came close even after a week. These last puppies were like little wild animals and could be tamed only in the way in which wild animals are usually tamed, by keeping them confined so that they could not run away and feeding them only by hand, so that they were continually forced into close human contact.

In still another situation (Fisher, 1955; Fuller, Clark, and Waller, 1960), in which puppies were kept isolated in individual pens 2 feet square, active escape responses never developed. When brought from his pen for the first time at 16 weeks of age, the puppy might crouch in a corner or adopt bizarre postures, or even show "fear biting," but he did not run away.

Playful fighting appears early in this period (see Fig. 4.2). At first the young puppies seem to be acting in slow motion, clumsily pawing and mouthing their litter mates without producing any real damage. As they grow older their teeth become longer, and a puppy which gets hold of a sensitive spot, such as the ear, may be answered by a yelp of pain. One indication that fighting is not serious at this early age comes from the dominance tests. At 5 weeks, less than one-fourth of the pairs of puppies showed complete dominance.

However, at about 7 weeks of age (the time when final weaning from the breast begins and mothers begin to threaten their offspring), puppies left with their mothers begin to attack each other in groups. The animal against whom the attack is directed is sometimes a small and weak individual, but it also may be a large and aggressive one. In most breeds this "ganging up" is temporary and playful. In the fox terrier breed, however, such group attacks are persistent and become so serious that the victim has to be removed in order to prevent serious injury (Fuller, 1953). In one litter of six animals there were three males and three females. The group began "ganging up" on the smallest female. When she was removed they began to attack another female, and when she was taken out they attacked the third. The final result was two separate groups of three, since one fox terrier can apparently stand off two but not three other animals. In larger groups, one puppy would get hold of the ears and another the tail, stretching their victim between them, while the third animal attacked in the middle. Not even a fox terrier can take much of this.

In other breeds the group attacks were never serious, and individual fights usually resulted in the formation of some sort of dominance relationship without either individual being seriously hurt. However, such relationships are not often completely formed until the next period of development.

Allelomimetic behavior.—At first the members of a litter are quite independent in their activities, wandering over the pen in different directions. Between 3 and 4 weeks of age they begin to follow each other around, and by 5 weeks we often see the puppies rushing toward the gate of their pen as a group (Table 4.4). This is the first appearance of the pack behavior of adults. By 7 weeks the puppies have begun the group attacks described above, which are a combination of allelomimetic and agonistic behavior.

TABLE 4.4

Time of Development of Allelomimetic Behavior as
Indicated by Co-ordinated Movement

Breed	Number of Animals	Day First Observed	Percentage on Day 35
Basenji.	33	20	70
Beagle.	39	31	44
Cocker spaniel.	42	20	74
Sheltie.	25	21	24
Fox terrier.	21	18	24
Total.	160	18	50

Sexual behavior.—Mounting and clasping may appear sometimes as early as 3 or 4 weeks of age. Supporting a puppy by a hand thrust between the forelegs and under the belly will sometimes stimulate thrusting movements. However, the sex organs themselves are still undeveloped, and such behavior is never complete. As our puppies grew older we sometimes observed the playful attitude which is the characteristic part of the adult courtship pattern.

TABLE 4.5

OCCURRENCE OF SEXUAL BEHAVIOR IN 160 PUREBRED PUPPIES (0–16 WEEKS OF AGE) DURING DAILY OBSERVATION PERIODS

BEHAVIOR	DAY FIRST OBSERVED		PER CENT OF ANIMALS IN WHICH SEXUAL BEHAVIOR OBSERVED	
	Male	Female	Male (83)	Female (77)
Nosing genitals..........	34	41	63	55
Mounting..............	22	27	41	23

Note that sexual behavior is seen slightly more often in males than in females, hence somewhat earlier. Results in the individual pure breeds were highly variable.

Other patterns of social behavior.—Epimeletic behavior is never seen except when the puppies scratch or lick themselves. Unlike primates, dogs show very litttle tendency toward mutual grooming, and the only situation which stimulates it is an open wound. The capacity for some of the patterns of maternal care may exist in older puppies, but under normal conditions the newborn puppies which might stimulate these patterns are never available. Shelter-seeking behavior continues but is likely to be lessened. The pups may sleep outside their nest boxes and sleep apart from each other more often as the period progresses.

Changes in basic capacities.—The retina of the eye is not completely formed until about 4 weeks of age; and the EEG, indicating the function of the visual portion of the cortex of the brain, does not assume the adult form until 7 or 8 weeks of age. This means that the puppy does not have sensory functions completely comparable to those of an adult until several weeks after the beginning of the period of socialization.

As to motor functions, the puppy assumes the adult forms of locomotion and chewing at the outset of the period but still has not reached anything like adult capacities by its end. A 7-week-old puppy is easy to catch even when running at top speed, and it can do very little damage with its immature teeth and jaws. Even at 12 weeks a puppy is a clumsy runner with little endurance.

With regard to learning capacities, the puppy has at the outset of the period the ability to make rapid associations between stimuli, and in this respect it is similar to an adult. However, its motor responses are still not completely developed, and this means that while the puppy may learn a great deal from the outside world, it is still not capable of learning complicated motor acts which require speed and good muscular control.

Conclusion.—During the period of socialization the puppy begins to show most of the adult patterns of behavior at least in a playful form. Ingestive behavior, which was most important in the neonatal period, is now reduced, and the most prominent aspect of behavior is social play. This is the time during which social relationships are easily developed, either with other puppies or with human beings. The behavior which initiates these relationships is social investigation, followed by playful fighting and sexual behavior, both involving bodily contact. These relationships can be developed with any individual who stays with the puppies. At the same time we see the appearance of behavior which limits the formation of relationships with casual strangers. The first reaction to a stranger is one of fear. As a puppy becomes older, fear and escape reactions become stronger and more difficult to overcome, so that a casual stranger never makes contact.

This is, therefore, a critical period for the formation of social relationships. We shall examine this concept more carefully in pages 110–12.

JUVENILE PERIOD

This period begins with the first long excursion away from the den or nest box and ends at sexual maturity. Thus it runs from approximately 12 weeks of age up to 6 months or later. The changes are not as striking as before and will not be treated in great detail.

Changes in basic capacities.—All the sense organs appear to be fully developed at the outset of the juvenile period. Permanent teeth begin to come in at about 16 weeks of age, and all are usually present by 6 months. Growth curves also begin to flatten out at 16 weeks. The period of rapid growth is over, and the puppy is approximately two-thirds of its adult size.

The development of motor capacities in this period consists of increases in strength and skill rather than the emergence of new patterns. Much depends on the environment. Puppies raised in open fields are more active and skillful than those raised in kennels, although any puppy at 4 months is still an awkward and gangling animal. By 6 months the most advanced pups are very similar to adults in size and motor capacities. In a wild species, we would judge them to be capable of existing independently from their parents. They continue to grow slowly thereafter and usually reach complete physical development about the age of two years.

As to basic learning capacities the puppy appears to be fully developed before the outset of the juvenile period. At about 4 months of age the speed of formation of conditioned reflexes begins to slow down. This is probably not because the nervous system deteriorates but rather because what the puppy has previously learned begins to interfere with new learning. As will be seen later, there is some evidence that the behavior of the puppy begins to reach a stable organization about this age; that is, he has established the foundation for what he will learn in the future. On the other hand, puppies still cannot be trained in difficult tasks, partly because of poor motor skill and partly because of a short attention span and ready emotional excitability.

Changes in social behavior patterns.—Puppies raised in a large field first began to move away from the nest area and explore the surrounding environment at about 12 weeks of age. This probably represents the beginning of the patterns of hunting behavior. We also have some evidence that puppies make a transition from one physical environment to another more easily at this stage than any other. Distress vocalization in response to a changed environment reaches a low level at 12 weeks (Elliot and Scott, 1961). Guide dog puppies usually make a successful adaptation when placed in homes at 12 weeks, but those kept in a kennel longer than 14 weeks show increasingly poor performance during their later training (Pfaffenberger and Scott, 1959).

The male pattern of eliminative behavior appears sometime during this period, varying considerably according to the speed of development of the animals. In dogs raised outside a kennel, Martins (1947) found that this first occurred between 5 and 8 months. This includes the leg-lifting reaction at scent posts and scratching after defecation.

Agonistic behavior has developed into a definite dominance-subordination pattern by 15 weeks, and the puppies concerned show the patterns of behavior typical of adults. This means that the occurrence of actual fighting is much reduced, although the growls and yelps of threat and subordination reactions may frequently be heard in a litter raised together.

The dominance order also limits contact with strange animals, as the puppies tend to attack strangers placed with them. The degree of tolerance toward strangers depends upon the breed as well as the state of development. Tinbergen (1958) observed that puppies growing up in an Eskimo village began to defend territories at the time of sexual maturity, which would be at the end of the juvenile period.

Allelomimetic behavior becomes more and more common, the litter tending to react as a group in many situations. When one animal moves to investigate a noise or movement, the rest usually follow. Playful sexual behavior continues, but the animals are usually distracted from this by any other sort of stimulation. Such incomplete sexual behavior is not usually seen unless the animals are completely unaware of the observer. This is very different from the concentrated attention on sexual behavior seen in adults. Estrus of the females and the emergence of the complete pattern of adult sexual behavior by both sexes marks the end of the juvenile period.

Conclusion.—Compared to the previous period, the juvenile period is one of gradual change, chiefly involving the maturing of motor capacities. It ends with the maturation of sexual capacities and the consequent ability to form complete sexual relationships.

Behavioral development continues throughout life. The next important changes follow the birth of pups and include the origins of epimeletic behavior. This behavior enables the development of the mother-offspring social relationship. A long and stable period of maturity follows, ending with the cessation of reproductive activity. This occurs about the age of 8 in most females and considerably later in males. Many dogs begin to show a decline of fertility and physical vigor after the age of 5, although they frequently live to three times this age.

CRITICAL PERIODS IN DEVELOPMENT

We have divided the development of the puppy into periods based on major changes in social relationships. Immediately after birth the puppy establishes the nursing relationship with its mother, marking the beginning of the neonatal period. The transition period consists of rapid changes in this relationship. At its end, the puppy is capable of forming a new type of social relationship which will persist into adult life. At this point the mother ceases her constant care of the puppies, so that the strongest relationships tend to be formed with the litter mates rather than with the mother. These relationships form the foundation of the typical social group of adults, the pack. The relationship with the mother is still further weakened when final weaning from the breast occurs, although in wild *Canidae* the parents continue to feed the young for some months by vomiting or by bringing them meat. The older animals, nevertheless may spend only a few minutes out of the entire day in company with their pups, spending the rest of the time hunting. The next major change occurs with the formation of sexual relationships as the puppies become adults.

These facts suggest a major hypothesis: that the period at which each new relationship is formed is a critical one for determining both the nature of the relationship and the identity of the individuals with which the relationship is formed. For example, the neonatal period is critical, for unless the puppy nurses successfully, it will die. The period of socialization is critical, since it determines what species and individuals will become the chief adult relatives of the puppy. A puppy taken from its litter early in development and raised by hand will form its paramount relationships with people, becoming an "almost human" dog and paying little attention to its own kind. Removed a little later in the period, it forms strong relationships with both dogs and human beings. Still later it has already formed strong relationships with dogs and its ties with human beings tend to be relatively weak.

This is the most important critical period in the life of the animal. In addition to the determination of social relationships, the emotional sensitivity and still undeveloped motor and intellectual capacities of the puppy suggest that this may also be a critical period for possible psychological damage. Emotional sensitivity is apparently a necessary part of the socialization process, and this automatically makes the animal susceptible to psychological damage as well.

Still another critical period begins with sexual maturity, when the mating relationship is determined, along with poor or good adjustment in sexual behavior. Both of these adjustments may be greatly influenced by what has already happened, particularly in the period of socialization.

We have defined the dividing lines between periods as major changes in social relationships. The periods themselves are characterized by certain processes. That in the neonatal period is the process of neonatal nutrition, or nursing. The same process goes on in the transition period, but there is also a new major process going on, this time a biological one. The greatest maturational changes in basic capacities take place in the transition period. At its end the puppy can move independently of its mother and also begin to take solid food. Further maturational changes continue to take place, but nothing as sudden and drastic as these.

The major process of the next period is that of socialization, the formation of lasting social relationships. This again is an ongoing process which probably never stops entirely in the life of the animal, but the period of socialization is the point at which the biggest changes occur and also the time when it is easiest to initiate such relationships.

Another major process which takes place during this period of development is that of localization, in the sense of a process comparable to socialization. The puppy becomes strongly attached to a place and is seriously upset if he is moved away from it. Our data show that this process reaches its peak at 6 to 7 weeks of age but that the emotional responses to being moved continue for some time.

The results of socialization and localization are so similar that we wonder whether they may represent the same process applied to different objects. This would mean that the puppy becomes attached to both the living and non-living parts of its environment at this age.

SUMMARY AND CONCLUSION

A puppy comes into the world with its behavior organized for life in the neonatal period, centering around the function of obtaining nourishment from the mother. It is highly protected from outside psychological stimulation both by maternal care and its own immaturity. The neonatal puppy is blind and deaf, and it is doubtful if even the olfactory sense is developed. Its motor capacities are so limited that it can move only within a radius of a few feet. Its capacities for learning are highly limited compared with those of older puppies, and it shows a limited number of emotional reactions.

Between 2 and 3 weeks of age a profound reorganization of behavioral capacities occurs. By 3 weeks, the puppy is able to move and eat by adult methods. All sense organs are functional, and it is capable of making quick and easy associations between outside events. Many of the patterns of adult social behavior and their accompanying emotions appear at the same time. The acquisition of these capacities is followed by a rapid organization, through learning, of behavior in relation to other animals, people, and places. Whatever happens here sets a general pattern which will affect almost everything in later life, because by the end of the period the puppy has formed patterns of responding to the major influences in any sort of future existence. There will be other periods of rapid organization of behavior in later life, at the time of sexual maturity and birth of the young, but their effects will be more limited.

As research workers, we can conclude that the puppy is in many ways an ideal animal for studying the effects of early experience. The periods of development are so well divided and easily recognized that it should be possible to give puppies the same experience at different ages and come to definite conclusions regarding the relationship between experience and maturation. Puppy development is a happy compromise between extremely short development, which permits very little effect of experience, and extremely long development, which may require a lifetime to analyze.

These studies on the development of behavior lead to two general principles regarding our original problem of the effect of heredity upon behavior. One of these is that genetic differences in behavior must be produced through developmental processes. We cannot think of behavior as being something relatively fixed and unchanged such as hair color, but rather something which is rapidly developing and changing throughout early life and which continues to change as the animal grows older.

Hereditary factors must then act by affecting developmental processes, such as the growth and differentiation of sense organs and organ systems. Heredity may also affect ongoing physiological processes throughout life, one of which is the process of learning itself. This in turn is the basis for other major processes: the foundation of social relationships and relationships with the inanimate environment.

A second principle is that the majority of hereditary differences in behavior are expressed as components of social relationships, either with other dogs or with people. Even when a dog is completely alone, he may react to the absence of others, and in fact usually does so, showing every evidence of emotional distress. This complicates the problem of measurement of hereditary differences, because any social relationship is an interaction between two individuals, and is therefore affected by the heredity of both. However, it is just these relationships which characterize the organization of adult behavior and form what we ordinarily call individuality or personality in a human being.

HUMAN COMPARISONS AND APPLICATIONS

When we try to compare the development of a puppy with a human infant (Scott, 1963a), we find that much of the information which we would like to have is missing. Child psychologists have generally studied children in schools, where they are easy to get at, and very seldom in homes, where their most basic social relationships are formed. This means that, in spite of its theoretical importance to later behavior, the study of the period of early infancy has been relatively neglected.

TABLE 4.6

NATURAL PERIODS OF DEVELOPMENT IN DOG AND MAN

Dog

PERIOD	INITIAL POINT	MAJOR PROCESS	CHANGE IN SOCIAL RELATIONSHIPS
Neonatal.......	Birth	Neonatal nutrition	Begin
Transition.....	Eyes open (2 weeks)	Transition to adult sensory, motor, and psychological capacities	Increasing complexity based on new behavior patterns
Socialization....	Rapid, stable conditioning, startle to sound, tail wagging (3 weeks)	Formation of primary social relationships	Partial independence of mother, increasing responses to litter mates
Juvenile.......	Distant exploration (12 weeks)	Rapid growth, development of motor skills	Increasing independence of mother

Man

Neonatal.......	Birth	Neonatal nutrition	Begin
Primary socialization..	Stable conditioning to sound stimuli, smile to mother (5 weeks)	Formation of primary social relationships	Responds at a distance
Transition (1)..	First tooth, crawl (7.5 months)	Transition to adult methods of eating and locomotion	Weaning from breast or bottle likely
Transition (2).	Walk, first words (15 months)	Transition to adult method of communication and social control	Escape and voluntary approach possible
Verbal socialization	Use and understand sentences (27 months)	Formation of verbal relationships: communication	New type of social control possible, both by and of infant

We do know enough to make at least some general comparisons. Like that of the newborn puppy, the behavior of the neonatal infant is primarily concerned with neonatal nutrition. The infant has better developed sense organs and responds to both sound and light. However, vision of the human neonate is not perfectly developed, as the EEG reaches its adult form at about 8 years instead of 8 weeks as in the puppy.

In motor capacities a human neonate is less developed than the puppy, not even being able to crawl. Its major transition to these adult capacities comes between approximately 7 and 14 months, beginning when the baby develops its first teeth and begins to crawl, and is ending when it is later able to walk. In this respect the 14-month-old baby is quite similar to the puppy of three weeks.

Here we come to a major difference. While the capacity for social responsiveness in the puppy develops after its transition to the adult form of locomotion, the baby develops the capacity for rapid conditioning by one or two months and begins to smile in response to social stimulation long before its general motor capacities develop. From 2 until 6 months, the baby responds indiscriminately to all human faces but by the end of this time is beginning to develop fear responses to strangers and by 8 months of age shows what is often called "8-months anxiety." It is only then that the average baby begins to crawl and develop its other motor capacities.

Thus, the period of human primary socialization precedes the development of the adult capacities of locomotion instead of following them, as in the puppy. This means that the baby develops its strongest relationships with the person or persons who take care of it at this time, usually the mother, but also the father and siblings. This also produces an emphasis on the development of strong relationships between older and younger individuals rather than the canine emphasis on strong relationships between contemporaries.

We can also see that in different species of mammals differences are not simply a matter of longer or shorter development; the order of certain major developmental changes is actually reversed in the two species we have compared. This, of course, raises the question of how much variation of this sort can take place within as well as between species.

The similarities between human and puppy development suggest the existence of a critical period for primary human socialization having great practical significance for the problems produced by the loss of parental care and changes produced by adoption. While the evidence is somewhat unclear due to the undesirability of making deliberate experiments with such serious consequences on human beings, all of the evidence indicates that such a period exists sometime during the first year.

This conclusion is bolstered by the evidence from other species of animals. Every highly social species of animal which has been studied so far has a short period early in life when primary social relationships are formed. In birds, with their very short periods of development (a song sparrow progresses from hatching to adult flying in 4 weeks), the process of socialization takes place extremely rapidly and so has received the name "imprinting." In the bird, the critical period for socialization may be a matter of a few days or even hours;

in the puppy a matter of weeks; and in the human infant a matter of months. These longer periods permit the development of a wider and less specific set of relationships, but the basic phenomenon appears to be the same.

When we realized the theoretical and practical importance of these discoveries, we did two things. One was to call these results to the attention of a conference of scientists interested in comparing human and animal behavior. (Scott, 1952). We realized that such a general principle as that of critical periods could be established only through the co-operation of many different scientists and institutions working on many different species of animals, and comparing the results with observations on children. The other was to institute a research program at our own laboratory, in order to verify the observations we have described in this chapter through experimental means, and also to discover the physiological and psychological causes of these phenomena. The results of this program are described in the next chapter.

CHAPTER 5

THE CRITICAL PERIOD

As we have seen in the preceding chapter, our studies of the development of behavior show that a puppy enters into a period of great change and sensitivity with regard to social relationships at approximately 3 weeks of age, and that his experiences at this time determine which animals and human beings will become his closest social relatives. This is a time of major decision affecting all the rest of his life, and the period is therefore a critical one.

In order to translate this into a workable scientific theory, we must sharpen up the concept of critical periods. By a critical period, we mean a special time in life when a small amount of experience will produce a great effect on later behavior. To draw an analogy, it is a great deal like pulling the trigger on a high-powered rifle. A very small amount of effort causes the bullet to travel at high speed and produce a smashing impact at a great distance.

This is the kind of scientific discovery which can have immense practical value if it turns out to be correct. A great part of scientific effort is directed toward discovering precisely this sort of causal relationships. It is not enough to know a cause, for if the effort required to manipulate it is equal to or greater than its result, the discovery has only limited usefulness.

Going back to the idea of the critical period, we can see that it is a relative rather than an absolute concept. The difference between the amount of effort needed to produce the same effect at different periods determines just how critical the period is. In the case of the puppy, it looks as if a small amount of contact shortly after 3 weeks of age will produce a strong social relationship which can be duplicated only by hours or weeks of patient effort at later periods in life —if, indeed, it can be duplicated at all.

It might be supposed that the effect of this sort of contact is inversely proportional to age, i.e., the younger the animal the greater the effect. If this were so we could not properly speak of a critical period. In order to define a period there must be definite and important changes bounding its beginning and end. One important scientific problem is determining the nature of these limits, for this information may in turn lead to the discovery of how to modify or prolong the period.

There may be more than one critical period in the life of an individual. The period we have described in the puppy is a critical one for the determination of primary social relationships, but this need not be the only one. There could be a variety of critical periods in development for different events. For example, we have stated the hypothesis that the time in which any important social relationship is begun is a critical one for determining the nature of that relationship. The times of the first mating and the birth of offspring should also be critical periods for the formation of sexual and parent-offspring relationships.

On the other hand, different sorts of critical periods might coincide. The emotional sensitivity of the puppy during the period of socialization suggests that this period might also be a critical one for permanent psychological damage. If the nature and importance of critical periods can be established first in animals and eventually in human development, the concept has immense practical possibilities for the improvement of mental health. If we knew that a relatively small amount of effort at the right time in development would start a child toward being a basically happy and productive individual, we could make vast strides in the improvement in human behavior within a generation. At the present time we know that disorganized family conditions appear to have a bad effect on young children. If the critical-period hypothesis turns out to be correct, we can in the future say to a parent: "If you do thus and so at a particular time, you will be more likely to have a well-adjusted child than if you do it at an earlier or later time." Meanwhile, we can get some idea of the needful scientific information from our studies on the dog.

BOUNDARIES OF THE CRITICAL PERIOD

As we have seen in the preceding chapter, it is easy to establish a beginning point for the critical period of primary socialization in the puppy. Between 2 and 3 weeks of age a very large number of changes take place in rapid succession, and all of these modify the capacity for forming a close social relationship. The period of change or transition begins with the opening of the eyes and includes the opening of the ears and the startle response to sound. At its end the puppy is capable of walking, so that it can either approach or avoid another individual. It begins to eat solid food and hence starts to be independent of the mother. It develops the ability to make rapid associations between outside events and unpleasant feelings. This last change is probably the most important one with regard to establishing a boundary for the critical period for forming primary social relationships, because the puppy is now capable of discriminating between individuals. This capacity is difficult to determine by observation and can be measured accurately only by a series of elaborate experiments. Fortunately there is a change in behavior which seems to be closely correlated with it, and which is easy to determine —the appearance of the startle response to sound. We can arbitrarily use this as a marker to indicate the beginning of the critical period.

How is this developmental change related to age? We can record the date on which the startle response occurs and calculate the age of each animal. There will obviously be some variation in age, reflecting differences in the state of development, and one important practical problem is that of determining the extent of this variation in normal dog populations.

Normal variation in development.—In the first place there is variation in the rate of prenatal development. The average gestation period of the dog is usually reckoned at 9 weeks, or 63 days after the onset of receptivity. This itself is only an estimate, because ovulation takes place approximately 72 hours before the end of receptivity, and the total length of the period of receptivity varies a great deal in different animals. Embryonic development in the dog is unusual in many respects which are correlated with its life as a hunting animal. The fertilized egg develops very slowly and does not become implanted in the uterus for 3 weeks. Even after this, it begins rapid development only when the placenta has been established. This means that the embryo grows very slowly at first and very rapidly in the last 4 or 5 weeks of pregnancy, so that the bitch has a relatively short period when she is heavy and incapacitated for active life. Both the time in which the eggs become implanted and their position in the uterus could produce variation in the state of development. We have no evidence on the dog, but studies on the embryology of the guinea pig show that embryos of the same chronological age from conception can vary at least a day or two in state of development without any obvious sign of abnormality (Scott, 1937). More than this, there is some variation in the times at which puppies are born, some litters being obviously more mature at birth than others. This probably adds at least another day or two of developmental variation. Finally, there is the accuracy of the measurement of age itself. When we recorded two litters as born on the same date, they actually could have been born 24 hours apart, since no observations were made at night. All this would add up to a range of 5 or 6 days' variation in development for animals recorded as having been born on the same day to different mothers.

Let us now look at the actual figures. We examined and tested the dogs once per week in order to avoid excessive handling in the early stages of development. Hence we must estimate variation on the basis of these weekly figures rather than on daily ones. The first change is the opening of the eyes (Table 5.1). Out of 195 purebred

TABLE 5.1

Estimated Variation in Developmental Events Associated with the Beginning of the Periods of Transition and Socialization in Purebred Puppies

Event	Beginning of	Mean, days	Standard Deviation
Eyes completely open.........	Transition period	13.0	2.3
Ears: startle to sound.........	Socialization period	19.5	2.3
First teeth erupted...........	Socialization period	20.8	2.9

puppies, only one cocker spaniel had its eyes completely open at one week of age. There are large differences in development between the different breeds, but weighting the breeds equally we find that 59 per cent have the eyes completely open at 2 weeks and 100 per cent at 3 weeks. Assuming that the time of opening the eyes is distributed in a normal curve, we can estimate that the average time for this event is 13 days with a standard deviation of 2.3 days. The standard deviation measures the point of inflexion of the normal curve, and this means that two-thirds of the puppies would open their eyes between 10.7 and 15.3 days. One-sixth of them might open their eyes sooner and one-sixth later. This result fits very well with our theoretical expectation that while the average might fall on a particular day, we might expect a range of variation of approximately a week in these developmental events (Scott, 1958).

A more important event for marking the critical period is the date on which the animals give the startle response to sound. Two animals, or 1 per cent, gave this reaction at 14 days, and 74 per cent at 3 weeks. The estimated average would be 19.5 days, with a standard deviation of 2.3 days. Again, this would mean that one-sixth of the animals should show this response before 17.2 days and one-sixth (the slow developers) after 21.8 days.

These figures are important to remember in interpreting any experimental results. Of animals given experimental treatment between 14 and 21 days of age, three-quarters of them would have passed into the critical period sometime during the week, and only about one in a hundred might have spent the entire week in this period. We would therefore expect the experimental results from this period to be quite variable.

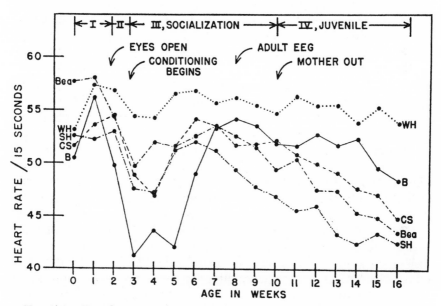

Fig. 5.1.—Development of the heart rate. Note that all breeds show a depression of the heart rate during the early part of the period of socialization.

We can now attempt to define the end of the critical period. This is not as easy to do, and our measurements are much less exact. The data on the development of the brain waves (based on a sample of only 10 puppies) indicates that the mature condition is reached sometime between 7 and 8 weeks (Charles and Fuller, 1956). Mothers begin to completely wean their puppies from the breast at about 7 weeks of age. They may or may not be influenced by the development of the pups, but weaning does indicate a change in social relationships. Our measurements of fearful reactions to human handling indicate that the maximum reduction of this behavior occurs before 7 weeks of age (Fig. 5.7). However, our best objective evidence of change at this time comes from the measurement of heart rates. As can be seen in Figure 5.1, all breeds show a reduction in the heart rate between 1 and 3 weeks of age, and in two of the breeds this change begins between 1 and 2 weeks. A low point occurs at 3 and 4 weeks, this being extended to 5 weeks in the basenji. A high point is reached again at 6 weeks in three of the breeds, 7 weeks in another, and 8 weeks in another. This change in heart rate may reflect either a change in the ability of this response to be conditioned, thus bringing it under the adult type of emotional control, or it may simply reflect a change in the emotional responses to the handlers. At any rate, it is the most objective measure of developmental change which we have available, and it led us to estimate that the end of the critical period occurred at approximately 7 weeks.

We have one more piece of confirming evidence: the vocal reaction of a group of 32 beagle puppies to isolation in a strange room at different ages. The maximum vocalization was elicited between 6 and 7 weeks, with a rapid decrease thereafter. All these data indicate that important developmental changes take place at 7 weeks of age or shortly thereafter, and we tentatively set the end of the critical period at this age. This date can be decreased or increased by individual variability of at least a week, since other data indicates that individual variability increases with age. As further experimental work shows, this estimate was actually too early by several weeks.

Developing a technique for testing the critical period hypothesis. —A healthy litter of 6 beagle puppies was being raised in the normal testing program in the "School for Dogs" when Dr. Emil Fredericson joined our staff. He performed a pilot experiment on these animals by taking them away from the mother and litter mates at different ages and giving them close contact with human beings (Scott, Fuller, and Fredericson, 1951). The first puppy was taken at birth and raised on a bottle until 2 weeks of age. Aside from difficulties in getting it to nurse, it showed no behavioral disturbance, although it was not as well nourished as its litter mates. The second puppy was taken away between 2 and 4 weeks of age. It readily adjusted to the nursing bottle and showed no disturbance of behavior either when taken away or when returned. According to our theoretical calculations, this experience with people included a week or more of the early part of the critical period. The third puppy, removed from the mother at 4 weeks of age, showed excessive emotional disturbance and at one point yelped continuously for 24 hours. However, it soon became adjusted to the family and home in which it was kept and began to act like a pet dog. The final experimental puppy was taken away from 6 to 8 weeks of age. Because the cries of the previous puppy had upset the human family, this puppy was kept at the laboratory and given extensive contact with people only during office hours. It showed the same sort of emotional disturbance as the previous puppy, but not so intensely. The remaining two animals were left with the mother at all times and used as controls.

After this experience, all the puppies were put back with the mother and litter mates and given the regular testing routine. Within a very short time the behavior of one was indistinguishable from another. In this situation any effects of their early experience were no greater than the normal variability in the litter.

We now attempted to find out whether the early experience had some less obvious effect, and we tried to duplicate the original experience by taking each animal to a strange private home and keeping it overnight. Table 5.2 shows the general results. In its over-all

TABLE 5.2

RELATIVE ADJUSTMENT OF PUPPIES AT 1 YEAR OF AGE

Test	AGE REMOVED FROM MOTHER, WEEKS					
	0–2, ♂	2–4, ♂	4–6, ♀	6–8, ♀	Control ♂	Control ♂
Tail up on walk	4.5	4.5	2.0	1.0	4.5	4.5
Tail wag on walk	4.0	4.0	4.0	1.0	4.0	4.0
Tail up in house	4.5	4.5	1.5	4.5	1.5	4.5
Tail wag to experimenter	4.5	4.5	1.5	3.0	1.5	6.0
Tail wag to stranger	4.0	5.5	2.5	2.5	1.0	5.5
Eat from hand	3.0	5.0	1.0	2.0	5.0	5.0
Sum of ranks	24.5	28.0	12.5	14.0	17.5	29.5
No. of tests ranking above 3	0	0	5	4	3	0

score, the animal which had been removed between 4 and 6 weeks made the quickest and best adjustment to the situation, confirming our expectation regarding the extent of the critical period. On the other hand, one of the controls did almost as well, and we concluded that the experience of socialization in the ordinary laboratory conditions was sufficient to produce almost as great an effect on some animals as did the intensive home treatment. One of us (Fuller, 1961) has since found that semi-isolated puppies can be successfully socialized with as little human contact as two 20-minute periods per week. We therefore began to look for another technique which would produce maximum differences.

In the previous chapter we have described the results of observing dogs living in large fields apart from human beings. The experiment included raising puppies in the same fields, and one obvious result was that these puppies became extremely wild. We took one home at the age of 12 weeks and attempted to make a pet out of it. It was extremely fearful at first, but by confinement and hand feeding we eventually calmed it down and got it to accept close contacts with people. However, it was always difficult to control and timid with strangers, and whenever there was a choice between human and dog contacts it chose the dog. It looked as if rearing in large fields produced the maximum effect for which we were looking. Beagles and fox terriers became equally wild, the only difference being that the latter could not be trusted not to kill each other.

Still another experiment gave us an idea of the limits within which we could work. Miss Barbara Arndt, a summer student, raised a litter of six beagles in a large field, making direct contact with them only once a week, when they were weighed, and every two weeks, when she gave them a handling test. She left the parents a dish of food once each day, immediately going away to make observations from the gate, where she stood about 20 or 30 feet away from the puppies with only her head and shoulders showing. Two of the puppies were taken away from the parents at 3 weeks, raised by hand for 2 weeks, and returned. Under these conditions, none of the puppies became excessively timid, although all six eventually showed avoidance scores above the mean of those raised in the laboratory with daily contact with people. The hand-reared puppies were the least timid when returned to the pen, but eventually grew more timid and much like the others. Five of the six puppies were much more responsive to the observer than to anyone else, and none of them acted like the wild dogs of earlier tests. It began to look as if a primary social relationship with a human being could be formed with a very small amount of contact during the critical period and, in fact, that it might be formed merely by having the person in sight at frequent intervals. However, the puppies could still have associated the observer with being fed, and become socialized in that way.

The "wild dog" experiment.—With these results in mind, Freedman, King, and Elliot (1961) designed an experiment based on large numbers of puppies. They used the large outdoor fields, introducing food and water through a hole in the fence so that there was almost no contact with the experimenters. They brought in each experimental puppy for only one week and gave it standardized contact with the experimenters each day.

From the evidence cited above, the experimenters supposed that the critical period extended roughly from 3 to 7 weeks of age, and they attempted to socialize puppies at 2, 3, 5, 7, and 9 weeks, leaving certain puppies continuously alone until 14 weeks. At this time all the puppies were brought into the laboratory and retested.

The results can first be analyzed in terms of initial effects. The puppies were given the handling test immediately after being brought in and again after a week of human contact. Figure 5.2

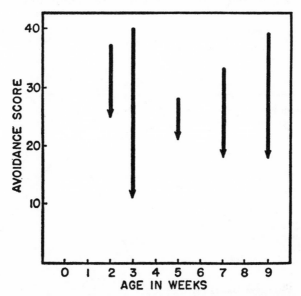

Fig. 5.2.—Effect of socialization at different ages upon avoidance of an active handler. Note consistent rise of avoidance scores after 5 weeks of age. A large number of the responses at 2 and 3 weeks of age are of the "no response" category; hence scores at these ages are not completely comparable to those at later ages. Length of lines denotes the change resulting from contact; top of line indicates initial score.

shows that the puppies brought in at 5 weeks of age showed the least initial avoidance of human handlers, while those brought in earlier or later showed larger numbers of fear responses. Nevertheless, after a week of contact with the experimenters, the puppies taken at 3 weeks of age showed the least fear, and those taken at 5 weeks were relatively timid. Thus, the maximum *reduction* of fear was produced at 3 weeks, and the minimum at 5.

In another test (the "Passive Handler Test"), the experimenter sat quietly in the room with the puppy each day for 10 minutes and recorded how long it took the puppy to approach and remain in contact. At 2 weeks, the puppies were so immature that it was 4 days before they began to spend the full time with the experimenters. At

3 and 5 weeks this occurred on the first day, but it was two days before the 7-week-old puppies passed the test, and three days before
this occurred with the 9-week-old puppies.

Thus we have measures of several different kinds of behavior. The
first is the positive response of making contact with a passive and
hence non-frightening new individual in the environment. This be-

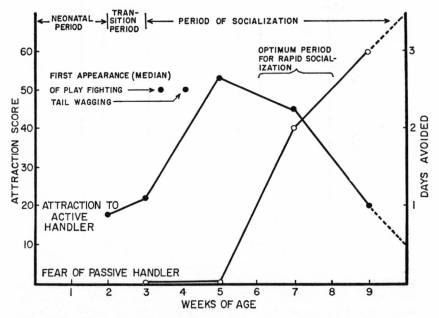

Fig. 5.3.—Timing mechanisms limiting the process of socialization in puppies. Initially, the puppies are unable to respond to an active person but show
little fear. Later, the developing fear response limits the capacity to be attracted.

havior is most obvious from 3 to 5 weeks of age and declines thereafter. Second, there is the negative, or fear response, to new and
active individuals, which continuously rises after 5 weeks of age and
probably causes the decline of the earlier approach behavior. Third,
there is the response of recovery from initial fears, which is almost
instantaneous at 3 to 5 weeks of age and becomes progressively
slower thereafter. All these behavioral processes affect the formation
of the young puppy's first social relationships during the critical
period. Their net result is that the puppy makes rapid contact with a
completely strange individual only within the relatively short period
from 3 to 5 weeks of age.

We may now consider the later effects of these experiences. The puppies were all removed from the large fields at 14 weeks of age, brought into the laboratory rearing rooms, and subjected to a series of tests, including new ones as well as those previously given. One test whose results showed the largest differences was that of leash-control. A puppy which has never been on a leash becomes alarmed and fearful, both in reaction to restraint and to being led into strange places. The test consisted of leading the puppy through the laboratory building and up the stairs (the most frightening part of the trip) and is described in detail in chapter 9. The results are shown in Figure 5.4. Those animals taken at 5, 7, and 9 weeks made the best

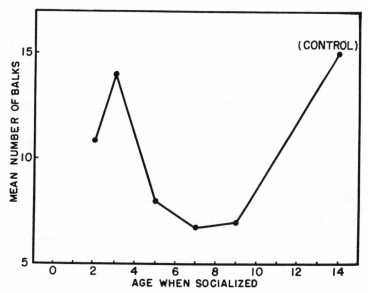

FIG. 5.4.—Performance in the leash-control test. Animals socialized at 5, 7, and 9 weeks of age balked fewer times when led into strange situations.

scores on the over-all test. Animals brought in for the first time at 14 weeks and those taken at 2 and 3 weeks did much worse. One 14-week animal was given intensive human contact and training for over a month and showed only slight improvement.

Part of the test included offering a food reward at the end of the route. Severely frightened animals refused to eat, and the best response was made by those animals taken at 5 and 7 weeks of age (Fig. 5.5). When we take all ages together and rank them according to performance on each test (Table 5.3), we see a definite trend. The

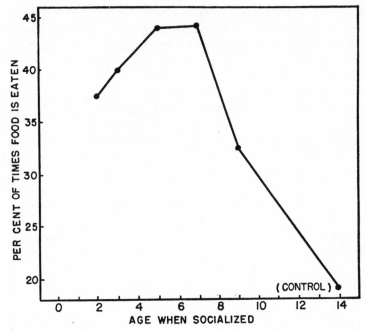

Fig. 5.5.—Eating during the leash-control test. Animals socialized at 5 and 7 weeks of age ate more often than the rest.

controls (14-week group) gave the poorest performance in 4 out of 6 tests, while the animals taken at 7 weeks were best in 5 out of 6. Next best were those taken at 5 weeks, confirming our assumptions regarding the length of the critical period. However, the relatively good performance of animals taken at 9 weeks shows that there is no

TABLE 5.3

Rank Order of Puppies on Tests Given After 14 Weeks of Age
(After Freedman, King, and Elliot, 1961)

	Age Socialized, Weeks					
Test	2–3	3–4	5–6	7–8	9–10	Controls
Handling:						
Initial attraction to handler............	5	4*	3*	2*	1*	6
Leash-Control:						
Eating in strange situation............	4	3	1.5*	1.5*	5	6
Fewer balks........................	4	5	3*	1*	2*	6
Reactivity Test:						
Total activity......................	2	4	3	1	6	5
Heart rate..........................	6	3	4	1	2	5
Vocalization, panting, and tail wagging..	2	3	4	1	5	6

* Distinctly superior ranks.

sharp cut-off point after 7 weeks. Rather, this seems to be the peak from which a decline in the capacity for socialization begins. The end of the period of socialization must be placed in the neighborhood of 12 to 14 weeks.

The results of this "wild dog" experiment also lead to the conclusion that the process of socialization is a complex one, involving several behavioral mechanisms. Similar experience at different ages will produce different results, but the final effects also differ with the kind of behavior tested.

BEHAVIORAL MECHANISMS
IN EARLY SOCIALIZATION

The effects of transferring the care of a young animal of one species to that of another during or before the normal period of socialization are so striking that one is likely to be highly impressed with the effect of what seems to be a simple behavioral factor. This is, however, a complex phenomenon involving many different behavioral mechanisms. We obtained some clues as to what these might be from the following experiments.

Many years ago we took a female lamb away from the mother at birth and raised it on a bottle for 10 days (Scott, 1945). We then returned her to the field in which the parent flock lived. She approached the other sheep curiously, but the females butted her, and she made no further approaches. Three years later this animal was still following an independent course around the field, not joining the flock, even though she had been mated and had borne lambs of her own. We then tried the same experiment with a young male which had been rejected by its mother and raised on the bottle. During the first two or three days he had some contact with other sheep, so that the experiments were not exactly comparable. The male acted in the same way as the female when he was young, but as he grew older and became sexually mature he became interested in the females and began to associate with the flock more and more. This showed that one of the positive mechanisms producing social relationships in the flock is the sexual behavior of males, although this factor is unimportant in females.

Later, Collias (1956) performed a series of experiments at the Cornell University Behavior Farm which clearly brought out some of the behavioral mechanisms of primary socialization in the sheep. If a lamb is taken from its mother at birth and returned at any time within four hours, she will accept it and allow it to nurse. After this time she rejects it, along with any other strange lambs. Thus the critical period for socialization in sheep is short and sharply limited, but limited by the behavior of the mothers rather than that of the lambs.

Meanwhile, we did an experiment with a young female puppy, similar to the one we had done with the newborn lambs, and expected to get a somewhat similar result. We took a hybrid pup (which soon became known as Christy) and raised her from birth completely away from other dogs, with the enthusiastic help of the entire summer student colony at the Jackson Laboratory. As might have been expected, she became a very friendly puppy, normal in health and appearance. At 9 weeks of age we introduced her to other dogs for the first time. She was somewhat fearful at the start, but not for long. Although adult animals did some growling at her, they made no real attacks. When she was placed with her own litter mates they reacted with playful aggressiveness, to which she rapidly responded. Within 4 days, she could not be told apart from the others except that she was somewhat more responsive to human beings. This indicated that one of the positive responses which leads to socialization is the playful aggressiveness of young puppies. It also indicated that rejection of strange young animals by adults is not as important in limiting socialization in the dog as it is in the sheep.

We can now begin to classify some of the behavioral mechanisms which enter into the process of socialization. There are positive mechanisms which tend to produce close contact during the initiation of a lasting relationship. On the other hand, there are negative mechanisms which prevent such a relationship from being set up with all animals. Under natural conditions these negative mechanisms prevent the formation of relationships with any other species and usually limit contacts to a few animals of the same species. As shown in the sheep, positive and negative mechanisms of behavior are found both in the young individual and in other members of the social group.

Individual mechanisms.—The positive response of any young puppy to a new animal or object in its pen is to approach it and investigate it with its nose, perhaps also biting and chewing it if this is possible. When the new animal is another puppy this may develop into playful fighting, with the puppies pawing and chewing on each other. As they grow older we also observe playful sexual behavior with clasping and mounting. Fairly early in the critical period we see the tendency toward allelomimetic behavior, with puppies following each other and reacting as a group. Social investigation, allelomimetic behavior, playful fighting, and playful sexual behavior all bring the puppies into contact and keep them together.

Ingestive behavior is probably not as important as the above in drawing the puppies together, since the puppies tend to growl over the possession of food even at a very early age, but food does enter into the relationship with the mother. The appearance of the mother is a signal for the puppies to crowd around her, either attempting to nurse or waiting for her to vomit food. This may be a vestige of the wolf "greeting ceremony," which involves tail wagging, licking, and muzzle biting in a stereotyped fashion (Ginsburg, 1963).

The most important negative mechanism is agonistic behavior. The immediate response of a puppy to anything new in its environment is to run away or show some sort of startle or freezing behavior. In a free situation, like that in the wild dog experiment, the capacity for escape develops rapidly, so that by the time a puppy is 4 months of age it is practically uncatchable. This fear response could probably be partially overcome, but it would take weeks of patient effort, just as it does when one attempts to make contact with a wild animal. Along with escape, the puppy develops the capacity for more serious fighting. Our experiments indicated that most of the puppies had developed dominance relationships by 11 weeks of age, and that these were well stabilized by 15 weeks. Thus the behavior of an older pup toward a stranger is likely to be a serious attack rather than playful fighting. This effectively prevents the formation of a social bond. Figures 5.7 to 5.9 show that the development of these behaviors is gradual rather than instantaneous.

Social mechanisms.—As well as reacting to its social environment, the puppy is acted upon by other animals. The first impact of these comes from the care-giving behavior of the mother. This is most intense during the neonatal period when the puppies lack the capacity to form a complex relationship. During the period of early socialization the mother still approaches the puppies, allows them to nurse, cleans them if they are dirty, and vomits food for them. However, her behavior during this period is much less intense and lessens as the puppies grow older. The mother acts as if her attachment to the puppies was strongest during the neonatal period, growing weaker thereafter.

Another positive mechanism acting on the puppy is the playful behavior of siblings. During the period of socialization a puppy is constantly approached by his litter mates and stimulated to this kind of behavior as well as initiating it himself.

Under normal circumstances there is very little opportunity for negative mechanisms to act during early life. The neonatal puppy lacks the ability to make contact with strange mothers, and even if it is placed with a strange female, the latter does not show the strong pattern of rejection seen in sheep. Many mothers will accept strange puppies without difficulty, although others have been known to kill strange pups offered for adoption. When older, the puppies can approach strangers on their own power. Adult dogs usually growl at strange puppies, although they seldom if ever actually hurt them. Such contacts usually happen long after the primary social relationships have been established and result in the establishment of dominance rather than separation.

A second negative factor is the developing aggressive behavior of puppies. This was brought out clearly in an experiment by Fisher (1955). He took litters of four from fox terrier mothers and raised them in boxes which isolated them from the sight of other dogs and people. Two from each litter were raised together so that each had the constant company of another dog. Another was taken out at regular periods for contact with people. The fourth animal was given no animal or human contacts until it was removed from the pen at 16 weeks of age. At this time Fisher attempted to put the litter back together. The puppy which had been completely isolated was strikingly different from the others, standing around in what appeared to be stupefied astonishment. The reaction of the other pups was to attack it. While the isolated pup fought back, it was ineffectual, and it always came out on the bottom of a dominance order. This meant that the first contact of the isolated puppy with other dogs was associated with being attacked and threatened, effectively preventing the development of a positive relationship. The experience of such puppies was like that of a bottle lamb attempting to rejoin the flock.

Fisher finally put four of these isolated animals together in the same pen and found that they lived peaceably together. However, they never formed a close positive relationship and played independently of each other. If they had been placed in a large field, it is likely that they would have behaved a good deal like the orphan lamb which stayed away from the flock.

Still further observations on these isolated puppies throw more light on their behavior. In connection with another project we raised a large number of beagles in isolation until 16 weeks of age (Fuller, 1961). When first placed with other puppies, they gave no reaction and the others either ignored them or growled at them. However, when we took these same puppies and began to handle and play with them, we elicited the same sort of playful fighting behavior which is typical of pups at a younger age. Within a few days they were reacting normally to people.

By 16 weeks these isolated puppies had lost the capacity to exhibit playful behavior spontaneously, although it could be elicited by external stimulation. By this age, normal puppies have concentrated this behavior in special relationships and no longer offer it to strangers. Thus another factor entering into the limitation of the capacity for socialization is the loss of the capacity for initiating playful behavior toward strangers.

GENETIC DIFFERENCES AFFECTING THE SOCIALIZATION PROCESS

One of our best objective measures of socialization in puppies is the handling test. In this test we do all the things which people ordinarily do to puppies. First we take them outside their home room and allow them a few minutes to become accustomed to the holding cage. Then the handler takes them back into the home room, one at a time. First he places the puppy a foot or two away, stands perfectly still, and notes any reactions which occur in the next 15 seconds. Then he moves slowly away, turns, and suddenly walks toward the puppy from a distance of 5 or 6 feet. This is the most frightening part of the test. The handler then squats down and holds out a hand toward the puppy. If it comes up and touches the hand, the experimenter lifts his hand into the air so that the puppy can jump up and touch it if it likes. Then the handler repeats the same behavior but calls the puppy at the same time. He then strokes the puppy gently five or six times and follows this by patting. At this point, the handler picks the puppy up and observes its reactions. Then he puts it down again and repeats the stroking and patting technique. The final part of the test is another 30-second period in which the handler stands perfectly still and allows the puppy free activity.

In recording data we mark on a check list whatever the puppy does as an immediate response to each type of stimulation. No attempt is made to estimate the intensity or duration of the response; only its occurrence is recorded. Thus the results give us the number of times in which a particular kind of behavior has been stimulated or, in other words, the number of times the threshold of response has been reached.

The different parts of the test can be thought of as stimuli of different intensity. Walking toward the animal is most likely to stimulate fearful reactions. Squatting and calling the puppy stimulates an approach, whereas patting and stroking are likely to bring out playful fighting, at least in older puppies.

Escape and avoidance behavior.—A very timid puppy will run away to the corner of the room, crouch, and give a high-pitched yelp which is unmistakably fearful in tone. When we measure this be-

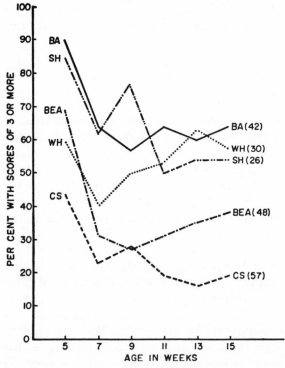

Fig. 5.6.—Changes in avoidance and vocalization in response to handling. In a well-socialized animal (here defined as one with a score less than 30), fearful responses are reduced to a minimum. The graph shows the relative numbers of incompletely socialized puppies at different ages.

havior, we find that the basenjis as a breed are much more fearful than the others at 5 weeks of age but show a great change by 7 weeks in response to the frequent handling received in daily testing between those times. Thereafter the amount of escape behavior stays quite level (Figs. 5.6 and 5.7). The other breeds show much less fear at 5 weeks and relatively little change at subsequent testing periods. Under the standard conditions of rearing up to 5 weeks of age, the puppies get little contact with people except when food is brought in once each day. This amount of contact is enough to prevent the development of fearful behavior in most of the breeds, but it is not enough for the basenjis.

Puppies raised under other conditions may not show these breed differences. For example, a puppy raised in a home from birth will show almost no fearful behavior toward people at 5 weeks, no matter what the breed; and our 3 home-reared animals showed a minimum amount of escape behavior in the handling test, their scores being located very near the lower end of the scale in their respective

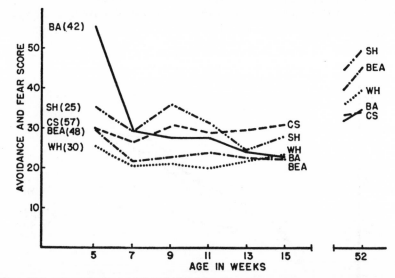

FIG. 5.7.—Avoidance and fearful behavior in response to handling. Basenjis have much higher scores at the outset, and the scores of all breeds fall during the next 2 weeks. At one year of age dogs respond to this test as they would to catching, and the scores of all breeds are markedly higher except for cockers.

breeds. Under these conditions, breed differences tend to disappear.

Breed differences also disappear when puppies are given the maximum opportunity to develop fear responses in the large fields. From our experiences in the laboratory we had expected that cocker spaniels and beagles would become naturally friendly despite their lack of contact with human beings. Such was not the case. Even cockers became as wild and fearful as other breeds, the only difference being that when caught they did not bite quite as hard, probably because of selection in the cockers for a "soft mouth," useful in retrieving.

Our results indicated that the basenjis had a greater capacity for developing escape behavior, probably because, as African village dogs, such behavior had survival value, whereas in other breeds this trait is considered undesirable by most dog owners. We can also conclude that such fear responses can be greatly modified in either direction by training and experience, and that genetic differences are less important at either extreme.

Aggressive behavior.—Breed differences in aggressive behavior can be measured in the handling test, but we never actually observed any serious attacks or even threats. All reactions were strictly playful, and the most extreme ones consisted of pawing and gently biting the experimenter's hands or clothing. As seen in Figure 5.8,

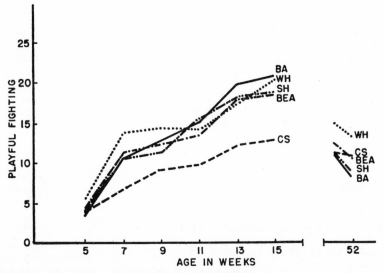

FIG. 5.8.—Agonistic behavior (playful fighting) in response to handling. This score consists chiefly of playful biting and pawing. Note that cockers are consistently lower than other breeds and change the least at 1 year of age.

this behavior reaches its height about 13 to 15 weeks of age. We should remember, however, that most of it is not spontaneous but stimulated by the active approach of the human handler. There appear to be definite breed differences, with wire-haired terriers at the top and cocker spaniels at the bottom of the scale. The results generally agree with the reputation of the breeds for fighting ability, except that the beagles show a surprisingly high rank, and this is a breed which shows almost no tendency toward real fighting. It appears that playful aggressiveness and serious aggressiveness are not necessarily correlated.

We also have confirming evidence from dominance tests. The best indication of aggressiveness is the existence of "complete dominance," in which one puppy always takes command over a bone presented to a pair. The wire-haired terriers develop a large number of these relationships before 15 weeks, other breeds being slower (Fig. 6.1). By one year of age shelties and basenjis have nearly caught up, but beagles and cocker spaniels never develop a large number of these relationships. We can conclude that the terriers are the most aggressive, the basenjis and shelties next, and the other two breeds much the least. We can also suppose that the socialization of a strange puppy to a group of fox terriers would be prevented by aggressiveness before 11 weeks of age (which corresponds to our observations of the severe fighting that breaks out in this breed as early as 7 weeks), and that the socialization of a stranger to a basenji group would be inhibited somewhat later, perhaps by 15 weeks. In the more peaceful breeds, this factor might be quite unimportant.

Social investigation and attraction.—One of the primary reactions involved in socialization is the puppy's tendency to come toward a handler and investigate him, usually wagging its tail at the same time. This kind of behavior often appears in the handling test and reaches its height about 7 weeks of age, staying at a fairly constant level thereafter (Fig. 5.9). Such activity may be divided into initial attraction and subsequent investigatory behavior. Considering attraction alone, the basenjis appear to be less attracted to people at all ages (Fig. 5.10). Shelties are low at certain ages, and cockers are consistently high.

A somewhat different picture is presented by the social investigative behavior shown after the puppies reach the experimenter and begin nosing his hands and clothes (Fig. 5.9). The beagles exhibited more investigative behavior than other breeds, although cocker spaniels were slightly higher at 5 weeks, just as they were in the attraction score.

The development of tail wagging.—The pattern of rapid, horizontal tail wagging is one of the most commonly noticed behaviors in domestic dogs. It begins at a very early age and continues as long as the dog lives, appearing most often in situations in which the dog is friendly and submissive. It has no function other than a social one, and in all these respects is much like the human smile.

Its development can be charted from the records of our daily ob-

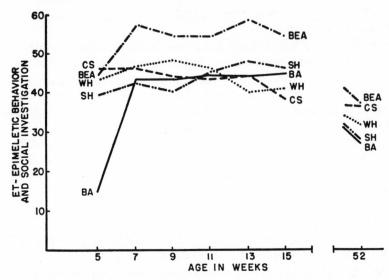

FIG. 5.9.—Et-epimeletic behavior and social investigation in response to handling. The bulk of this score is made up of tail wagging, often accompanied by nosing. Note that beagles are consistently higher than other breeds. Cockers again show little change at 1 year, indicating that this breed shows persistently immature social behavior.

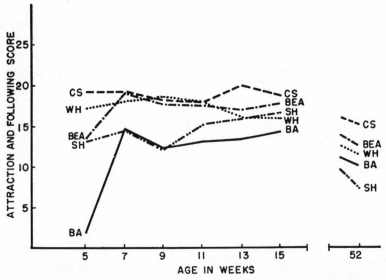

FIG. 5.10.—Attraction and following in response to handling. Basenjis are consistently low in this score compared to other breeds, but shelties show the greatest decrease at 1 year.

servations. A research assistant watched each litter for 10 minutes a day and recorded everything that the puppies did, later checking each item on a standard list. In this way we have a record of the first day on which each puppy was seen to wag its tail.

Tail wagging first appeared at 17 days of age (Table 5.4*a*). By 30 days one-half of all the purebred puppies had been seen wagging their tails. This behavior thus begins early in the period of socialization and rapidly increases in frequency.

TABLE 5.4*a*

DEVELOPMENT OF TAIL WAGGING

BREED	TOTAL NUMBER		DAY FIRST OBSERVED		MEDIAN, DAYS		LAST ANIMAL OBSERVED, DAY	
	Litters	Animals	♂	♀	♂	♀	♂	♀
Basenji...........	6	33	28	28	43	34	88	89
Beagle...........	10	39	21	19	30	30	47	68
Cocker...........	9	42	17	17	23	23	37	35
Sheltie...........	6	25	17	19	27	31	38	42
Fox terrier........	5	21	18	18	36	31	64	58
All pure breeds....	36	160	17	17	30	30	88	89

When we compare the different breeds we see certain obvious differences. By 18 days of age one-fourth of the cocker spaniels were seen wagging their tails, whereas it was not until 30 days of age that basenjis were observed doing this. At this latter age 83 per cent of the cockers had been seen tail wagging. The rest of the breeds were very similar to each other, with about 50 per cent tail wagging by 30 days of age (Table 5.4*b*).

TABLE 5.4*b*

DEVELOPMENT OF TAIL WAGGING: OCCURRENCE AT 30 DAYS OR EARLIER

Breed	Litters	Total Number of Litters	Individuals	Total Number Individuals	Proportion of Individuals
Basenji.............	3	6	9	33	.27
Beagle.............	7	10	19	39	.49
Cocker.............	9	9	35	42	.83
Sheltie.............	5	6	14	25	.56
Fox terrier.........	3	5	9	21	.42
All pure breeds......	27	36	86	160	.54

The differences between cockers and basenjis are highly significant, and both breeds also depart widely from the average of all breeds. We therefore have an important breed difference in the development of a form of social behavior closely connected with the process of socialization. However, the difference between basenjis and cockers is probably one of thresholds of stimulation rather than rate of development, since there is almost no difference in tail wagging when both breeds are strongly stimulated, as they are when they are being given their weekly weighings. Cockers apparently wag their tails in response to a slight stimulus, while basenjis require stronger stimulation.

These results show that genetic factors have important effects on the process of socialization. Heredity affects both the individual reactions which limit socialization from within, such as fear responses of the basenjis, and the group reactions limiting socialization from without, such as the aggressiveness of the fox terriers. In addition, genetic factors can affect the positive behavior of approach and social investigation.

One possible hypothesis is that these differences are the result of a single trait of fearfulness which could, of course, affect all the others. This, however, will not explain the results. The basenjis are the most fearful at first and show the least attraction. However, they are not the lowest in social investigation and are actually second highest in playful aggressiveness. We are obviously dealing here with a number of separate traits with different genetic bases.

This separation of traits also gives us a clue toward understanding the organization of behavior. Our usual assumption is that fearfulness and aggressiveness are the opposite ends of a single behavioral scale. This is not the case with dogs. A breed may have the capacity to become both highly fearful and highly aggressive, as in the basenji; or highly aggressive, but not fearful, as in the fox terrier; or showing neither extreme, as in our other three breeds. We also see that even aggressiveness is not a unitary trait, for a breed like the beagles can show both a high degree of playful aggressiveness and a very low amount of serious aggressiveness.

In general, the behavior of domestic dogs shows the result of selection against the development of the escape and avoidance behavior normally found in wolves. This means that the development of one of the factors limiting the period of socialization has been delayed, so that successful socialization can be achieved at a much later age. Murie (1944) achieved good socialization with a wolf puppy taken before the eyes opened. Crisler (1958) apparently was somewhat less successful with puppies taken at a slightly later age, although the situation was not strictly comparable, in that she raised a pair rather than a single individual.

Ginsburg (1963) has recently demonstrated that wolves captured as adults can be successfully socialized. The process takes many days and the wolves must be kept in a pen so they cannot escape. The experimenter enters the cage and maintains a passive role until the wolf makes its first positive advance. At this point he must neither threaten nor withdraw from the wolf, but must respond appropriately and positively. If there is successful mutual adjustment, the wolf will thereafter accept not only the experimenter but other human beings as well. Such results indicate that the capacity for forming new social relationships is not lost in an adult but is ordinarily prevented from functioning by escape and fear reactions.

There has been selection for decreased aggressiveness within many dog breeds in order to make them easier to handle. In the case of the terriers, however, selection for aggressiveness has proceeded in the opposite direction. This does not seem to have limited the capacity for socialization, since one of the positive mechanisms of socialization is playful fighting, but rather to have limited the capacity for group living. There is no indication that the tendency toward social investigation has been prolonged in any breed, except by the elimination of fearful behavior which interferes with it. As long as they are not afraid, members of all breeds act like typical friendly puppies until at least 15 weeks of age. Some breeds show more playful behavior than others, but there is no true prolongation of an immature social state.

THE CRITICAL PERIOD IN OTHER SPECIES

The fundamental technique for testing the existence of the process of socialization is to foster a young animal on another species. With higher animals it is often easiest to do this by hand rearing, but the living requirements of others such as the social insects may be so difficult to duplicate that they can only be fostered on similar animals. If the fostered animal transfers its social relationships to the new species, we can conclude that socialization has taken place. The next step is to attempt this at varying ages in order to test for the existence of a critical period for the process. Such experiments have so far been done with only a limited number of species, but results are generally consistent.

Insects.—Ant colonies usually attack strange species and often give the same reaction to strange members of the same species. If young ants are experimentally placed in a strange colony they become a permanent part of it, being cared for by the resident workers.

This regularly takes place in the slave-making ants, which raid colonies of other species and bring back young. These captured individuals then grow up and take care of the young of their masters. No attempt has been made to find a critical period, but it must end soon after the adults emerge from their pupal cases, if not before.

Honeybees also recognize members of their own colony. As with ants, recognition is based on scent and has been shown to be an acquired rather than an inherited reaction. Furthermore, bees of two colonies can be brought together without fighting provided they have been fed on identical diets for a sufficient length of time (Ribbands, 1953). There is no direct evidence as to when the scent of the whole home colony is learned, but it must be early in life because the workers tend to take up guard duties before they become foragers. Before becoming guards, the young bees work around the hive, cleaning the cells, so that there is ample opportunity to learn the scent of the hive. Thus, socialization does take place among social insects, and probably during a short period early in life.

Birds.—The great importance of the process of primary socialization was first recognized in birds (Lorenz, 1935). The naturalist Heinroth had attempted, with characteristic German thoroughness, to hand rear all of the common European species of birds in order to learn more about their development and behavior. Some of the results were spectacular examples of transference of social relationships from one species to another. In one case, although the adult bird mated with a member of its own kind, it was attracted away by its foster parent whenever he appeared. Konrad Lorenz extended these observations and made more specific studies of two species, the graylag goose and the jackdaw. Geese are precocious, being able to walk and swim a few hours after hatching, whereas jackdaws, which are related to the crow family, are hatched in an immature state. Lorenz found that newly hatched geese formed their social relationships within the first few hours, whereas jackdaws did this more gradually at a later stage of development. Once these relationships were formed, it was difficult if not impossible to restore the natural relationships. Lorenz called the process of forming this primary social relationship *Prägung*, which has been translated as "imprinting." The word also means impress, and this might have been a better translation. At any rate, the young birds seem to be deeply "impressed" or "imprinted" by a limited experience early in life.

The fundamental importance of these results in relation to theories of psychiatry regarding early experience was soon recognized. Fabricius (1951), Collias (1952), Hess (1959), and others followed up Lorenz' work with detailed analytical studies. In precocious birds like ducks or chickens, there is indeed a critical period. It is easiest to demonstrate imprinting in the chick at about 13 to 14 hours after hatching, and difficult if not impossible by 30 hours. The primary behavioral mechanisms which limit the period are, at the beginning, the development of the ability to move toward a strange object or person and, at the end, the development of a fear response which keeps the birds from approaching strange objects. Still other mechanisms are involved; as Hess (1957) has shown, a duckling becomes more firmly imprinted if it has to make an effort to follow a model.

Comparable studies have not been made with the more slowly developing birds, except for the very excellent descriptive studies of Margaret Nice on song sparrows (1943). It is evident that such birds are much more like the slowly developing mammals than are chicks and ducks, and that comparable periods of development exist (Scott, 1962b).

Mammals.—Many species of mammals other than dogs and wolves have been experimentally hand reared, and most of these show a spectacular attachment to their human foster parents. Chimpanzees, for example, readily become members of human families when raised in homes (Kellogg & Kellogg, 1933; Hayes, 1951). There have, however, been almost no detailed analytical studies of the critical period for socialization except in dogs, sheep, and guinea pigs (Gray, 1958). These last two species are highly precocious and show a short critical period very early in development. There are indications of a critical period in rhesus monkeys during which contact with other young animals of the same species is necessary for the development of both sexual and maternal behavior (Harlow *et al.*, 1963).

Mammals in general differ a great deal in their speed of development, some being born in a relatively mature state, and others being highly immature. There may be a great deal of variation even in closely related animals. For example, rhesus monkeys show rapid and precocious development, chimpanzees and other great apes are much slower, and human infants are slowest of all.

In summary, we can state that all highly social animals which have been so far studied show a critical period for socialization early in development. The more precocious the animal, the shorter and earlier the period is likely to be. The behavioral mechanisms which limit the period differ from species to species and cannot be predicted in advance. However, a developing fear reaction is a common mechanism.

It would be surprising if human beings did not show at least the vestiges of a similar critical period. This conclusion will receive greater support if the generality of the phenomenon of critical periods can be extended by studies on a wider variety of vertebrate and invertebrate species.

BASIC NATURE OF THE PRIMARY SOCIALIZATION PROCESS

We have so far talked about the behavioral mechanisms which facilitate or limit socialization but little about the process itself. When Lorenz (1935) first wrote about the importance of imprinting, he stated that the process was quite different from that of conditioning, in that it occurred very rapidly and the results seemed to be permanent. Psychologists were quick to point out that conditioning can also occur very rapidly, "one-trial learning" being commonplace in higher animals. Furthermore, Pavlov had emphasized the long persistence of conditioned reflexes. Even when a reflex had been extinguished by non-reinforcement, it always returned spontaneously after an extended period of rest.

An attractive and simple alternate theory was that the emotional bond between mother and offspring was produced by feeding. A child associated its mother with the pleasure of eating, and this formed the basis of the attachment which was later more generalized. This idea was implicit in Freud's description of the oral stage in human development and has since been elaborated as the theory of acquired drives (Miller and Dollard, 1941).

Brodbeck (1954) was the first to test these ideas on the dog. He raised a litter of cocker spaniels and a litter of beagles, taking them away from the mothers at 3 weeks of age. He fed half of them by hand, and fed the other half by a system of ropes and pulleys so that the puppies never saw who fed them. He then gave both groups the same opportunity to have direct contact with him and later tested their reactions as he sat quietly in the same room. Both sets of puppies were highly reactive toward him, and those of one group stayed with him just as closely as those from the other. We can conclude that feeding is not a necessary part of the development of the social bond. Stanley (1962) and Feider later did a similar experiment with two litters of beagle-terrier hybrid puppies. The hand-fed puppies would vocalize more at the sight of the experimenter, but this was the only important difference between the two.

Elliot and King (1960) did the opposite sort of experiment in that they hand fed a group of puppies but did very little else with them. Half of the puppies were given all they could eat and the other half were underfed. When these puppies were given a handling test, the underfed group showed more positive responses to the handlers, but both groups showed more timidity than other puppies which had been raised in the regular "school for dogs" program in which they received regular handling. From this experiment we can conclude that feeding by itself does not produce a highly socialized animal.

Stanley and Elliot (1962) conducted a series of experiments in which they painstakingly attempted to discover the factors which make a human being attractive to young puppies. The basic experiment is a modification of the passive-handler test. Beginning at about 6 weeks of age the puppies are weaned, never being fed by the experimenters. They are then taken one at a time to a different room, placed in a small box, and allowed to run toward an experimenter. This is repeated daily for several weeks. The puppies become highly socialized to human beings, even if the person toward whom they run does nothing more than sit passively. Basenji puppies appear to be more attracted by a person who is completely passive than one who attempts to pet them. All puppies become closely attached to human beings with no more than the daily contact involved in the experiment.

Fisher (1955), in his experiment with isolated fox terrier puppies, treated one group by punishing them whenever they made any positive approach. These puppies were reared in special boxes from 3 weeks of age and were never fed by hand. As long as he continued to punish them, the fox terriers stayed away; but once he stopped they almost immediately overcame their fear and came toward him. In fact, they paid much more attention to him than a comparable group of puppies which he had treated with uniform kindness. The experiment shows not only that food rewards are unnecessary in getting puppies to become attached to people, but that they will form an attachment in spite of considerable punishment. We must remember that these were fox terrier puppies which have been selected to take a good deal of punishment in fighting, and the results might have been different with a more sensitive breed.

Meanwhile, Harlow (1958) had been performing an interesting experiment with rhesus monkeys. These are quite precocious animals as primates go, and normally cling to their mothers immediately after birth and continue to be carried by their mothers for several months. Harlow took the baby monkeys away from their mothers at birth and offered them a choice of two imitation mothers. One was deliberately intended to be uncomfortable, being made of wire mesh hardware cloth, and the other was covered with soft terry cloth. A nursing bottle was inserted in the hardware cloth mother, and the infant got all its food in this way. Harlow found that the baby monkeys spent all their time with the comfortable mother, only going briefly to the other in order to nurse. The baby monkeys gave every evidence of being profoundly attached to the terry-cloth mother, and Harlow concluded that the theory that the attachment of the infant to the mother was an acquired drive based on food rewards could be abandoned. He was inclined to consider "contact-comfort" as the essential element in the process.

Igel and Calvin (1960) wondered what would have happened if the monkeys could have chosen between two comfortable mothers, one of which produced milk and one which did not. Not having monkeys available, they did a corresponding experiment with young puppies and found that, if the puppies were equally comfortable, they spent more time with the mother that produced milk. In short, food rewards will still affect the behavior of a young animal, even if these are not the primary element in the process of socialization.

Meanwhile, other psychologists had been experimenting with the imprinting of baby chicks on various sorts of models. At first it was thought that the model must move to be effective, but later James (1959) showed that chicks became imprinted on a model which was merely exposed to flickering light. Finally, Gray (1960) did an experiment in which baby chicks were exposed to a motionless model which they could see but not touch, and their subsequent behavior indicated that they had become imprinted. The baby chicks therefore became attached to an object which neither rewarded nor actively stimulated them in any way.

Taken together, the evidence from these different species of mammals and birds leads to the conclusion that the process of forming an emotional attachment to members of the parent species is largely independent of outside circumstances. Whether rewarded, punished, or treated indifferently, the young animal of the proper age proceeds to form an emotional attachment to whatever is present in the environment at that time. The essential mechanism appears to be an internal process acting on the external environment. In this way it is indeed quite different from conditioning, which is directly dependent on outside circumstances.

This hypothesis fits much of the experimental and observational evidence. To state it more clearly: a young animal automatically becomes attached to individuals and objects with which it comes into contact during the critical period. The capacity to do this need not be lost at later ages, but the process can be slowed down or prevented by the development of interfering behavior, particularly fear responses. Likewise, the process can be prevented at later ages by the decline of positive behavior which normally brings the young animal into contact with others.

HUMAN APPLICATIONS

We have reasoned that there is a high probability that a critical period for primary socialization exists in human development. However, this conclusion can be verified only by direct observation and experiment on human infants. One result is the finding that, as we pointed out in the last chaper, the sequence of developmental processes in man and the dog occur in a somewhat different order (Table 4.6). The period of primary socialization in human infants extends from approximately 6 weeks to 6 months, thus preceding the period of transition to the adult methods of feeding and locomotion rather than following it. This means that the baby can form its primary relationships only with those persons who take care of it, as it is unable to make contacts on its own; and it also means that the first and probably the deepest relationship will be formed with the mother rather than with the father or siblings. All these conclusions are based on descriptive evidence. Experimental work of the kind done with lower animals is difficult if not impossible with babies, and all we have to go by are the results of various accidents and variations in human behavior (Gray, 1958).

There have been a few isolated cases of children who have been confined by parents or caretakers in social isolation comparable to that of a puppy raised in a small box. The best known of these cases is that of Kaspar Hauser, a youth who turned up in Nuremberg, Germany in the year 1828 (Singh and Zingg, 1939). No one was able to verify what his actual previous experience had been, but he apparently had been given an elementary education while kept in close confinement from babyhood until nearly an adult. Since this was before the development of the science of psychology, the reports concerning him were made by a judge rather than a trained scientist. He described the boy as good natured and socially responsive but completely naïve about the outside world. There was no evidence of unusual fearfulness. For what it is worth, the description of Kaspar Hauser's behavior sounds somewhat more normal than that of puppies reared in isolation.

Wolf children.—Deliberately fostering a human baby on another species is completely impossible from a humanitarian viewpoint, but there are occasional news reports of this occurring by accident. These somewhat legendary accounts usually concern "wolf children" in India, and have been fictionalized by Kipling in his jungle books, but one such report has a more scientific basis. In the 1920's a Reverend Singh was said to have found two "wolf children" in India and brought them to an orphanage, where he attempted to educate them (Singh and Zingg, 1939). According to his account, he found the two children living in a wolf den from which wolves also emerged. This, of course, would not prove that the children had been reared by wolves but simply that wolves and children were found together. Ogburn and Bose (1959) later made an intensive effort to check up on this story but were able to verify very little except that there had actually been a Reverend Singh and that there were newspaper stories about wolf children at that time. The accounts of living persons who remembered the incident were highly contradictory.

Apart from the dubious nature of the evidence, the chances that a human baby could actually be reared by wolves are so small that we can discount such reports as fantasy unless some trained observer actually sees a baby being mothered by a wolf. Wolves, like dogs, nurse their offspring for approximately 7 weeks, after which milk is no longer available. As early as 3 weeks, they begin feeding their offspring on vomited food and meat which is often partially spoiled, particularly in warm climates. Even in a primitive human society, a baby has to be nursed for two years or so in order to survive, and there is usually a high infant mortality rate from intestinal infections. It is conceivable that a wolf might adopt a human baby but not that the baby could survive with only wolf care.

It is significant that these reports of wolf children come from areas of great poverty where child neglect and desertion are common. It is quite possible that children in such areas have been abandoned by their parents considerably later than early infancy and have been able to survive and "run wild." Judging from Ginsburg's success in socializing adult wolves, it would even be possible for such a child to be accepted and tolerated by wolves. As scientific evidence, however, the reports concerning "wolf children" have little value because it is impossible to check what actually happened to them before they were found. For what it is worth, such children are reported to be wild and fearful, like the puppies raised entirely apart from human beings.

Adoption.—A less drastic type of fostering frequently takes place in the human practice of adoption. For one reason or another babies may be taken from their natural parents and either raised in orphanages or adopted by other individuals. From our knowledge of human development, we would expect that a baby could be transferred from one parent to another during the first two months without the baby's noticing what had been done as long as maternal care was sufficient. From 2 to 5 months of age, we would expect that the baby would be emotionally disturbed by the change but would make a relatively easy adjustment to the new parent, just as puppies seem to do during the period of primary socialization. Beyond this age the baby would have formed definite relationships which would be broken off with considerable emotional disturbance, making adoption more difficult, both because the baby would not readily form new relationships with strangers and because the foster parents would be upset by the emotional behavior of the child.

The actual studies of adopted children are not impressive as scientific experiments, but people who work with children agree on two things. One is that children adopted before 6 months of age get along better than others, and the other is that prolonged experience in an orphanage produces bad results on many children. Another conclusion is that the results are not always bad, since many children turn out well in spite of these early circumstances. This variation, of course, could be caused by differences in genetic constitution of the individuals involved. Some children may be more resistant to emotional distress.

Another kind of "accidental experiment" is produced by hospitalization of very young children. Here the baby may be taken away from its parents suddenly and put with strangers for days or weeks. In addition, it may be suffering from illness and physical pain in these strange surroundings. Bowlby (1951) and others have found that such children show every evidence of severe emotional disturbance at the time, despite efforts by nurses to keep them reasonably comfortable and happy. Bowlby also found, working in the other direction, that maladjusted children frequently had suffered such drastic early experiences. He concluded that such children were sometimes unable to develop deep emotional relationships in later life. If true, this would mean that the process of socialization had been definitely disturbed.

Looking at these experiences in another way, we can see that for a young social mammal living under natural conditions, a situation in which it is separated from its parents in strange surroundings is acutely dangerous. Such an animal would be unlikely to survive unless it gave a strong emotional reaction which might attract the attention of its parents. Young puppies certainly react in this way, and it may also be true of babies. Why a prolonged separation and the resulting emotional disturbance would disturb the relationship between mother and child is a matter of conjecture, but we can suppose that a baby might conclude in a primitive way that the parents themselves were the cause of his emotional discomfort and anticipate that they might abandon him again and produce the same painful feeling.

These are, of course, human phenomena into which subjective reactions always enter. Our animal experiments do give support to current ideas regarding the desirability of early adoption and the keeping of children in families rather than in orphanages. Late adoption is, of course, still the only practical course of action in many cases, but it must be done with a great deal of care and patience, including the realization that the formation of an emotional attachment by the older child will be a much slower and less automatic response than in a young infant. By adding to our understanding of the behavioral mechanism involved in primary socialization and adoption, the animal experiments also suggest ways in which psychological damage may be avoided when children are placed in such difficult situations by accident or necessity.

CHAPTER 6

THE DEVELOPMENT OF SOCIAL
RELATIONSHIPS

INTRODUCTION

A social relationship may be defined as regular and predictable behavior occurring between two or more individuals. This definition applies not only to relationships in animal societies but to those in human societies as well, where relationship systems form a fundamental type of social organization. In human societies a relationship between two relatives such as father and son consists of both the observed behavior between them and a system of verbal rules which may or may not correspond to the actual behavior. In an animal society the verbal element is, of course, missing.

Psychological origin of a relationship.—When two puppies meet for the first time, each represents a problem to the other. If one puppy initiates playful fighting, will the other respond in kind, or by passive submission, or by running away? At the first meeting the puppies may try out a variety of solutions to the problem. With subsequent experience, they work out some sort of solution, often on a basis of trial and error, and begin to reduce it to a habit, omitting much of the original behavior. Thus a social relationship begins as a problem and ends as a simple habit.

From what we know about learning and habit formation, we can predict that habits are not invariable; in fact, there is a normal tendency for an animal to vary its behavior even in what appear to be identical situations. This is, of course, an essential part of the process of learning, for without it no improvement of adaptation would be possible. The amount of variation is reduced in frequently repeated behavior but never entirely eliminated. We also know that habits are not unbreakable. A habit which once led to successful adjustment, but no longer does so, may be repeated for a short while but soon dies out. Therefore a social relationship is neither invariable, nor incapable of change. However, social relationships under the proper conditions can become extremely stable and invariable.

The differentiation of behavior.—When we apply the idea of development to social relationships, we can see that there are two kinds of development. One is the rapid psychological development of a relationship which may occur within a few days or even hours, and the other is the long-term biological development of a relationship based on the slower processes of growth and biological change. These two kinds of development may both be described in terms of differentiation of behavior. In a relationship based on playful fighting, two puppies at first show much the same kind of behavior. At the end of the relationship, one may always attack and the other one always submit; their behavior is now differentiated on a psychological basis, and the process of learning may result in a very slight or a very high degree of differentiation.

This psychological process is also affected by the other type of differentiation of behavior. A newborn puppy has relatively few alternate behavior patterns. As it grows older, the capacities for a wider variety of behavior patterns appear. Whereas at first the puppy could only yelp in response to discomfort, it can now run away or attack the source of discomfort as well as vocalizing. Thus the puppy develops through biological processes the capability to differentiate its behavior in different situations.

More than this, genetic factors differentiate the behavior of one individual from another. In social mammals there are basic differences in behavioral capacities between males and females and between adults and young. Within each of these main types, individuals may be affected by differential heredity and so show different kinds of behavior. It is the importance of the latter phenomenon with which this chapter is mainly concerned.

Fundamental classification of social relationships.—In his field studies of primates, Carpenter (1934) developed a classification of observed relationships. With three types of individuals—males, females, and young—it is possible to have three relationships between different kinds of individuals (male-female, male-young, and female-young) and three more between the same kinds of individuals (male-male, female-female, and young-young). In the case of the dog, which regularly forms relationships with human beings, six basic kinds of individuals are involved. This means that there are fifteen relationships possible between different kinds of individuals and six between like types, making a total of twenty-one (see Table 6.1).

TABLE 6.1

Social Relationships* of Dog and Man

	Dog			Human		
	Male	Female	Young	Male	Female	Young
Dog						
Male............	♂♂					
Female..........	♂♀	♀♀				
Young..........	♂Y	♀Y	YY			
Human						
Male............	♂♂	♀♂	Y♂	♂♂		
Female..........	♂♀	♀♀	Y♀	♂♀	♀♀	
Young..........	♂Y	♀Y	YY	♂Y	♀Y	YY

* There are 21 possible relationships, using all combinations of age and sex: 9 dog-human relationships (lower left corner of table) and 6 each of dog-dog and human-human combinations (upper left and lower right corners).

In our experiment, the relationships formed were limited by the system of rearing. Among dogs, only two relationships were studied seriously, those between mother and young and those between young belonging to the same litter. The dog-human relationships were likewise limited. Since our experimenters were all adults, the chief relationships studied were the human male versus young dog and human female versus young dog.

Analysis of social relationships.—Since it is defined as the behavior exhibited by two individuals, a social relationship such as that between mother and offspring must consist of behavior patterns belonging to one or more of the nine important systems of behavior described in chapter 3. This makes it possible to analyze into various sorts of behavioral adjustments the relationship existing between any two individuals. Since there are nine systems of behavior, there are likewise nine possible relationships in which both individuals exhibit the same patterns of behavior. If two individuals respond to each other with unlike types of behavior, a total of thirty-six more relationships are possible. In actual practice only a few of these theoretical relationships appear to be important in any one species (Scott, 1953b).

In the mother-offspring relationships of dogs, the predominant behavior of the mother is epimeletic. The puppies exhibit et-epimeletic behavior (whining and yelping when in distress), ingestive behavior, and eliminative behavior. We can call this type of relationship *care-dependency*. Other kinds of relationships may be developed as the puppies grow older. In the process of weaning, the mother keeps the puppies from nursing by growling and threatening them, thus establishing a *dominance-subordination relationship.* Other systems of behavior are exhibited by the mother but have relatively little importance. Since the mother comes to feed the puppies while they remain in one place, there is little opportunity for a *leader-follower relationship* to develop between them, and the mother usually discourages any attempts at sexual play on the part of the puppies.

The most important relationship between litter mates is based on agonistic behavior. The puppies gradually develop dominance-subordination relationships which are mainly concerned with the distribution of food but which may also affect the distance between individuals as they occupy their living space. The puppies in a litter show a great deal of allelomimetic behavior with mutual following, but there is little or no indication of important leader-follower relationships. Usually, the first puppy to notice a strange object will run toward it and the rest will follow, and which puppy initiates such a movement is a matter of chance. As for sexual behavior, while puppies exhibit much playful behavior of this sort, the development of adult *sexual relationships* was not permitted by the design of our experiment. Puppies do very little grooming either of themselves or each other so that the *mutual care relationship* seen in primates has little importance. The puppies have the capacity for developing *mutual defense* and *coordinated-attack relationships* but again the conditions of rearing prevented this from developing, since contacts with strangers were not permitted.

As might be expected in a school for dogs, the dog-human relationship is more complex. The human handlers and caretakers take over the care-dependency relationship from the mother dog and maintain it throughout the lives of the puppies, so that in this relationship the domestic dog never becomes an adult. Likewise, the human handlers develop a dominance-subordination relationship with the puppies in which the handler is always dominant. By means of various control methods the handlers also develop something of a leader-follower relationship with the puppies.

Social control.—Any social relationship can be analyzed in terms of control. In most relationships, the animals obviously control each other and the concept becomes important only when control is unequal, as it is in a dominance-subordination relationship or a leader-follower relationship. The dog-human relationship is complicated by deliberate attempts to extend and increase the amount of social control. Two devices are important: one is an extension of the care-dependency relationship in which the dog's food is made conditional on his performing certain actions, and the other is an extension of the dominance-subordination relationship in which the dog is forced to perform certain acts such as following on a leash or sitting. As many observant writers have pointed out, these methods are very similar to those once employed in human slavery. From a psychological viewpoint they can be called *reward training* and *forced training*. In either case the balance of control is shifted strongly toward the human member of the relationship.

One of the obvious features of the dog-human relationship is its many resemblances to the human parent-child relationship. Its development therefore brings the elemental problems of child psychology into sharp focus. The successes and mistakes of parents with their children are clearly mirrored in the reactions of young puppies to their handlers.

THE DEVELOPMENT OF DOMINANCE RELATIONSHIPS BETWEEN LITTER MATES

The dominance test.—We deliberately set up our conditions of rearing so as to minimize competition. Our puppies were fed on dry food and milk. Dogs never fight over the possession of liquids and seldom attempt to defend a feeding dish in which dry dog food is always present. In order to study the development of dominance, we therefore had to introduce a competitive situation which we could control and observe. From 2 until 10 weeks of age each litter was given a weekly dominance training period. The mother was removed and a single bone placed in the pen for 10 minutes, during which time the reactions of the puppies were recorded.

Unlike liquid and dry dog food, a fresh bone is a strong stimulus to agonistic behavior. Even puppies of 2 weeks of age will accasionally growl and bark when given a meat-covered bone. This is one of the few cases in dogs of a specific primary stimulus producing a behavior pattern and is similar to the many cases of "releasers" in birds.

As the puppies grew older, there was more and more competition over the bone, and occasional fights broke out. At 5 weeks of age we gave a detailed dominance test instead of the usual training period and repeated this at 11 and 15 weeks of age.

In the dominance test all the puppies were taken out of the pen and brought back, one pair at a time. When they had quieted down, a single bone was brought in, shown to both puppies, and laid between them. An observer then recorded the behavior of both puppies for 10 minutes, timing the possession of the bone by each puppy and noting the occurrence of growls, barks, attacks, and other items of agonistic behavior.

All degrees of dominance appeared in the interactions between the puppies. Very rarely, two puppies would share the bone. Others appeared to take turns; i.e., as long as a puppy held the bone he was dominant. Still others struggled with each other continually, and in some pairs one puppy would immediately seize the bone and hold it for the entire 10-minute period. We decided to define dominance arbitrarily as a condition in which one puppy kept the bone for at least 8 out of 10 minutes. In order to prove that this was real dominance and not a situation in which the puppy was allowed to keep the bone simply because he got it first, we made a check test at the end of the period by taking the bone away and giving it to the apparently subordinate member of the pair. If the subordinate animal could keep it, we said that the first animal was incompletely dominant. If the bone was taken away, we said that the first puppy was completely dominant. In short, we defined a completely dominant animal as one that kept possession of the bone the majority of the time and was able to repossess it at will.

The results showed that little dominance had been developed at 5 weeks. Not more than 25 per cent of the tests showed complete dominance in any breed at this age. By 11 weeks, all breeds showed a large increase in the number of completely dominant relationships. Beyond this point only the wire-haired terriers showed a continuing increase, although the basenjis and shelties showed an increase between 15 weeks and one year (see Fig. 6-1).

It was obvious that the dominance tests themselves served as training periods for the development of dominance since actual fights occurred in many. The winning puppy afterwards became dominant over the loser. We also saw occasional fights among puppies at other times, arising from undetermined causes. Again, the winner of the fight tended to become dominant in the future. It should be added that these fights between young puppies were mostly a matter of noise and struggle, with very little actual damage inflicted. Thus dominance was settled early in development before adult capacities for more serious fighting had appeared.

Effect of dominance on the amount of fighting.—It is often stated that the development of a dominance order has the effect of decreasing the total amount of fighting, since once a definite relationship is set up there is no longer any need for fights. In most of the breeds there was not enough observed fighting so that relative percentages at different ages could be accurately calculated. However, in the basenji breed, fighting was fairly frequent and some conclusions can be drawn (Fig. 6.2). The total amount of fighting increases with age, but this does not hold for all relationships. Although it is true that fights between males are much more frequent at one year of age, presumably as a result of sexual maturity, the percentage of tests in which actual fights occur between females, or between males and females, remains constant between 10 and 20 per cent at any

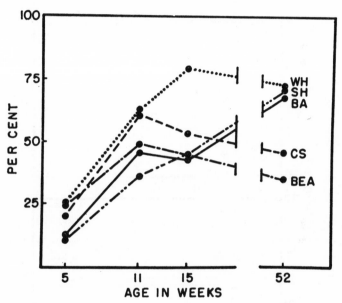

Fig. 6.1.—Percentage of occurrence of complete dominance in litters of purebred dogs.

age. Thus, in this particular breed, the development of dominance organization seems to keep fighting at a constant level, rather than decreasing it, and is unable to suppress fighting between males at sexual maturity. Of course, we have no figures on the amount of fighting which might take place between strange puppies not affected by a dominance order, and this would probably show a much higher figure.

In addition, in both male-female and female-female relationships, the number of *attacks* (in which one animal assaults another that does not fight back) tends to rise after 11 weeks of age. The number of attacks of one male basenji on another is higher at 11 and 15 weeks of age but decreases at one year, probably because these attacks are now resulting in actual fights and are entered as such on the records. This means that fighting by the dominant animal is not suppressed and may increase somewhat. Nevertheless, the total amount of fighting is probably much less than it would be if the subordinate animals fought back.

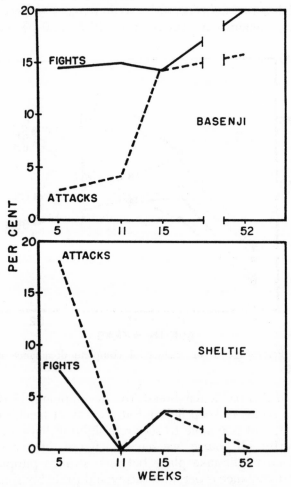

FIG. 6.2.—Occurrence of fights or one-sided attacks during dominance tests.

In Shetland sheep dogs the results are much closer to what we would expect (Fig. 6.2). There are more fights and attacks at 5 weeks of age than at any other time. Also, at 5 weeks of age, there are more attacks in male-female pairs than any other combination, this number being decreased to zero at 11 weeks of age.

In the fox terrier breed there were so few fights or attacks at any age that no percentages could be calculated. This is the breed in which the most complete dominance is seen. When serious fighting broke out in this breed, it was almost never between pairs but involved group attacks on one individual, and these, of course, could not take place in the paired dominance tests. In the other two breeds, beagles and cocker spaniels, there were likewise almost no observed fights, not because of a rigid dominance order, but because these breeds have a very low degree of aggressiveness.

Thus, the theory of the dominance order as a controlling agent in fighting is upheld in two breeds, fox terriers and shelties, but not in another, the basenjis. This finding may be partially explained by the fact that the basenjis have seasonal breeding and consequently more sudden changes in the level of the male hormone as they approach maturity. However, there is also more fighting in this breed at younger ages, and it is possible that all this is the result of forming less rigid habits of dominance and subordination. At any rate, there seem to be definite breed differences in the capacity to develop an effective control system over fights between individuals.

The effect of breed on dominance organization.—Knowing that the breeds have been selected for differences in aggressiveness, we would expect that the development of dominance relationships might show large differences between breeds, and this is indeed the case.

In wire-haired fox terriers, the number of dominance relationships was greatly reduced after 5 weeks of age by the group attacks which led to the separation of animals from their litters. However, the percentage of tests which resulted in complete dominance continued to rise (Fig. 6.1). When we (Pawlowski and Scott, 1956) analyzed this rise, we found that it consisted almost entirely of male-female relationships and that more and more males were becoming completely dominant over their sisters (Fig. 6.3).

The basenjis showed a different picture. The total number of complete dominance relationships rose between 5 and 11 weeks but stayed at almost the same level at 15, then rose sharply again at one year. The fighting behavior of the basenji thus tends to appear in later development, rather than in early puppyhood as it does in the wire-haired fox terriers. The dominance of males over females showed the same steady rise as in the wire-haired terriers, and at one year no females were completely dominant over males.

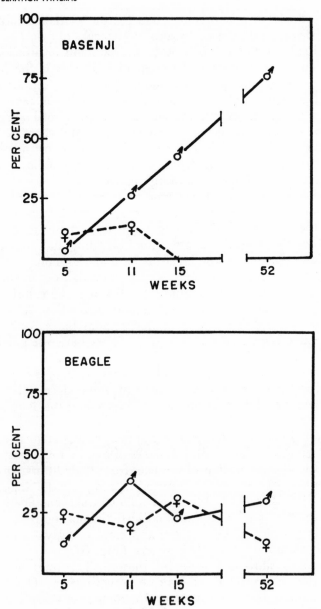

FIG. 6.3.—Occurrence of complete dominance in male-female pairs.

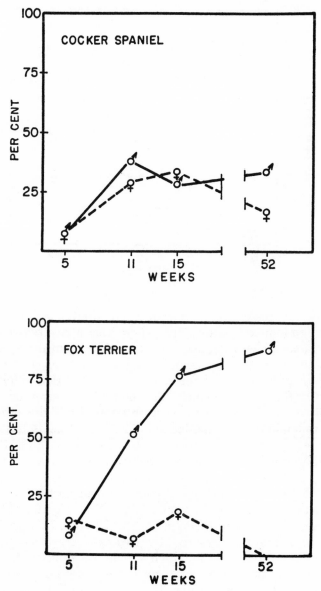

Fig. 6.3.—Continued.

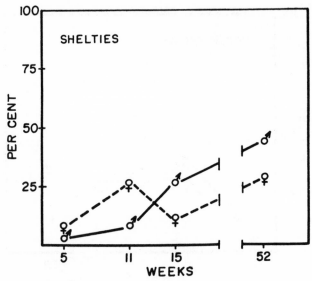

Fɪɢ. 6.3.—Continued.

Beagles and cocker spaniels were quite similar in their develop-ment of dominance relationships. There was a rise between 5 and 11 weeks and a slight decline after that time, continuing until one year. As adults, both cockers and beagles were likely to be indiffer-ent to possession of the bone and to pay attention solely to the observer, so that no dominance status could be determined. Further-more, these two breeds were very different from the others with respect to male-female dominance. There is little indication that more males are dominant than females, and if anything, there is a slight difference in favor of the females by one year of age.

Shetland sheep dogs, like beagles and cockers, showed a large number of cases of incomplete dominance and particularly of cases where no dominance could be determined. The majority of the latter cases turned out to be between males and females. In a small sample of 11 male-female pairs, there were two cases in which the female was dominant, three cases in which a male was dominant, and six in which there was no dominance. This is quite different from the other breeds. In cockers and beagles there are more cases of complete dominance between males and females than between like sexes, and in basenjis and fox terriers there is a strong tendency for males to dominate females. The shelties tend not to dominate females in con-flicts over food.

This breed difference may reflect a characteristic of the shelties, who are related to the Scottish collies. In the stories of Alfred Payson Terhune there are many references to the "chivalry" of the male collie toward the female. Certainly we observed nothing of the kind in the puppies of four of our five breeds, but the shelties may be an exception. However, these results with competition over a bone do not tell the whole story of the dominance relationships in this breed. As observed in their large outside pens, the shelties developed a strong dominance order based on space. Whenever an observer came near, the entire litter would rush out and start to bark, but almost immediately two or three of the animals would turn on their litter mates and drive them back toward their house, where the food and water were kept. In one litter, one female could never emerge from the house without getting chased and barked at. A small male was allowed to sit outside the house, while two large males could run up and down the pen, barking freely. In short, there appeared to be a strong dominance order based on space rather than food, with no deference paid to the weaker sex. Basenjis, on the other hand, seemed to be completely tolerant of one another with respect to space. A group would sit peaceably or run side by side without interference, but compete fiercely as soon as food was brought into the pen. In a separate experiment, in which basenjis were given special food in a dish, the dominant member immediately took charge of the dish and would allow none of the others to eat. One of the subordinate animals developed the technique of rushing in when the dominant animal's back was turned and tipping over the dish. He was then allowed to eat the food spilled on the ground, but not out of the dish itself.

Such differences in behavior indicate that dominance hierarchies are not universal but are developed in specific kinds of situations. Whether it would be possible for a dog to occupy a different position in a hierarchy under two different situations is a question on which we have no evidence.

The breed differences are also complex. The four hunting breeds are all strongly motivated by food but differ greatly in aggressiveness, the beagles and cocker spaniels being much more peaceful animals. The shelties, on the other hand, appear to have a rather high degree of aggressiveness but are relatively little motivated by food, so that they simply do not compete strongly for its possession. There has undoubtedly been selection in this breed for the lack of food motivation which would be likely to interfere with their training as working dogs.

Genetic effects on the dominance organization of the entire group. —The kind of test described above does not reflect the effect of several members of a group upon each other. However, we can express the total organization in terms of relationships between individual pairs. When we do this, we find that there is a greater tendency in the aggressive breeds to develop a straight-line dominance hierarchy with each animal dominant over those in the lower ranks. In the less aggressive breeds, a dominance diagram has a sprawling appearance with several individuals in the same rank, although we get few, if any, cases of the circular diagrams which sometimes occur in flocks of chickens.

The effects of sex and size.—With respect to sex, we had three kinds of relationships in our experiment: male-male, female-female, and male-female. When males and females met each other the males tended to dominate, particularly in cases of complete dominance. Also, there were more cases of complete dominance between males and females than there were between like sexes, either males or females. This reflects the fact that males are on the average larger and more aggressive than females.

Table 6.2 summarizes the results of a thorough analysis of the effect of size on dominance at 15 weeks of age. There is no indication that size has any effect on the dominance existing between female pairs, even in pairs which showed large differences in weight. In the male pairs, heavier animals tended to be dominant, provided there was a difference of at least 1 kilogram. This effect is more striking in the hybrids than in the purebreds, but if all hybrids and purebreds

TABLE 6.2

EFFECT OF SIZE ON DOMINANCE-SUBORDINATION RELATIONSHIPS
(HYBRIDS AND PURE BREEDS COMBINED)

	FEMALE-FEMALE	MALE-MALE	MALE-FEMALE
Differences less than 1 kg			
Dominant lighter......	71	58	62
Dominant heavier......	90	55	124**
No dominance.........	11	2	11
Total.............	172	115	197
Differences 1 kg or greater			
Dominant lighter......	27	23	67
Dominant heavier......	26	41*	172**
No dominance.........	9	4	16
Total.............	62	68	255

* $P < .05$
** $P < .01$

are added together, the difference between the actual figures and those which would be expected if there were no effect of size is statistically significant at the .03 level. With regard to the male-female contests, there is a uniform tendency for heavier animals to be dominant in a ratio of at least 2 to 1, but this, of course, reflects both the heavier weight and greater aggressiveness of males.

This difference in dominance relations developed by the two sexes is easily explainable by the observed differences in behavior. Females rarely get into actual fights with each other and seem to establish dominance largely on the basis of vocalization and threats. On the other hand, males often get into real fights, in which superior size is, of course, important in determining the outcome. Similarly, males are likely to make actual attacks on females rather than being intimidated by bluff, and the result is usually dominance by the male, particularly in the more aggressive basenji and fox terrier breeds. In contests between males and females, the superior size of the males as well as their greater aggressiveness undoubtedly plays a part in the outcome.

Genetics and the differentiation of behavior.—The most important conclusion from the above results is that we get extreme differentiation between the behavior of two individuals involved in a relationship only where there are large differences in the capacities of the two individuals, as there are between male and female puppies. More than this, both individuals must have the basic capacity to develop in a differential way. For example, beagles do not seem to possess the capacity to become highly aggressive in any of the circumstances which we studied. If the above generalization is correct, we would expect to get greater differentiation between like-sexed pairs in segregating hybrids (F_2's and backcrosses) than we would in nonsegregating hybrids (F_1's) or in pure breeds, and our experiment on genetics and behavior provides a good test of the hypothesis.

As Table 6.3 shows, this expectation is well borne out by the data. Not only are the like-sexed pairs more greatly differentiated in the segregating hybrids, but also the male-female pairs. The best comparison is between the F_1 hybrids and the segregating hybrids, and there is a larger proportion of pairs showing complete dominance in every segregating hybrid population. The same is true when the segregating hybrids are compared with the pure parent breeds, except for the female basenjis, which show slightly greater differentiation than do the cocker backcrosses or the F_2's. We may conclude that the differentiation of a social relationship is proportional to the difference in basic capacities of the two individuals concerned.

TABLE 6.3

Percentage of Complete Dominance in Basenjis, Cockers, and Their Hybrids at 15 Weeks

RELATIONSHIP	BASENJI	COCKER	F₁	TOTAL P's + F₁		BACKCROSS TO COCKER	BACKCROSS TO BASENJI	F₂	TOTAL BACKCROSSES + F₂	
				Number	Per Cent				Number	Per Cent
Female-Female..	44	15	17	67	24	33	48	41	77	42*
Male-Male......	50	54	39	83	37	66	68	64	64	66**
Male-Female....	42	43	35	114	45	55	72	67	163	66**

* $P < .05$
** $P < .01$

One of the basic reasons for studying genetics and behavior is to find out just how important genetic differences in behavioral capacities may be. In many instances these seem to have little importance in comparison to the effects of environmental factors. Here, in the development of social relationships, we have a case in which genetic factors are highly important determinants of behavior. This does not exclude other determinants, of which we have many examples even in dogs raised in a uniform environment. In the mother's relationship to the puppy, relative age is extremely important in the development of a dominance relationship, as mothers are almost invariably dominant over their puppies, although they do not always enforce this in competition over food.

In highly developed relationships the behavior of two individuals becomes differentiated with respect to each other, i.e., one becomes dominant and the other subordinate. At the same time, each individual retains the capacity to further differentiate his own behavior in relation to other individuals. The same puppy may occasionally threaten another in a relationship in which he is dominant and in complete control, frequently attack in a relationship in which dominance is unsettled, or be extremely submissive in a relationship in which he is subordinate. The same animal can thus be correctly described as moderately aggressive, extremely aggressive, or unaggressive.

What then becomes of the idea of a "personality" composed of simple behavior traits? We are likely to think of an animal or person as being generally aggressive or generally submissive. This obviously does not fit the facts. If we want to characterize the behavior of an individual, we shall have to do it in terms of his relationships, and in any practical situation this means selecting and analyzing the most important and general relationships out of many special and minor ones.

Is it then possible to make any generalizations about an individual? We can say that an individual has a basic capacity, but this capacity can only be developed in terms of relationships. One answer to the problem is that while there are certain kinds of relationships which are special and completely unique, there are others which apply to several individuals and can be called general relationships. Examples of these two kinds are easy to find in human situations. A child has only one mother (a special relationship), but several aunts, all of whom are treated more or less alike (a general relationship). Our puppy experiment, on the other hand, was set up to develop social relationships with a small group of experimenters, all of whom attempted to act alike. As will be seen from the results of the obedience test described in chapter 9, the puppies evidently developed a general relationship with all experimenters.

One answer to the question of the generality of behavior traits came from the fact that when we took a puppy out of the pen and then put him back in with the litter mates, they frequently threatened him, exhibiting considerable aggressive behavior. This looked like an opportunity to study the phenomenon of jealousy, as the other puppies appeared to be "jealous" of the one who had been removed. Dr. Theodore Zahn, a visiting investigator at Hamilton Station, systematically studied this phenomenon in several litters, taking each puppy out, returning it, and observing the results. He soon found that fighting was confined to certain individuals in each litter, observing only 10 cases of dominance in 61 relationships. When he compared his results with those of the dominance test over bones, he found that most of the animals which threatened each other had not previously shown consistent dominance. In other words, the manifestation of "jealousy" reflected the fact that certain animals had not established a definite dominance relationship between themselves.

In another sort of experiment King (1954) placed litters of adult basenjis and cockers in large fields and observed their reaction to the introduction of strange animals, up to the point when serious fighting began to take place. When he analyzed his data, he found that the dogs were exhibiting most aggressiveness toward animals like themselves. Female cockers attacked strange females more often than they did strange males and attacked strange cockers more than they did strange basenjis. This interesting finding indicates that in dogs, at least, there is no basic hostility toward unlike individuals. If they are basically hostile, it is toward strangers resembling themselves. One possible explanation is that they attempt to extend the dominance behavior which they normally exhibit in their own group to include animals of familiar appearance, but that this generalization is less likely toward animals showing an unfamiliar size and shape. There is also a tendency to react to animals of the opposite sex through sexual rather than agonistic behavior.

Most of our studies of social relationships have been done between the two members of a pair, partly because this is probably the most important kind of relationship, and partly because the technique is relatively simple. When one begins to examine relationships in terms of groups of three individuals instead of two, the problem of analysis becomes extremely complex. It is, however, an important theoretical and practical question whether the relationship between two individuals is the same when they are both part of a group. In studies of chickens and goats, there is every evidence that individual pair competitions give the same result as group contests. In these animals fighting always occurs between two individuals and there are no group attacks. But dogs are capable of group attacks on an individual, and the situation here may be quite different.

We have some observational evidence that the presence of a group modifies the behavior of one individual toward another, in this case toward a human handler. We often found when we entered a dog pen that some animals came boldly forward and others hung back. Yet when we tested all animals in an individual handling test, we found that some of the apparently shy animals were strongly attracted to people. They stood at the bottom of the dominance order among dogs and were being kept away from the human handlers by the more dominant members of the group. Dog buyers frequently select puppies on the basis of which puppy in a group comes most boldly forward. In so doing they may be selecting for aggressiveness against other dogs as well as against shyness of people.

More objective data on this point were obtained by measuring the catching time of individuals in a group during the weekly inspection. When we compared these times with attraction scores in the handling test, we found a relatively low correlation. Out of 22 litters analyzed (2 each from all breeds and hybrids), 16 showed a positive rank-order correlation, with an average R of .30.

In conclusion, the expression of behavior in a social relationship is a highly complex expression of genetic differences whose action is dependent upon the development of behavior between two individuals who mutually affect each other. Nevertheless we can express the results as a general law: *a relationship tends to be differentiated in proportion to the differential biological capacities of the two indi-*

viduals involved. The most important of these biological capacities are determined by sex (which is, of course, hereditary), by individual heredity, and by age (which reflects biological development). All of these determine the process of psychological differentiation of behavior which eventually freezes a relationship into a relatively stable form.

This generalization needs to be widely tested in order to be firmly established. Assuming that it is correct, all sorts of practical applications immediately suggest themselves. If we desire a highly differentiated relationship, as in a leader-follower relationship, this should be relatively easy to establish between quite unlike individuals, as between older and younger persons. It would be most difficult to establish leadership within a group of like age and sex, where the only basic differentiation would be that between individuals having markedly different heredity. It is no wonder that people looking at a boys' group are likely to conclude that one of them is a "natural" leader, since heredity is almost the only way in which one boy's behavior could be differentiated from the rest. On the other hand, it is also possible that his behavior could have been differentiated by unusual previous training. This last question has not been answered by our dog experiment, in which early environment was kept uniform. Experiments with inbred mice in which heredity was kept constant but certain individuals were given pre-training in fighting show that a relationship can also be determined by differential training (Scott, 1944). If we want to make our generalization more universal, we must add the factor of differential development of capacities by training and previous experience.

THE MOTHER-OFFSPRING RELATIONSHIP

From the description of development in the previous chapter we concluded that the puppy was incapable of forming any true social relationship until approximately 3 weeks of age. This conclusion does not apply to the mother, who can begin to form relationships as soon as the puppies are born. Thus the mother-offspring relationship is extremely one-sided in the first few weeks after birth. Furthermore, since the mother pays the greatest amount of attention to the puppies in these first few weeks and begins to leave them more and more as they grow older, the eventual relationship of puppies to their mother is weaker than that formed with their litter mates.

The general behavior of the mother toward her puppies at first includes a great deal of nursing, combined with much licking and cleaning. If disturbed by experimenters or caretakers she comes out of the nest box and investigates or barks at the intruders. She makes no attempt to defend the puppies against handling by familiar persons, but often seems nervous and discontinues her care while they are about. Consequently, it is essential to keep the mother and her litter in undisturbed surroundings for the first few weeks.

Individual differences in the attentiveness of mothers toward their offspring are common, and we wished to find out whether there were consistent differences between the different breeds in respect to maternal care which might affect the behavior of their offspring. We therefore collected objective data on the mother-offspring relationship and selected the ages of 1 week and 7 weeks for particular study, the former because maternal attention was close to its height at this point, and the latter because it is a time when all mothers are in the process of weaning their puppies and might exhibit differential behavior if it ever occurred.

Nursing behavior.—While we made the weekly observations and weighings, we also recorded what the mother was doing as each puppy was taken away from her and placed on the scales. As seen in Figure 6.4, the mothers continued to nurse their puppies very nearly

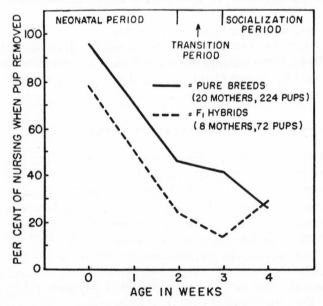

Fig. 6.4.—Decline of nursing behavior by mothers during the neonatal period. There are considerable variations between individual mothers, but a general downward trend is evident.

100 per cent of the time immediately after birth, but this figure steadily went downward in subsequent weeks, being approximately 30 per cent when the pups were 4 weeks old. There was almost no difference between cocker and basenji mothers, but the F_1 mothers apparently left their offspring in a greater number of cases. These hybrid mothers (F_1) had a very much better milk supply than the purebred ones and on the whole took excellent care of their offspring. It looks as if the amount of time spent nursing might not be directly dependent upon the milk supply but rather upon the responses of the puppies. That is, the stimulus which keeps mother with the puppies may be their sucking behavior. However, the numbers of animals concerned are too small to draw any definite conclusions except that the downward curves obviously confirm the impression that the mothers are much more attentive to the puppies in the early part of their lives. A similar conclusion can be drawn from observations on undisturbed mothers (Table 6.4), except that under these conditions there are indications of breed differences between cockers and basenjis. Similar results have been obtained by Rheingold (1963) with this and numerous other measures of maternal care.

TABLE 6.4

OCCURRENCE OF MOTHERS NURSING PUPPIES UNDER 2 WEEKS OF AGE DURING 10-MINUTE DAILY OBSERVATIONS

BREED	NUMBER OF LITTERS	PER CENT NURSING	
		Days 1–7	Days 8–14
Basenji..........	4	54	13
Beagle...........	5	58	24
Cocker..........	5	81	56
Sheltie..........	5	83	66
Fox terrier.......	5	63	40
Totals.......	24	69	41

Retrieving test.—When the puppies were 1 week old we took the mother out of the room for a few minutes, took the puppies from the nest box, and distributed them at equal intervals around the edges of the nursery room. Then we put the mother back and observed her for 15 minutes, recording which puppies were visited, the number of visits to each, and the time when each puppy was carried back to the nest. The mother usually went from puppy to puppy, touching each one with her nose. The puppies meanwhile usually made continuous whining noises. Finally, the mother would pick each one up and take it back to the nest, carefully carrying it in her jaws with the puppy's feet dangling down.

TABLE 6.5

OCCURRENCE OF RETRIEVING 1-WEEK-OLD PUPPIES
BY PUREBRED AND HYBRID MOTHERS

BREED OR HYBRID	NUMBER			PER CENT	
	Mothers	Litters	Pups	Mothers Retrieving	Pups Retrieved
Beagle	5	11	58	20*	2
Sheltie	3	8	28	66*	29
Fox terrier	2	5	23	50*	4
Basenji	8	19	103	75	44
Cocker	13	27	134	38	25
F_1	8	15	89	63	21

* Figures based on very small sample.

As Table 6.5 shows, the basenji mothers carried their young back to the nest more often than did the cockers. However, there were big individual differences, usually consistent from litter to litter. Some mothers almost always retrieved their puppies and other mothers never did so. Ten of the 13 cocker mothers did no retrieving at all, whereas 6 out of 8 of the basenji mothers retrieved at least some of their puppies. This is somewhat curious because the cocker spaniel breed has been selected for the ability to learn to retrieve and carry game. Apparently these two types of retrieving are not closely related, and we can conclude that the basenji mothers have been selected for better maternal care, probably because of less human assistance in their native African villages.

There was considerable variation in temperature between different tests, and one might suppose that the mothers would retrieve the puppies more promptly if the puppies were cold and making more noise. Analysis of the data showed that the temperature had no consistent effect, but rather that the characteristic behavior of the mother was the important factor.

Another interesting observation was that the mothers seemed to play no favorites. They visited each puppy about the same number of times, and when they began to bring them back they did so in what appeared to be a random order. Sometimes the mother would bring back some of the puppies and lie with them in the nest box, leaving the others outside. At other times the mother might stop and allow a puppy to nurse outside the nest. We did not, however, make the critical test of repeating the observation to see whether the puppies were collected in the same order.

These data confirm the impression that dog mothers make little if any distinction between members of the litter. This is quite different from the conditions of human motherhood, in which the necessarily different ages of children make differential treatment necessary. The dog mother is in the same situation as a human mother of twins, or perhaps quintuplets, all being the same age and very much alike. The only time a dog mother will show obvious differential treatment is in the case of a sick or dying puppy. Once the puppy begins to get cold and inactive, the mother pays very little attention to it.

Weaning test.—When the puppies were 7 weeks of age, we took the mother out of her pen for an hour, so that mother and puppies were separated for a considerable period. We then put her back with the puppies and recorded the ensuing activities on a check list. Usually the puppies would all rush toward the mother, attempting to nurse as she moved around the room. Their behavior was quite similar to the response given to a human handler when he holds out a hand to the young puppies. If the mother allowed them to nurse, she usually stood still for a few minutes while they leaped around, pawed, and sucked. While they were doing this she might vomit, whereupon they ate the new food. If they subsequently tried to nurse, she would turn on the offending puppies with a loud growl, opening her mouth as if to bite and rushing toward them. The puppies at once retreated, sometimes rolling on their backs and yelping, but the mother was never seen to actually bite one of them. At this age such behavior was typical of almost all mothers and their offspring.

Another test was to place a bone in the pen with the mother and puppies. In almost every case the mother did not compete with them, although she was always dominant over them when they attempted to nurse. In other situations where the mother and puppies were fed dishes of semi-solid food, we have observed mothers which would take possession, drive the puppies away, and bury the food with shavings from the floor. The normal behavior of mothers toward puppies of this age, however, seems to be to limit the puppies' nursing behavior but to allow them to eat other food at will without competition.

Following test.—The mothers were taken away from the puppies at 10 weeks of age. Neither appeared to be disturbed by the change. The mothers had only one contact with the puppies thereafter, at 18 weeks of age. At this time we began to train the puppies to walk on a leash. We soon found that if we began leash training without some preparation, the puppies became frightened and did very poorly on the test. Therefore, as an introduction to the leash, we first trained the puppies to follow without it.

On the first day of the test the puppies were fed a special dish of fish by the experimenter, and the same thing was repeated the next day. On the third day the experimenter went into the pen with the food dishes but instead of giving them to the puppies, she walked out of the pen and toward the laboratory through an experimental field. The puppies were fed after they had followed her into the laboratory. On the fourth day the puppies were tested one at a time. The experimenter did not carry the food but simply walked along, and instead of going directly through the field, she went around the edge of the fence. Thus, if the puppy was interested only in the food reward, it could run ahead of the experimenter. The amount of time which the puppy spent within a dog's length of the experimenter was recorded together with the total time spent walking from pen to laboratory. The experimenter also rated the puppy according to how closely it followed, providing a second measure of the tendency of a puppy to follow. On the final day of the experiment the same procedure was repeated, except that the experimenter led the mother of the pups on a leash, allowing the puppy to follow at will.

This is not a perfect experiment for determining whether the puppy is more attracted by the human handler or its mother. It should have been done in reverse order as well, to cancel the possible effect of training. As the experiment was actually run, almost every puppy spent more time close to the handler when the mother was also present, and often three times as much. We can at least conclude that the puppies were still strongly attracted by their mother (Fig. 6.5). On the other hand, the reaction of the mother to the puppies was either one of indifference or rejection. If the mother happened to be in heat, as was sometimes the case, she might be approached by a male puppy but would warn him off with a growl. Whenever this sort of behavior occurred, the mother was strongly dominant over the pup.

In conclusion, the mothers develop three types of relationships with their offspring. In the care-dependency relationship, development proceeds toward a decreasing amount of maternal care and increasing independence of the puppies. This change is brought about in two ways. The mother begins to leave the puppies more often and at the time of weaning actively rejects them. There is little or no differential treatment by the mother, who has the same general relationship with all the puppies in the litter. The phenomenon of sibling rivalry simply does not occur in ordinary dog development.

There is, however, competition over food, and the puppies, who never seem to compete with each other for maternal care, develop

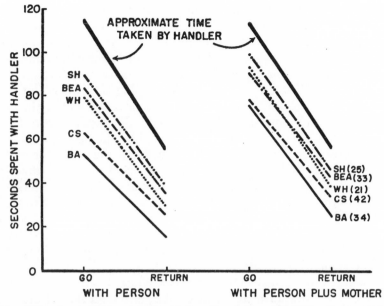

FIG. 6.5.—Average time spent with handler during the "following test." Note that the same order is maintained between breeds in every test, with one exception. All breeds spent more time with the handler on the second day, when the mother was also present.

strong dominance relationships with each other in connection with food. The dominance relationship between mother and puppies is quite different. The mother does not ordinarily exert dominance in the presence of a bone, although she is definitely dominant in other situations, particularly when weaned puppies attempt to nurse. Even this type of dominance is different from that between puppies, as it never seems to involve actual fighting, but is developed and enforced solely by means of threats.

The "following test" indicates that under the proper conditions a leader-follower relationship could be developed between mother and offspring, and it is possible that this may occur occasionally in the wild Canidae. However, in most cases wolf puppies do not stay with their parents but form new packs of their own. A well-developed leader-follower relationship is not a characteristic part of either dog or wolf societies.

THE DOG-HUMAN RELATIONSHIP

As we saw in the previous chapter, the relationships between dogs and human beings may develop in many directions. At one extreme, a puppy raised entirely apart from human beings may later react toward them with extreme fear and hostility. At the other, a puppy taken away from dogs at an early age transfers all its social relationships to the human species and becomes an "almost human" dog. Whatever this relationship may be, it is normally well formed by 7 weeks of age, although it goes on developing and changing long after that. We shall now discuss the further development of the dog-human relationship and the effect of genetic differences upon it.

Any relationship is, of course, determined by the behavior of both parties. Since we were interested primarily in the behavior of the dogs and not of their human handlers, we tried to standardize the human half of the relationship. We instructed all experimenters and handlers to treat the puppies with uniform kindness and not to make special pets. In test situations we made it a rule to give every dog an equal opportunity, without playing favorites. If any special treatment had to be given to one animal in a litter, the instructions were to do the same thing to all members of the group. While testing, all experimenters were instructed to wear white laboratory coats so that their general appearance was similar, although they differed in details of clothing.

Because of the good scientific training and co-operation of the experimenters, the effort to standardize human behavior was largely successful. From the viewpoint of analyzing the effects of heredity, the results were excellent. An unexpected result was that the quality of the social relationship which developed was somewhat shallow.

The development of the care-dependency relationship.—Our chief experimental measure here was the handling test, described in the previous chapter. This is essentially a test of social relationships, and its results are divided into several parts. The first includes initial responses by the puppy to threatening behavior by the experimenter, and these are scored as reactions of escape and avoidance. Second, as the puppies grow older they begin to react toward the handler in much the same way as they act toward the mother, running toward him and wagging their tails and licking his hands. Still later, the puppies begin to react as they would toward their own litter mates, with a great deal of playful fighting and occasional playful sexual behavior. The positive patterns of behavior tend to occur in combination with each other at this age, and their separation is perhaps somewhat artificial. In scoring the test, however, we divided the responses into fearful behavior, playful fighting, social investigation and et-epimeletic behavior, and attraction. The et-epimeletic behavior and attraction scores are perhaps most closely related to the care-depend-

The five pure breeds. *Left to right,* wire-haired fox terrier, American cocker spaniel, African basenji, Shetland sheep dog, and beagle.

Outside runs where the puppies lived from 16 to 52 weeks of age. In the background is the nursery wing of the Behavior Laboratory.

Nursery room interior. Shetland sheep dog female carries a puppy in a retrieving test. Puppies lived in rooms such as this one from birth until 16 weeks of age.

Behavior Laboratory and Staff

Back row: Daniel Reynolds, Pearl McFarland, Daniel G. Freedman, Donald Dickerson, John A. King, Douglas G. Anger, Duane Blume, Edna DuBuis, Frank Clark, Sheldon Ingalls, Howe Smith.

Middle row: Maxine E. Schnitzer, Marian Burns, Florence Smith, Margaret Charles Higgins.

Front row: Clarence C. Little, John Paul Scott. (John L. Fuller was away on leave when this picture was taken.)

Ancestral cocker spaniels 0414 ♀ and 0415 ♂ .

Some offspring of the ancestral co[cker]
spaniels used in the cross.

Breeding stock—BCS cross

Mated pair—basenji ♂ and cocker ♀ .

BCS F$_1$ hybrids— ♂ and ♀ pair.

Some offspring used in the cross. ♂ on the right.

Ancestral basenjis 739 ♂ and 1090 ♀ .

Breeding stock—CSB cross

Mated pair—cocker ♂ and basenji ♀ .

CSB F$_1$ hybrids— ♂ and ♀ pair.

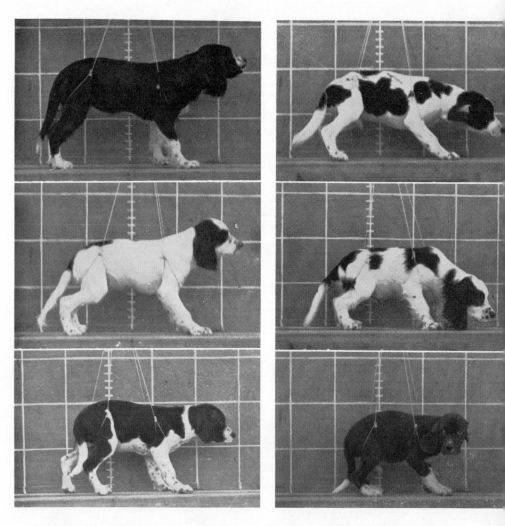

Backcross to the cocker spaniel, litter 2249–57. Note cocker-like ears and segregation of long and short hair. These and following pictures of hybrids were all taken at 16 weeks of age.

Backcross to the basenji, litter 3301–06. Note uniform short hair, the result of a dominant gene, and basenji-like appearance of some animals.

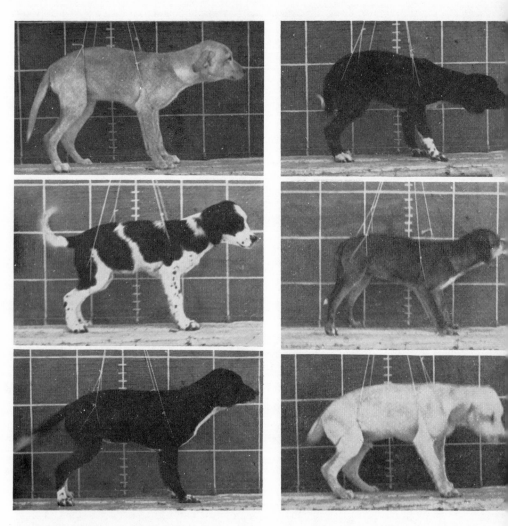

BCS F$_2$ litter 3193–3200. Note the great variety of form and color in these animals.

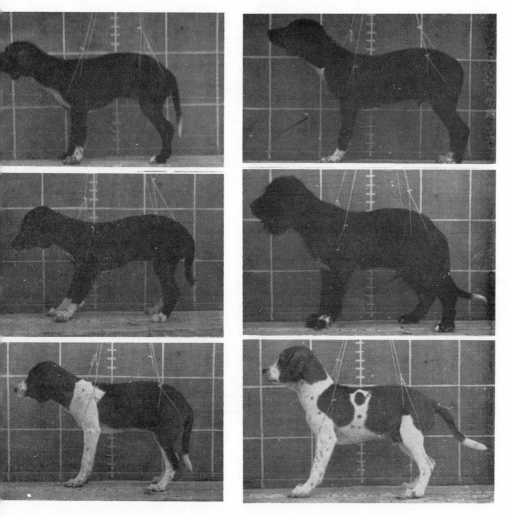

CSB F$_2$ litter 3394–3400. None of the 72 puppies in the total F$_2$ population showed an exact resemblance to either parent breed.

Pygmies returning from a hunting trip with a basenji. The nets are used to trap game such as the small antelopes carried by the hunters. The basenji runs with them as they beat the brush, and the bell around its neck helps to frighten the game (by permission of the American Museum of Natural History).

Hunting party at Ninevah, Iraq, 1952. Their three dogs belong to the breed from which the European. greyhounds were probably derived, being brought back by Crusaders (photo by R. T. Hatt). The saluki is a breed developed from the same source in modern times.

Left: Male purebred dingo in Australia (photo by Dr. N. W. G. McIntosh). Note general similarity to the basenji. *Right:* Kurdish guard dog, Iraq, 1952. Note strongly curled tail. Animals such as this were used in ancient times as war dogs, as well as for guarding flocks and property (photo by R. T. Hatt).

Historical records of land spaniels and hounds from Blome's *Gentleman's Recreation* (1686).

Use of spaniels in hawking. The spaniels crouch in response to a hand signal given by their master.

"Hunting the hare with deep-mouthed hounds." Modern beagles are related to hounds of this type.

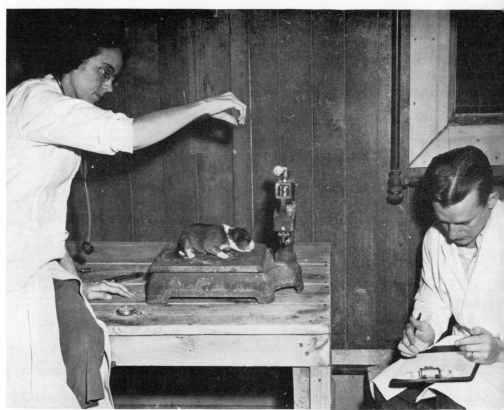

Forced training

Above, a puppy is taught to remain on the scales while weighed, starting at birth. The puppy in the picture is only two weeks of age.

Left, the leash control test. A balk at the stairs.

Basic reward training

Goal-orientation test. Upon release from the start box, the puppy learns to run to the dish of food in its own pen.

Advanced reward training

Motor-skill test. In this, the most difficult part of the test, the dog must climb the ramp to a dish of food on top of boxes 5 feet high.

Detour test (first barrier test). In this first problem-solving experience, the puppy must go away from the goal on the far side of the barrier in order to reach it.

Manipulation test. The puppy must get the dish out of the box with his paws or teeth in order to eat.

Maze test. A puppy approaches the exit and the food reward.

T-Maze and delayed-response test. Only one door leads to an exit. In the delayed-response test, the puppy must remember a previous signal.

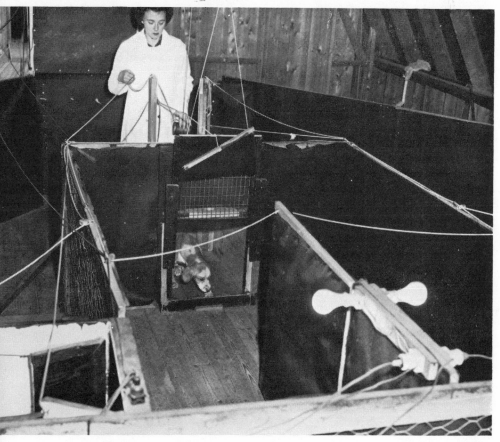

Trailing test. Puppy learns to follow an artificial trail. A single drop of fish juice is placed on each metal plate; the trail ends with a small dish of fish.

Spatial-orientation test. Only the righthand ramp leads to the goal. Arrangement of the goal box and ramps is altered in each part of the test.

ency relationship. As seen in Figures 5.9 and 5.10, these scores rise quite rapidly and then level off. By one year of age they have decreased considerably, indicating that the animals are now more independent.

Another measure of dependency is the "following test," already described in connection with the mother-offspring relationship. The time spent close to a human handler varied a great deal among individual puppies, but there are average breed differences. Cockers, for example, spend more time close to the handler than do basenjis on the average, although some of the individual dogs spending the most time with the handler were basenjis (Fig. 6.5).

For most of the dogs, being removed from the pens and tested was a rewarding experience. Whenever a handler appeared at the gate, the puppies would swarm toward him, attempting to get out. In most of the tests they obtained both a food reward and the opportunity to move into a new environment, and they presumably associated the experimenters with these rewarding experiences. We made certain special experiments in an attempt to analyze this relationship in more detail.

The first of these experiments was a study of the effect of feeding. Food rewards, which can be used so effectively to motivate learned behavior, might have a powerful effect on social relationships and the attraction of an animal toward those who fed it. The facts do not bear this out. Most of the regular feeding was done by animal caretakers. The puppies were intensely interested in them at feeding time but lost all interest as soon as feeding was over and paid little attention to them thereafter. And, as we saw in the last chapter, puppies which are machine fed become just as attached to people as those which are hand fed. In short, a puppy which is merely fed regularly by hand and given no other attention develops little more relationship with the feeder than he would with a feeding machine.

We can conclude that feeding itself is a minor part of the care-dependency relationship. This does not mean that food could not become an important part, particularly if food was made conditional on certain acts, and sometimes given and sometimes withheld. But it does look as if the mere act of feeding does not produce a strong emotional reaction. In everyday language, a puppy does not automatically love you because you feed it.

However, Elliot and King (1960) did a different sort of experiment in which they regularly hand fed a group of puppies but made no other contact with them. Half the puppies were underfed and half were overfed. The puppies were fed twice a day after being weaned at 4 weeks of age, and soon became strongly interested in food as their regular mealtimes approached, running to the gate and whining repeatedly. At 5 weeks of age, the underfed and overfed dogs showed no differences when given the handling test. At 6 and 7 weeks of age, however, the underfed puppies showed less avoidance and more attraction, the differences being significant at the .01 level. Both groups were more afraid of a strange handler than of the person who regularly fed them but exhibited the same differences in reaction to either handler. As they grew older the same average differences persisted but were no longer statistically significant. We can interpret these results as meaning that making the puppies hungry speeds up the process of primary socialization (i.e., the formation of a social relationship with human handlers), although the overfed puppies tend to eventually catch up with the others without the special impetus of hunger.

All the puppies in this experiment exhibited another interesting aspect of behavior, in that neither experimentals nor controls developed a social relationship with handlers as rapidly as puppies in the regular "school for dogs," where they were given much human contact in addition to feeding. We may conclude that hand feeding has a definite effect on the development of the social relationship between puppies and people, particularly if the puppies are unusually hungry, but that the relationship will also develop normally without feeding being involved. This agrees with our knowledge that strong relationships develop between litter mates in spite of the fact that one litter mate never feeds another but instead often competes with him for food.

Relationships of home-reared dogs.—More light is thrown on the origin of primary social relationships by our experiment with home reared dogs. This could be done on only a very small number of animals, so that general conclusions are impossible. Each puppy was taken from the litter at 4 weeks of age and raised in the home of a staff member. Each home was somewhat different from the other two. Silver, the Shetland sheep dog, lived in a house with two small boys and a tomcat. George, the beagle, lived with a young married couple who had no children but did have a cat. Gyp, the basenji, went to a home where there was already an adult female dog and several small children. All the families were fond of pets but differed somewhat in their ideas of dog-rearing and discipline.

The puppies lived the ordinary lives of household pets, except that they were frequently taken back to the laboratory and tested. While there, they were kept apart from other dogs and might spend all day by themselves before they were returned to their homes. As we saw from other experiments, isolation is an unpleasant experience for a puppy, and over and over again they had the experience of being placed in an unpleasant emotional situation and eventually rescued from it. We would expect that these puppies would have strong emotional reactions associated with human beings.

In general these puppies were more free and confident with human beings than the kennel-reared dogs. On the tests they sometimes did better, particularly when wider experience might be of some assistance, but they also did worse in some tests which were based on the kennel experience.

The most striking result was the difference in adult behavior. Two of the puppies became extremely dependent and highly attached to people. One of the usual problems of raising a pet beagle is its tendency to go off hunting for hours at a time, never returning to the house. George, the beagle, would scarcely leave his owners' yard. Silver, the Shetland sheep dog, was instantly responsive to human behavior and easily trained, either positively or negatively. He defended his yard against all strange dogs, both male and female. He became sexually responsive to the cat with which he was brought up, while the beagle became attached to the cloth bag of a vacuum cleaner. On the other hand, Gyp, the basenji, became more and more independent, roaming the streets and fighting with other dogs. Perfectly behaved in the laboratory, he became almost a canine delinquent at home and eventually became uncontrollable.

Although the environment was not uniform for these three animals, the same general treatment produced in two of the dogs a condition of overdependency and a lack of maturity, but had relatively little effect on the other. The basenji became, as a dog, a normally independent animal; as a pet, he was overly independent and difficult to control. The beagle and sheltie developed a much richer and deeper relationship with human beings than did the kennel dogs, whereas the basenji was much less affected, possibly because of contact with another dog in the home as well as his different heredity.

The effect of home rearing on the behavior of these three dogs can be determined by comparing their test records with those of other members of the same breed as distributed on the stanine scales. As might be expected, the most obvious effect was on the relationship to a human handler. The home-reared dogs were all less fearful and showed more playful aggressive behavior and more evidence of attraction during the handling tests. Gyp, the basenji, was an unusually confident dog at 5 weeks of age; and Silver, the Shetland sheep dog, showed an unusual amount of attraction to human handlers at 13 to 15 weeks of age. Both these scores were more than two times the standard deviation away from the average of their respective breeds.

Similar effects showed up in the reactivity tests. The pets were more active, stood more erect, and showed more tendency to investigate their surroundings. All these results are consistent with the interpretation of an animal which is confident in strange surroundings. One peculiarity was that all of them showed more muscular tremor than the average in reaction to restraint in a strange place. On the performance tests, all the animals did better on the first barrier test given at 3 weeks of age, where confidence in a strange situation undoubtedly has a strong effect on the outcome. In over-all performance, the basenji scored better than average on five performance tests, poorer than average in one, and average in four more. He did outstandingly better than other basenjis in the trailing test.

The beagle was better in three tests, worse in two, and average in five, confirming the impression of his owners that he was a dog of quite average intelligence. He did outstandingly well on the leash-control test, but this probably reflects previous home training.

The Shetland sheep dog did better on four tests, poorer in one, and average in five, turning in outstanding performances on the first barrier test and on the trailing test. The behavior of all dogs on the last two tests were apparently greatly affected by their reactions to strange situations. The sheep dog showed a slower than average performance on speed scores in two of the tests, reflecting the fact that this animal became obese as it grew older and could not run as fast as the others.

The home environment chiefly affected the relationship of the dogs with people and their confidence in strange situations, and this enabled them to do better than the average in certain tests in which this was important. Their additional outside training and experience with richer environments benefited them in some other tests but did not make them into "super dogs."

CONCLUSIONS

The most important conclusion that can be drawn from these experiments is that heredity does have an important effect upon the development of social relationships. Involving, as it does, a process of mutual adaptation and learning, the expression of heredity in a social relationship is far removed from the primary biochemical action of a gene. Nevertheless, the end result can be expressed in a simple statement that the differentiation of behavior in a social relationship is proportional to the differences in the capacities of the two individuals involved. In the case of the dominance-subordination relationship developed between young puppies, the outcome is affected by both sex and breed differences in the capacity to develop aggressive behavior. These experiments have many other implications, and they will be discussed in greater detail in the final chapter.

While we were able to work with only a limited number of mothers, there were obvious individual differences in their behavior toward their offspring. Very possibly a larger sample would show that there were important breed differences in maternal behavior. One interesting fact was that the hybrid mothers, which gave more milk than purebred ones, spent less time nursing their offspring, suggesting that the mothers tended to allow their offspring to nurse as long as the puppies kept sucking. There is a great decline in the amount of time spent in nursing during the neonatal period; so that by the time the puppies enter the period of socialization, nursing is a quick and casual affair.

As well as differences in maternal care, there are breed differences in the reactions of the offspring. This suggests the possibility that part of the behavior which we consider characteristic of a breed may be a reflection of the social environment, i.e., the behavior of mothers and litter mates. In the case of social behavior, genetic factors may in part create their own environment which in turn modifies their development and expression.

In contrast to the dog-dog relationships, which were allowed to develop freely, the development of dog-human relationships was more strictly controlled. As far as possible, the human handlers treated each puppy in the same way. They controlled and stimulated the behavior of the dogs but responded to them in only a limited way, so that the puppies could control human behavior to only a very small extent. The result was a relatively shallow social relationship compared to the deeper and more complicated relationship which house dogs ordinarily develop with their masters.

We were able to compare the effects of normal home rearing in the case of three dogs. The principal effect was to make them more confident in the presence of people, and this in turn affected their performance where confidence was needed. Basenjis have a tendency to be afraid of strange objects and apparatus, and the more confident home-reared basenji did much better in the trailing test, in which such fears were a real handicap to performance. Otherwise, home rearing seemed to help the animals in tests where wider experience was of value and to handicap them where kennel experience was important. We also had the impression that the three home-reared dogs had more highly differentiated behavior as individuals than did those reared in a kennel. This undoubtedly resulted in part from differential amounts of training but also suggests the possibility that part of what we consider the characteristic behavior of dog breeds is developed and exaggerated by differential treatment.

Thus we reach the general conclusion that the course of development of social relationships is determined by genetic differences between the individuals involved and by the nature of the social environment. Social relationships in turn influence almost all other behavior and are the framework within which all tests of performance take place.

Preceding chapters have dealt with basic information: the origin of dogs, their basic patterns of behavior, and the broad outlines of behavioral development. The next section will emphasize the development and analysis of breed differences in performance, including the capacity to accept training and to solve problems, and the dependence of each of these upon basic physiological, emotional, and anatomical characteristics.

PART II

THE DEVELOPMENT
AND EXPRESSION
OF BREED DIFFERENCES

CHAPTER 7

ANALYSIS OF GENETIC DIFFERENCES

Before presenting the results of the genetic studies it is necessary to explain scoring systems and methods of statistical analysis. For the most part, our procedures were conventional, but some explanation of the genetic significance of our results is necessary for persons more familiar with behavior than with biometrical genetics. It will become clear that, although statistical analysis is essential for an interpretation of our findings, it is not sufficient in itself. To a great extent, the internal consistency of the data must be considered in arriving at conclusions. In an experiment involving so many variables, numerous breed differences that are significant at the conventional 5 per cent level would be expected by chance. The 1 per cent level of significance is probably a safer boundary in multivariate comparisons.

Furthermore, quantitative estimates of the heritability of a trait depend upon clear separation of genetic and environmental sources of variance. Pure breeds of dogs are appreciably heterogeneous in many respects; hence, the within-breed component of variance is a reflection of genetic as well as environmental differences among members of the same breed. Frequently the variance between breeds was considerably greater than that within breeds; and on occasion we used the between-breed variance component as an estimate of genetic effects. Heterogeneity in the breeds causes the between-breed variance to give underestimates of the total effect of heredity.

The use of between-breed variance as a measure of genetic effects assumes the essential similarity of the environments of the different breeds. In practice there were systematic differences. For example, basenji pups developed in a basenji uterus, and for 70 days after birth were nursed and cared for by a basenji mother. For their entire period of observation they lived with their basenji litter mates, not with cocker spaniels or beagles. By cross-fostering, by rearing in mixed-breed groups, or by rearing in isolation, it would have been possible to isolate the effects of differential social environment at various stages of life. Pilot studies of each type were, in fact, carried out, but not on a scale large enough to permit detailed statistical analysis. The outcomes of all such experiments were consistent; breed characteristics persisted in cross-fostered pups, in isolation-reared pups, and in pups transferred after some weeks or months to a litter of a different breed. Because of these consistent results, we believe that genetic contributions to breed differences overshadow environmental contributions.

Genetic heterogeneity within parent stocks would tend to lower *estimates* of heritability, while systematic environmental effects would raise them. Because the two potential sources of error work in opposite directions, we may hope that a rough balance was attained. Before proceeding with a more detailed discussion of our methods of analysis, we will describe the stanine system of scaling quantitative scores (Guilford, 1950).

STANINE SCORING SYSTEM

The scores of our subjects were expressed in diverse units such as errors, running times in seconds, or arbitrary numerical ratings derived from a descriptive scale. These scores have meaning only insofar as they are related to a particular testing procedure. In order to make different sets of scores comparable, we have converted many of the test results to stanines. In this system all scores are divided into nine groups based upon rank ordering and each group of scores is represented by a single digit. The conversion is chosen so that the mean of the scores for the reference population is close to 5.0 and the standard deviation is about 2.0. Thus a score of 5.0 indicates that the subject was between limits of ±0.25 standard deviations from the mean of the transformed scores. The mean of the transformed scores, of course, corresponds to the median of the original scores.

Transformation of scores has a number of advantages. Intervals on the stanine scale are, in a sense, uniform throughout the range of scores within a particular population. This satement is not true for many of the original scores. For example, in one test involving running-time subjects running at full speed could cover the course in 2 seconds. An animal averaging 4 seconds was clearly running much below maximum speed, and the difference between 2-second and 4-second subjects was significant. Some dogs remained within the apparatus long after the gate had been opened, and their recorded running times averaged 20 seconds or more. Under these conditions, the difference between 20 and 22 seconds was not comparable to that between 2 and 4 seconds. Conversions to reciprocals or to logarithms might have been satisfactory, but we selected a transformation based on rank ordering as more generally applicable to all types of scores.

The stanine conversion makes it possible to compare performances in quite different tests. A stanine score of 7, for example, always means that the subject was between 0.75 and 1.25 standard deviations above the mean on the transformed scale. The original scores may have been running times or ratings of emotionality, but such differences do not alter the procedure of ranking which is the basis of the transformation. Stanines are also advantageous for computations, particularly when desk calculators must be employed. A variety of analyses can be conducted more easily on scores expressed as single digits.

There are also certain drawbacks. The disadvantage of transforming data in this way is that it may obscure or distort information which is obvious on the original scale, or which could be brought out more plainly by using another transformation. As Mather (1949) has pointed out, there is no a priori reason for assuming that one scale will appropriately measure the effects of every gene. Consequently the stanine scale may in certain cases obscure precise measurements of behavioral differences and indications of genetic segregation. The advantages of the method, however, outweigh its drawbacks, particularly if many computations must be made.

The use of stanines carries the implication that the traits measured by our scores are essentially continuous rather than discontinuous, and that the base sample used to compute the transformation tables was a representative sample of the population in which we were interested. Specifically, this means that we assume that dogs are not divided into bright and dull or emotional and non-emotional classes, but that intelligence and emotionality as measured by our tests are quantitative characters with the majority of the population intermediate and relatively few subjects at the extremes. Further, we have assumed that the five breeds chosen for study represent a fair cross-section of dogs. Although these assumptions were not strictly

proved, the distributions of the test scores are in general agreement
with them. Stanine conversions were not made on scores whose distributions obviously contradicted these assumptions.

The procedure for transforming original scores into stanines is
illustrated by data from the motivation test. The original score was
the total time in seconds a dog required to run ten trials in one day
through an L-shaped course. Each dog was tested on three separate
days, so that three scores were available for every dog. We chose a
sample of 100 dogs, 10 males and 10 females from each of the five
pure breeds, selecting at random within each breed and sex. The
distribution of the 300 scores is shown in Figure 7.1. The curve is

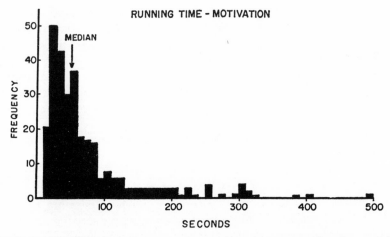

FIG. 7.1.—Distribution of 296 running times (4 scores off scale to right)
from 100 dogs given 3 trials each. The sampling procedure is described in
the text.

markedly skewed, but there is no evidence of bimodality which
would justify the classification of subjects into fast-running and slow-
running dogs.

The original scores were arranged in rank order and cut-off points
were selected at points corresponding to cumulative percentages for
segments of a normal distribution, one-half a standard deviation
wide (Table 7.1). Note that the cumulation of scores was started
from the slow-running end, because we wished to express our results in terms of speed and the fastest subjects have the lowest time
scores. The transformation to stanines shifts the distribution of scores
to an approximately normal form as in Figure 7.2. We can, there-

TABLE 7.1

CONVERSION TABLE—ORIGINAL SCORE TO STANINE

Stanine	Cumulative Percentage, Upper Limit	Group Limits in Standard Deviations	Range of Original Scores (Motivation Test) in Seconds
1	4.0	Below −1.75	Over 275
2	10.6	−1.75 to −1.25	171–275
3	22.7	−1.25 to −0.75	96–170
4	40.1	−0.75 to −0.25	61–95
5	59.9	−0.25 to 0.25	43–60
6	77.3	0.25 to 0.75	30–42
7	89.4	0.75 to 1.25	23–29
8	96.0	1.25 to 1.75	20–22
9	100.0	Over 1.75	Under 20

fore, by the use of stanines, essentially change the basis of measurement of an individual's behavior from a scale based on absolute physical units to one based on performance of a standard population.

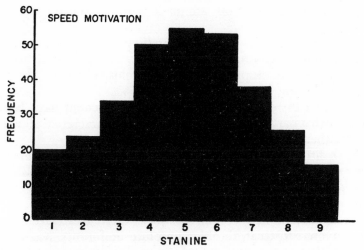

FIG. 7.2.—Distribution of 300 motivation-speed scores after conversion to stanines.

ANALYSIS OF VARIANCE

Analysis of variance separates the effects of two or more factors upon a set of measurements. In our experiments, the major interest lay in estimation of the contribution of heredity to total variance. Total variance is readily estimated by standard procedures, and to determine heritability it is only necessary to find some means of calculating the genetic component. In practice, this is somewhat complex. We shall describe the situation in the experiment comparing five pure breeds.

In this experiment, potentially genetic contributions can appear in three forms:

1. A systematic effect between breeds may appear because of differences in gene frequency between the several breeds. We shall label this variance σ^2_{HB}.

2. Some genetic variation may be attributable to differences in genotypes between the various mated pairs in each breed. We shall label this variance σ^2_{HM}.

3. Some of the variation among members of the same sibship is attributable to random segregation in the offspring of heterozygous parents. We call this variance σ^2_h.

The design of the experiment does not permit separation of σ^2_h from environmental variance, but σ^2_{HB} and σ^2_{HM} can be estimated, provided certain assumptions are made with respect to environmental effects.

We can also subdivide environmentally produced variance.

1. A systematic environmental effect could have been produced by the association of puppies almost exclusively with their own breed. This variance we shall describe as σ^2_{EB}.

2. A similar systematic environmental effect could have been associated with a particular mother and would characterize all matings of this mother. This variance we shall call σ^2_{EM}.

3. Each litter had a particular and unique life history characterized by season of birth, exposure to accidental stimuli, testing by a particular research worker, etc. The variance attributable to these common litter effects we will call σ^2_{EL}.

4. Each subject had certain unique experiences. Even though we endeavored to treat all alike, absolute uniformity was not possible, and litter mates were not, of course, bound by any experimental design in their treatment of other individuals in the litter. Variance dependent upon these individual experiential differences we denote as σ^2_e.

The complexity of sources of variation is somewhat discouraging at first. In a large experiment, it would theoretically be possible to isolate each component, although biological limitations on the number of litters which a single female could produce would interfere with even the best possible design. Our experiment was too small to permit direct estimates of all components. Hence, we were forced to look for defensible simplifications.

In the first place, cross-fostering failed to demonstrate major effects, and individuals reared in mixed litters did not differ markedly from their own breed. We concluded that in dogs, σ^2_{EB} and σ^2_{EM} were too small for measurement in small samples; hence they were neglected in most analyses. In effect, such an simplification assumes that cultural differences between breeds are not important determinants of behavioral variation, although we later found evidence that "breed environment" has an effect on some of the variables (see chap. 14).

The importance of σ^2_{EL} can be tested by determining whether litters of the same mating differ significantly from each other. When the question is put to the data the answer is not straightforward; they do on some tests but not on others. We conclude that the control of environment from litter to litter was imperfect in spite of our best efforts, and that in analysis we should take into account between-litter variation.

The separation between σ^2_h and σ^2_e, both contributing to within-litter variance, is not possible with the data from the pure breeds. An estimate of the distribution of the within-litter variance was, however, possible in the hybridization experiment. We can be certain here that the F_2 and backcross hybrids were genetically more diverse than the F_1's, yet the within-litter variances of behavioral measures were not uniformly smaller in the F_1's. The genetic diversity of the purebred litters was certainly less than in the F_2 hybrids, and the genetic contribution to the variance within such litters must have been less. Thus, when we assume that the within-litter variance of the pure breeds is environmental, we may underestimate total genetic effects for characters of high heritability, but the error is probably too small to be detectable except in very large-scale studies.

Finally we must consider σ^2_{HB} and σ^2_{HM}, although our primary interest is in σ^2_{HB}. By deriving our purebred stocks from a few closely related individuals, we endeavored to reduce σ^2_{HM} to a low value. We can omit separate calculation of between-mating variance and include it with between-litter variance. In so doing, some genetic components will be incorrectly classed as environmental, and the effect will be to reduce the probability of finding significant breed differences and so lead to a conservative estimate. In anticipation we shall note that this analysis proved suitable for the breed comparisons. Among the hybrids (see chaps. 12 and 14) rather large differences were found among matings of the same type.

In summary, a survey of the sources of variation in our experiment leads to the following simplified schema of analysis for a sample of b breeds, each made up of f litters of k individuals (Table 7.2).

TABLE 7.2

Simplified Interpretation of Analysis of Variance
in Breed Comparison Experiment

Source of Variation	Individuals per Subclass	Degrees of Freedom	Mean Square Is Estimate of . . .
Breed.........	fk	$b-1$	$\sigma^2_h + \sigma^2_e + k\sigma^2_{EL} + fk\sigma^2_{HB}$
Litter.........	k	$b(f-1)$	$\sigma^2_h + \sigma^2_e + k\sigma^2_{EL}$
Individual......	1	$bf(k-1)$	$\sigma^2_h + \sigma^2_e$

Practical computations must take into account inequalities in the various subgroups.

In this schema, three hierarchial mean squares are computed between breeds, litters, and individuals. The variance between individuals of the same litter is a measure of the effect of non-controlled environmental and genetic factors. The mean square for litters also includes a component which we interpret as predominantly the effect of specific-litter life histories. Any genetic differences in matings of the same breed will also be included. If between-litter effects are non-significant, then the within-litter mean square (estimate of $\sigma^2_h + \sigma^2_e$) is the proper term for the F test of breed differences.

If the between-litter effect is significant, then the proper denominator for the F test of breed effects is the mean square for litters (estimate of $\sigma^2_e + \sigma^2_h + k\sigma^2_{EL}$). Obviously, this test is more rigorous than a simple comparison of breed differences which neglect specific-litter effects, and it is more difficult to establish significance. Because the procedure is conservative, more faith can be placed on it.

The ratio of σ^2_{HB} to total variance $(\sigma^2_h + \sigma^2_e + \sigma^2_{EL} + \sigma^2_{HB})$ is known as the *intraclass correlation coefficient* and is perhaps the most generally useful index of the magnitude of the effects of heredity. A correlation of .25, for example, means that one-fourth of the total variance of a population drawn at random from the total sample would be due to breed differences. Another intraclass correlation can be computed for the litter effect $(\sigma^2_{EL}/\sigma^2_e + \sigma^2_h + \sigma^2_{EL})$ which is the proportion of variance within a breed attributable to specific-litter histories. Both measures have proved useful in our analyses.

The most general applications of this basic theory are presented in chapter 14 in connection with an over-all analysis of variation in the pure breeds and hybrids. Chapters 8 to 10 are concerned with the analysis of breed differences and the relative ranks of the pure breeds in various capacities, whereas chapters 11, 12, and 13 are concerned with the analysis of the results with hybrids.

More detailed statistical techniques are described and their application discussed in several of these chapters. The analysis of variance method can be extended to include detailed analysis of environmental effects. Elliot carried out an elaborate analysis of this sort upon the second barrier (or maze) test, which is described in chapter 10.

In chapter 11, methods are described for estimating the numbers of genetic factors involved in a particular trait. Chapter 12 includes methods for using offspring-parent regressions and for the analysis of half-sib and full-sib families. The design of our crossbreeding experiment was such that the same purebred female was bred to two different males, one of the other pure breed, and one being her own son. This same son was bred to two genetically different females, the mother and a full sister. This makes it possible to compare animals having one parent in common and thus to estimate the effects of the parents. This method, in contrast to the more general one described above, makes it possible to analyze the effects of heredity without assuming homogeneity of the parent breeds.

Finally, the well-known technique of factor analysis was applied to the results on three different occasions, and the findings are described in chapters 13 and 14. The assumptions and limitations discussed above apply to all procedures employed. Despite the complexity of the factors affecting behavior and the elaborate nature of the statistical techniques required for their analysis, one basic result clearly emerged: large effects attributable to heredity were demonstrated in almost every test we administered.

CHAPTER 8

EMOTIONAL
REACTIVITY

The work of Hall (1941), Searle (1949), and others on the genetics of rat behavior indicated not only that emotional behavior was affected by heredity, but that emotionality had an important effect on performance tests. In our dogs we therefore studied the early development of such emotional reactions as distress vocalization and tail wagging, and particularly the changes in emotional reactions in response to human beings during the process of socialization. Important breed differences were found in all these characteristics.

In addition to their overt expression, emotions also include internal responses which can be measured by well-known physiological techniques. These techniques are most easily applied to larger animals under restraint, and we therefore planned a major test of the emotional and physiological reactions of older puppies and grown dogs as they stood on a Pavlov stand. We hoped to define the effect of heredity upon external and internal emotional reactivity.

Reactivity is a broad term which may have several meanings. We defined reactivity as the intensity of all responses, external and internal, made to a stimulus change. Such a broad definition creates some problems of measurement, since the various possible responses do not necessarily vary together. Nevertheless, the concept seems to have validity from the point of view of genetics.

DEVELOPMENT OF A REACTIVITY TEST

Apparatus.—In order to reduce the reactivity measurements to manageable proportions and to facilitate physiological measure-

194

ments, subjects were restrained upon a Pavlov stand. Here they were exposed to a standard set of situations designed to elicit an emotional response. Since we were interested in observing physiological correlates of reactivity, we connected our subjects to an oscillograph, using modified battery clips as surface electrodes. To prevent them from pulling the clips off or tangling the wires, dogs were lightly restrained by leg loops which were tied over their backs and connected to a steel bar about 4.5 feet above the floor. The supporting frame for this bar was built into a large cage $8 \times 8 \times 7$ feet which was covered with screen wire for electrical shielding. Observers and recording equipment were stationed just outside the cage in the same room.

Behavior in the test situation.—Within this cage we observed the responses of the purebred and hybrid dogs to a series of social and non-social episodes. The responses were obviously modified by the conditions of testing, and reactivity measured in another situation might have taken very different forms. The majority of subjects tolerated the attachment of electrodes and the fitting of leg loops with a nominal display of excitement. Good visual observations and oscillograph records could be obtained from such animals. A few subjects reacted so violently to simple restriction of movement that testing had to be discontinued or carried out in a modified form. Even more common were dogs which slumped down in the harness and were so inhibited throughout that little differentiation of reactivity was detectable from episode to episode. These two types corresponded to Pavlov's (1928–41) "excitable" and "inhibited" animals, which he tested under similar conditions. Between these two classes, there was an enormous range of variability. The extreme patterns appeared to be more a reaction to the test situation as a whole than to its specific components.

Age of testing; observer variation.—As a standard procedure each subject was tested at three ages: 17, 34, and 51 weeks. At the time of the first test the dogs had just been placed in an outside yard; at 34 weeks they had completed the major behavioral tests with the exception of spatial orientation. The 51-week test came after approximately 12 weeks of living in the outside yards with a minimum of human handling. In these three tests the effects of changes in age, previous test experience, and current conditions of maintenance and handling are combined. It would have been possible to design an experiment to separate these effects, but this would have involved more subjects than were available and would not necessarily have given more information regarding breed differences, since all breeds were treated alike.

Over the course of the experiment changes in personnel occurred, thus introducing problems of variation between the ratings of individual observers. To reduce this source of error, each new observer was trained by the principal investigator until ratings were judged to be reasonably uniform. A sample of correlations between observer ratings on the same subject varied between .85 and .95.

Episodes in the reactivity test.—The reactivity test was divided into ten episodes: (1) *Preparation.* During this period of variable duration the experimenter attached electrodes and adjusted the restraining loops. (2) *Control I.* The subject was left alone for one minute in an electrically shielded cubicle, lighted from the inside. An observer viewed the subject from the side through a screen which blurred but did not obscure the dog's view to the outside. (3) *Quieting.* An experimenter entered, spoke softly to the subject while extending his hand, walked to the rear of the cubicle, and left quietly after 30 seconds. (4) *Control II.* A repetition of Control I. (5) *Bell.* A doorbell attached to the cubicle wall was sounded for 30 seconds, followed by (6) *Control III.* Another repetition of Control I. (7) *Shock.* Four single shocks from an induction coil were delivered to the dog's left foreleg at 5-second intervals and observations were continued for 60 seconds following the first shock. (8) *Threatening.* An experimenter entered the cubicle, grasped the dog's muzzle while speaking in a loud harsh voice, and forced the subject's head from side to side. In addition, it was often necessary to untangle loops or re-attach electrodes which were dislodged during the shock period. All such handling was done in a brusque manner. (9) *Control IV.* Repetition of Control I. (10) *Release.* After releasing the dog from the leg loops, the experimenter remained standing in the cubicle for 30 seconds, then stepped outside and waited by the door for an additional 0.5 minute. Latency of the dog's leaving the cubicle was recorded to the nearest second.

Rating scales.—Ratings of behavior were made in each of the 10 episodes with a standard checklist. Each specific behavior pattern was rated on a 5-point scale, the higher values corresponding to greater energy expenditure. Exceptions were made for such categories as aggression (biting) which was scored no lower than 4 or 5, since aggression appeared only when subjects were more active than average. Some items, such as posture and tail carriage, were rated in each episode; others, such as tail wagging or vocalization, only in those episodes in which such a response was observed. With this rating system, the total score is a measure of the average vigor of all forms of response over 10 episodes. Details of scoring are available in Scott and Fuller (1950).

In addition, graphic records of respiration, electrocardiogram, and electromyogram (recorded from the thigh) were available for all except the preparation and release episodes. The respiratory rate data have not yet been analyzed completely. Heart rate changes from episode to episode were noted, as well as the heart rates under control conditions and an index of sinus arrhythmia. An arrhythmic heart would slow down and seem to almost stop, then speed up for a few beats. The electromyograms were rated on a 5-point scale by comparing them to a set of standard records ranked from extreme relaxation to violent tremor.

When possible, ratings were converted to stanines to facilitate group comparisons, but some distributions were too skewed to permit transformation, so that comparisons were made in terms of the proportion of each population responding at a level arbitrarily called high. Eventually we used 18 specific-response ratings (Table 8.1). In addition, composite scores (sum of scores on all overt be-

TABLE 8.1

SPECIFIC RESPONSE RATINGS

Converted to Stanine Scale	Unsuitable for Stanine Scale
Body posture	Lip licking
Tail posture	Vocalization
Tremor	Panting
Investigation	Tail wagging
Attention to observer	Resistance to forced movement
Escape behavior	Biting
Heart rate in control periods	Latency of exit
Heart rate change to quieting	Elimination
Heart rate change to bell	
Sinus arrhythmia	

havior) were computed for the control episodes alone, for reactions to the bell, and for the shock episode. A handler-effect score was calculated based on the net change in total score between stimulation periods and control periods. Some dogs were more active when alone than when stimulated; in others, the reverse was true. We were looking for evidence that such responses could be affected by heredity.

RESULTS

Reactivity changes associated with age.—Changes in total scores from one age to another were not spectacular. As shown in Figure 8.1, terriers, beagles, and basenjis are consistently more emotionally reactive than shelties and cockers at all three ages. Specific ratings

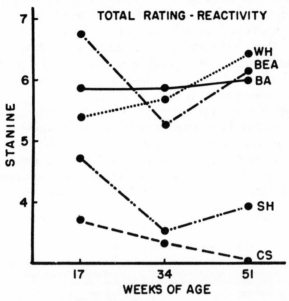

Fig. 8.1.—Total reactivity ratings for the five pure breeds. Note that three of the breeds rank definitely higher than the other two.

showed similar results. Terriers were high on tremor and panting at all ages and low on vocalization. In fact, the means of each breed were remarkably consistent in view of the relatively low correlations between ratings on successive tests. Tetrachoric correlations between ratings at 17 and 34 weeks ranged from .063 to .818 with a median of .333. Corresponding figures for scores at 34 and 52 weeks were from .219 to .692 with a median of .532. Although most correlations were statistically significant, they were too low to be of great value for prediction. Actually the mean of its breed is often a better predictor of an individual's score than its score on a previous test.

A few scores changed significantly between the three tests. Except in terriers, panting was much more prevalent at 17 and 34 weeks.

Lip licking declined in all breeds except basenjis. Furthermore, younger dogs tended to be more active when an experimenter was present in the cubicle; older dogs were more active when the experimenter was outside.

TABLE 8.2

"RESTING" HEART RATE (BEATS/MINUTE)

Age (Weeks)	Basenji	Beagle	Cocker Spaniel	Shetland Sheep Dog	Wire-Haired Terrier	Intraclass Correlation
17	179	150	143	142	173	.247**
34	163	134	121	122	161	.656**
51	138	137	113	125	127	.412**

** $P < .01$

In all breeds heart rates declined with age, from approximately 220 beats per minute at one week of age to 130 at one year (see Fig. 5.1). The relative positions of the breeds remained fairly constant from 17 weeks throughout the rest of the first year. The intraclass correlation for control period heart rate was highest at 34 weeks, indicating a somewhat greater effect of genetic differences at this age than with either younger or older animals (Table 8.2).

In Table 8.3 are *changes* in heart rate accompanying the entrance

TABLE 8.3

CHANGE IN HEART RATE (BEATS/MINUTE) DURING QUIETING

Age (Weeks)	Basenji	Beagle	Cocker Spaniel	Shetland Sheep Dog	Wire-Haired Terrier	Intraclass Correlation
17	+1.2	0.0	−24.8	−9.7	+26.5	.413**
34	+22.4	+17.6	−11.5	+10.2	+3.2	.215**
51	+21.4	+10.2	−10.1	−8.4	−3.8	.245**

** $P < .01$

of an experimenter for Episode 3, quieting. In these data the breed differences are greatest at 17 weeks, when the cocker spaniels showed a marked drop and the wire-haired terriers a similar increase. On the later tests basenjis and beagles consistently showed cardioacceleration, while spaniels continued to show deceleration under the same conditions. In contrast, the heart rate responses of the sheep dogs and terriers are inconsistent at different ages and within breeds. The small average change for terriers does not indicate that all members of this breed maintained a constant heart rate when the stimulus was changed. Both marked acceleration and deceleration were found in a few individuals, but the mean was close to zero.

TABLE 8.4

Mean Stanine Scores of Five Pure Breeds on Ten Behavioral
and Four Heart Rate Measures at Three Ages

Measure	Age (Weeks)	Basenji (23)	Beagle (26)	Cocker (29)	Sheep Dog (20)	Terrier (20)	P of Difference	Intraclass Correlation
Total behavior rating	17	5.8	6.8	3.7	4.7	5.4	.001	.350
	34	5.8	5.3	3.3	3.5	5.7	.001	.366
	51	6.0	6.2	3.0	3.9	6.4	.001	.446
Handler quieting effect	17	4.9	4.6	3.6	4.1	3.2	.05	.064
	34	7.0	5.6	4.9	4.8	5.3	.01	.212
	51	6.3	5.5	5.2	4.8	5.4	N.S.	.050
Bell response (all categories)	17	4.4	6.4	4.3	4.7	4.4	.01	.137
	34	4.8	4.9	5.0	4.3	4.3	N.S.	.001
	51	4.7	6.2	3.7	4.8	6.7	.001	.297
Shock response (all categories)	17	5.1	5.9	4.0	4.0	4.9	.001	.158
	34	6.1	5.0	4.2	4.3	5.2	.001	.270
	51	6.1	6.2	4.7	5.1	5.9	.01	.106
Escape activity	17	5.5	6.5	3.7	5.5	5.0	.001	.167
	34	5.7	4.8	3.2	4.5	4.8	.001	.168
	51	6.1	5.8	3.6	5.0	5.0	.001	.244
Heart rate, isolation	17	7.3	5.6	5.1	5.0	7.0	.001	.247
	34	6.4	4.4	3.4	3.4	6.2	.001	.656
	51	4.7	4.6	2.6	3.7	6.4	.001	.412
Change in heart rate, quieting	17	4.7	4.6	2.7	3.7	6.4	.001	.413
	34	6.2	5.9	3.6	5.3	4.9	.001	.215
	51	6.1	5.4	3.7	3.8	4.3	.001	.245
Change in heart rate, bell	17	3.3	4.9	4.1	4.4	3.7	.05	.060
	34	5.2	5.8	5.4	5.3	5.1	N.S.	.000
	51	4.3	6.0	5.5	5.2	5.5	.05	.070
Arrhythmia index	17	5.2	3.4	5.1	4.5	2.9	.001	.209
	34	6.2	4.1	6.6	5.9	4.1	.001	.369
	51	6.9	3.6	7.0	5.2	3.2	.001	.614
Erect body posture	17	6.5	4.6	3.6	4.9	4.8	.001	.320
	34	7.5	4.9	3.3	5.6	4.8	.001	.498
	51	6.7	5.0	4.7	5.6	4.6	.001	.161
High tail carriage	17	6.2	5.7	5.8	5.2	4.5	.001	.204
	34	6.2	4.3	5.6	4.4	3.8	.001	.249
	51	6.4	3.6	5.9	4.5	4.1	.001	.402
Tremor (read from myogram)	17	4.6	4.8	3.5	4.7	6.8	.001	.309
	34	5.1	4.2	4.3	4.2	7.3	.001	.323
	51	4.8	5.0	4.1	5.2	8.0	.001	.382
Investigation	17	7.2	5.8	4.4	4.7	5.5	.001	.347
	34	6.5	5.4	4.0	4.4	5.6	.01	.257
	51	5.4	5.0	3.0	4.7	4.4	.01	.281
Attention to observer	17	6.0	7.0	3.1	4.8	4.4	.001	.522
	34	5.0	6.1	3.6	4.1	5.0	.001	.318
	51	5.4	6.4	3.1	4.5	5.1	.001	.352

Breed comparisons.—Tables 8.4 and 8.5 summarize reactivity test scores for the five pure breeds. In Table 8.4 are measures which were convertible into stanines, together with the results of analysis of variance of the transformed scores. Readers interested in comparing the significance of differences between pairs of breeds can use the approximation that differences of 1.0 stanine units will occur in about 5 per cent of comparisons in a random sample. For example, the first four breeds are separated from each other by a difference of one or more on the total behavior score at 17 weeks. Fox terriers are intermediate between basenjis and shelties and thus are not significantly different from these two breeds. This means that 8 out of a possible 10 interbreed comparisons are significantly different at 5 per cent or better.

Perhaps the most striking feature of this table is the almost universal occurrence of highly significant differences between the five breeds. Thirty-one differences out of 42 are significant at the .001 level. Only heart rate change during bell ringing and the handler-effect score fail as discriminators of breed differences. A preponderance of significant effects of heredity has not been achieved by discarding measures which failed to discriminate between breeds. All of the measures on which reliable data were collected have been included. It is much more difficult to find scores which are not affected by breed differences than to find those which are affected.

In Table 8.5 are set forth the results of measurements which were not suitable for stanine conversion because of skewness of distributions—usually the occurrence of many zero scores. Evaluation of the significance of differences between breeds was accomplished by the chi-square method, and contingency correlations were calculated in order to compare degrees of hereditary influence. As with the more normally distributed scores, differences between breeds were highly significant on all measures except that of elimination, where the incidence was too low for a satisfactory statistical comparison.

It is instructive to compare the values of the intraclass correlations at each age. Over two-thirds of these correlations fall between .15 and .45. The median values at successive ages are .23, .27, and .29. Thus we inferred that the proportionate effect of heredity on all scores considered together is much the same between 17 and 51 weeks, although correlations for individual measures such as the heart rate may rise and fall. There is some possibility that genetic differences are expressed more definitely in the one-year phenotype, as high intraclass and contingency correlations occur twice as often at 51 weeks as at either of the other two ages.

TABLE 8.5

PERCENTAGES OF SUBJECTS EXCEEDING ARBITRARY THRESHOLDS
ON EIGHT MEASURES IN REACTIVITY TEST

Measure	Age (Weeks)	Basenji	Beagle	Cocker	Sheep Dog	Terrier	P of Difference	Contingency Correlation
Lip licking*	17	87	65	38	47	70	.001	.341
	34	48	38	3	0	54	.001	.410
	51	74	62	7	0	41	.001	.522
Vocalization	17	48	81	45	47	20	.001	.359
	34	87	85	55	55	40	.001	.355
	51	65	89	59	68	6	.001	.458
Panting	17	35	73	79	79	80	.001	.346
	34	49	100	72	45	85	.001	.420
	51	0	54	11	16	77	.001	.522
Tail wagging	17	26	81	34	37	25	.001	.394
	34	35	77	55	55	20	.001	.365
	51	13	61	34	32	18	.001	.345
Resistance to forced movement*	17	48	42	69	79	80	.001	.303
	34	22	12	21	30	30	N.S.	...
	51	22	23	7	47	53	.001	.353
Biting	17	78	62	35	53	60	.001	.477
	34	78	23	10	25	50	.001	.454
	51	83	42	7	26	53	.001	.475
Elimination	17	17	15	31	32	15	...	...
	34	22	8	14	10	5	...	...
	51	9	19	10	16	0	...	...
Latency of emergence*	17	74	46	93	89	45	.001	.412
	34	17	23	69	60	50	.001	.387
	51	9	19	34	42	12	.05	.297

* Thresholds for grouping scores on lip licking, resistance to forced movement, and latency of emergence were not the same at all ages because of developmental shifts in average frequency of occurrence.

Since a substantial portion of the observed variation in reactivity was not correlated with breeds, we may inquire into causes other than heredity. Had the pure breeds been highly inbred so as to be nearly homozygous at all loci, we could conclude on the basis of the intraclass correlations that 55 to 85 per cent of the variations in the test measures was attributable to environmental factors acting to differentiate individuals. Each breed, however, was actually fairly diverse in genotype so that the calculated correlations are minimal estimates of genetic influences. Environmental factors were probably not as important as the above figures would imply.

Still another approach is to investigate the effects of sex and of litter influences on the test. Among the pure breeds, total reactivity scores of males averaged 4.9 on the stanine scale; females averaged

5.0. The two sexes gave almost identical results, indicating that this form of hereditary variation has little effect on the scores. On the other hand, there are significant differences between litters when all three tests are grouped together. Our experimental design does not permit allocation of the source of this effect, but some or all of the following may be involved: specific genes from a particular set of parents, maternal care differences, climatic changes, changes in observers—in fact any variable which might be more alike for members of litters than for dogs in general. Removing the litter effect, however, does not in this instance decrease the breed differences. The intraclass correlation for breeds (based now upon litter means), is .61, denoting remarkably large hereditary effects.

Interpretation.—After emphasizing differences between breeds in the form and intensity of emotional expression, we wish to caution the reader against accepting the idea of a breed stereotype. Typically the range of scores for a breed extended over 5 or 6 points, and occasionally over the entire 9 points of the stanine scale. Basenjis and cocker spaniels had mean stanines of 6.00 and 3.03, respectively, at one year; but 65 per cent of the basenjis and 41 per cent of the spaniels overlapped in the middle range of 4 to 6.

No one-year-old terrier was rated below 6 on tremor although all other breeds had individuals with low scores. Although terriers did not have scores in the low zone, the other breeds were not restricted in this way, and rating of 7 and 8 were common in four. The least amount of overlap was found in biting at one year; 83 per cent of basenjis were biters and 93 per cent of cocker spaniels were non-biters.

The most general interpretation of these differences between breeds is that they are related to differential responses to inhibitory training. This is particularly evident in Figure 8.1, in which the cockers and shelties ranked consistently below the other breeds. Both these breeds are easily trained to repress any activity, although the responses take quite different forms. Any sort of threat, but particularly a hand motion, will make a cocker cease all activity, but the effect is momentary and the animal shows no signs of emotional disturbance. Inhibition of a sheltie is long lasting and appears to be accompanied by prolonged emotional arousal.

This brings up the problem of whether the reactivity test measures a single underlying trait, or a composite of genetically influenced capacities. Many of the scores are partially correlated with each other, and two of the measures, heart rate and sinus arrhythmia, are highly interdependent. A slow heart rate is arrhythmic and, conversely, a faster heart rate is regular. A factorial analysis, as will be seen in chapter 14, resulted in splitting the various reactivity measures among several factors, heart rate being one, and a general factor for confidence-timidity being another.

SUMMARY

The emotional responses characteristic of the different breeds continue to change and develop throughout the first year of life so that breed differences on the reactivity test were present and highly significant in a statistical sense at all ages. The test situations seem to accentuate emotional differences by stimulating distinct alternate responses, such as passivity or attempts to escape. Selection in the past has obviously been highly successful in separating the five breeds with respect to a large number of emotional responses. So much variation still remains, however, that it would probably be possible to select cocker spaniels for a few generations and produce offspring like terriers or beagles—at least with respect to particular responses. To synthesize by this method a total pattern of responses similar to another breed would be a more difficult if not impossible task.

All this strongly supports the conclusion that heredity greatly affects the expression of emotional behavior and also that differences in emotional behavior form a prominent part of the characteristic behavior of breeds and individuals. Comparable differences have been reported in human individuals by Lacey and Lacey (1958). Whether or not such differences are related in turn to the performance of breeds in training and problem-solving tests is a question which will recur in future chapters.

CHAPTER 9

EXPERIMENTS
ON TRAINABILITY

A number of our procedures involved modification of a dog's behavior toward a pattern specified by the experimenter. Such learning differs from problem solving by requiring a more stereotyped response from the animal, although the fundamental process of selective reinforcement of successful responses operates in both types of tests. It is therefore convenient to separate such tests as quieting, leash training, retrieving, motor skills, and obedience from problem-solving tests in the more usual sense.

Training tests can in turn be divided into forced and rewarded categories. In forced training, the trainee is punished for a deviant response or prevented from carrying it out. In reward training he is rewarded when he makes the correct response.

Training to follow approved cultural modes is an important part of child development and is likewise imposed upon a puppy adopted into a human family. Remaining quiet, walking on a lead, and sitting quietly on command are part of the usual regimen of pet dogs. Our procedures were simply standardized forms of such training, coupled with an objective rating scale for performance. In such forced training, the experimenters established in advance a pattern of behavior which was to serve as a model and systematically punished deviations. Analogies with certain aspects of child rearing are fairly clear. Training was continued for a specified period, and comparisons between genetic groups were made at various stages. Reward training is more closely related to ordinary problem solving since the subject is freer to adopt any individual mode of responding which is successful. The tasks in retrieving and motor skills, however, involve a minimum of sensory discrimination; instead, the experimenter encourages and rewards the subject to perform relatively simple acts. Parents similarly train a child to throw a ball or climb stairs.

That there are individual differences in trainability is, of course, common knowledge. Our interest was in determining the portion of variability attributable to genetics, and in the relationship between trainability and other characteristics of the subjects. In particular, we were interested in whether trainability was correlated more closely with other measures of learning or with measures of emotional response.

FORCED TRAINING

Quieting.—The earliest forced training arose out of attempts to weigh puppies accurately. It was necessary to keep them quiet on the scales without actually touching them in order to take a reading. We developed a technique of placing a subject on the scale platform, holding our hands near but not touching him unless he started to wriggle or step off. The best possible performance was for a puppy to remain inactive for one minute of weighing and observation. Ratings were given on each trial as follows: 3, active; 2, partly active; and 1, quiet.

Figure 9.1 shows the results at different ages. None of the very

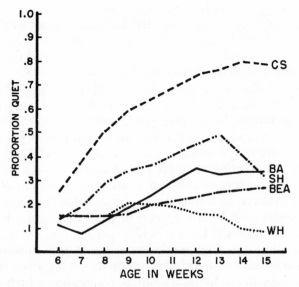

Fig. 9.1.—Proportion of animals rated as quiet during weighing. The curves are smoothed by taking 3-point moving averages.

young animals remained quiet for a whole minute, but beginning at about 5 weeks of age, some individuals in some breeds began to do this. Gradually more and more were completely trained. At the same time the breeds began to differentiate, so that by 16 weeks nearly 70 per cent of the cocker spaniels were remaining quiet but only 10 per cent of the wire-haired fox terriers. Shelties were the second most easily trained, and basenjis and beagles were close to the mean of the fox terriers.

Leash training.—Training to walk on a lead was begun at 19 weeks of age. The puppies were given five trials (one per day) over a course leading from their outside pens to the laboratory. Beginning on the third day, they were led through a portion of the laboratory and up a flight of stairs. The puppies wore a choke collar attached to a chain leash. An animal which balked three times so firmly that it could not be induced to continue walking was lifted and carried by the experimenter over the course. The subjects were led back over the same path, but no records were made of their responses on the return journey. A second series of five trials was given during the twenty-second week when they were led to the laboratory for discrimination training.

We wanted to train the dogs to walk on a slack lead at the left of the experimenter without vocalization or bodily contact. Demerits were recorded for the following faults, with an arbitrary maximum score on one day of three demerits in each category.

Balks on the outdoor course
Balks at doors or gates
Fighting or biting leash
Dragging behind or running ahead (position errors)
Interference with the experimenter (body contacts)
Vocalization

A convenient way of measuring the effectiveness of leash training is the progressive reduction of the total number of demerits. Of equal interest is the frequency of occurrence of the several types of demerits over the whole course of training. Such measures proved to be particularly suitable for characterizing patterns of responses for the five breeds.

In the leash test, unlike most of our battery, there was no significant increase in reliability as training proceeded. This is well shown by the tetrachoric correlations between the three pairs of tests recorded in Table 9.1. The correlations were: $r_{tet1,2} = .72$; $r_{tet4,5} = .76$; $r_{tet9,10} = .79$. Although almost all subjects were receiving very few demerits at the end of the training period, they were no more predictable from day to day than they were at the start of training. The correlation between the sums of odd and even trials was .86, which is high when one considers that handlers were changed on alternate trials.

The effects of training on the numbers of demerits in the five pure breeds are shown in Table 9.1. For simplicity, only six of the ten days are summarized: the first two, the fourth and fifth, and the last two days. These were chosen to be representative of the beginning, middle, and late stages of training.

Inspection of Table 9.1 shows that the means of the breeds come

TABLE 9.1

Mean Number of Demerits in Leash Training

BREED	NUMBER	DAYS					
		1	2	4	5	9	10
Basenji.......	34	9.4	7.0	5.9	4.8	3.3	3.4
Beagle........	33	9.8	6.2	3.5	2.1	1.7	1.2
Spaniel.......	41	6.3	5.0	4.2	3.2	1.4	1.1
Sheep dog.....	24	12.0	7.9	5.8	5.3	3.3	2.7
Terrier........	21	6.9	5.1	3.5	2.7	1.4	1.3

closer together as training proceeds. Thus, the range of breed means on Day 1 was 5.8 demerits, and on Day 10, 2.1 demerits. One might infer from this that the effect of training was to diminish the importance of hereditary differences upon behavior, but the conclusion would be incorrect. When the proportion of total variance attributable to breed differences is computed for Day 1 and for Days 6–10 combined, the respective values are 53.7 and 52.4 per cent, a negligible difference. The last five days were pooled in this calculation in order to make the mean number of demerits more nearly equal for both sets of scores. In general, the dogs made about as many errors on the first day as they did during the whole second week.

Training for the rigid criteria demanded by this test does not eliminate nor even diminish the relative contribution of heredity to behavioral variation. Of course, if the process could be continued to the point where no demerits were assigned, all individual and breed variation would vanish, and hereditary effects would disappear. In this test, over 50 per cent of the variation in the total score could be attributed to breed differences. The proportion is unusually high, and therefore it is interesting to analyze this case further.

One obvious course is to compare the breeds on the different types of demerits separately in order to see whether the groups differ in number of demerits only, or in their categories as well. By conversion to stanines, it is possible to compare each type of demerit on the same scale.

Figure 9.2 shows the mean stanine scores for each demerit category. It is clear that breed differences in numbers of balks and posi-

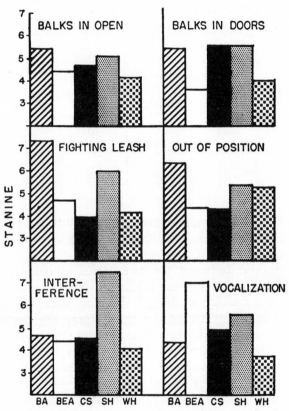

FIG. 9.2.—Types of demerits given during leash-control training (stanine scale).

tion errors are less marked than in fighting the leash, interference, and vocalization. Basenjis were outstanding in their vigorous resistance to the restraint of a collar and lead; Shetland sheep dogs similarly interfered excessively with the handler by leaping on him and winding between his legs; beagles were the foremost emitters of howls and wails during initial stages of training.

Applying an analysis of variance, we can compute for the several demerit categories the proportion of variation attributable to breed differences. These values are: balks in the open, 1.9 per cent (not significant); balks at doors, 28 per cent; fighting the leash, 72 per cent; out of position, 39 per cent; interference, 55 per cent; and vocalization, 40 per cent. All differences except the first are significant at better than the .01 level.

The pattern of intercorrelations between the six types of demerits shows some striking relationships. These have been arranged in Table 9.2 to bring the larger coefficients of correlation along the main

TABLE 9.2

INTERCORRELATIONS OF TYPES OF DEMERITS DURING LEASH TRAINING

	Balks at Doors	Balks in Open	Out of Position	Vocalizes	Fights Leash
Interferes..........	−.45	−.28	.00	−.12	−.14
Balks at doors......	...	.53	.51	.12	.14
Balks in open.......	.53	...	.55	.31	.21
Out of position.....	.51	.55	...	.25	.23
Vocalizes...........	.12	.31	.25	...	.35
Fights leash........	.14	.21	.23	.35	...

diagonal. Such an arrangement, according to Guttman (1955), shows the degree to which the several kinds of demerits can be considered as expressions of common factors.

Intercorrelations between the two types of balks and position errors are the highest and seem to define the central characteristics of the model toward which training is directed, that is, a dog walking freely on the leash by the side of the trainer. The demerits which most strongly separate the breeds are peripheral to this core characteristic and may be interpreted as alternative forms of emotional response to the situation. Demerits for interference correlate negatively with all other types. By adding interference ratings to obtain a total demerit score, we tended to understate breed differences.

Although training reduces the number of demerits in all categories, it does not do so equally. The proportion of balks and vocalization falls, but the proportion of position errors and interference increases. Details are given in Table 9.3. Thus the hereditary differences at the beginning, middle, and end of training are expressed in different behavior patterns. If only the total scores are compared, this interesting fact is concealed.

We conclude that each breed is characterized by a pattern of responses which are differentially affected by training. As training proceeds, the expression of hereditary differences shifts, but the pro-

TABLE 9.3

TYPES OF DEMERITS IN EARLY, MIDDLE, AND LATE PHASES OF LEASH-CONTROL TEST

TYPE OF DEMERIT	DAY 1		DAY 5		DAY 10	
	Mean	Per Cent	Mean	Per Cent	Mean	Per Cent
Balking..............	2.4	26.8	0.9	24.6	0.2	12.1
Fighting leash.........	1.5	17.1	0.4	11.0	0.3	14.9
Out of position........	2.2	25.7	1.1	30.7	0.6	34.5
Interference..........	1.2	13.5	1.0	27.7	0.6	32.0
Vocalization..........	1.5	16.9	0.2	6.0	0.1	6.5
Total............	8.8		3.6		1.8	

portion of variance attributable to genetics changes little or not at all. Within a few days, subjects are transformed from puppies who pull off at wild angles or drag their feet, emitting bloodcurdling yelps, to docile creatures walking freely by their handlers. But the hereditary contribution to individuality is detectable at all stages of training.

The obedience test.—A third major procedure designed to train subjects to achieve a specified criterion of behavior was the obedience test. Animals were trained over a 3-day period to remain on a stand for 30 seconds and then to jump down on command. During the early stages each subject wore a choke collar to which a lead was attached. As training progressed restraints were gradually removed, and the handler on successive trials moved away to distances of 1.5, 3, 6, and 12 feet, and finally behind a screen 14 feet from the stand.

If the subject leaped prematurely during a test, the handler moved closer to the stand for the next trial, and started a new series of increasing-distance trials. A measure of the level of training attained on each day was obtained by arbitrarily assigning a score to the subject-trainer distance and adding the scores of a day's trials. Assigned values were: choke collar (0); 6 inches (1); 1.5 feet (2); 3 feet (3); 6 feet (4); 12 feet and behind screen (5). Subjects who failed to leap at command within 10 seconds were gently pushed from the stand by the trainer.

The training procedure was designed to bring individuals to a common standard as quickly as possible. Equalization was promoted by counting two choke-collar trials as equivalent to a single hand-control trial in scoring. Thus the more recalcitrant subjects received additional training. In addition, training was discontinued on any day when the subject stood for 30 seconds with the handler at the 12-foot mark. This was done to avoid overtraining subjects and thereby increasing the range of individual variation.

On Day 4 and, for most subjects, again on Day 5, a standardized set of five trials, one each at 0.5, 1.5, 6, and 12 feet and behind the screen, was repeated twice. Procedure was the same as in training, except that no punishment was given for premature leaping. A composite score for the test was obtained by adding the time spent on the stand, plus 30 for each trial in which the subject jumped on command, plus 60 for each trial in which the subject delayed for 10 seconds following the command to jump. Scores could range from zero for a subject who always leaped from the stand immediately upon release, to 450 for a subject who failed to jump on command in all five trials. The optimum "obedience" would be a score of 300. However, the results made less sense when arranged according to a scale of "obedience" than when arranged according to a scale of inhibition of movement. In our system, therefore, high scores denote inhibited subjects which remained immobile on the stand; low scores denote hyperactive subjects not well controlled by the handler; and middle-range scores were obtained by dogs which were readily controlled by the handler. Results were converted to the stanine system.

In addition, two other ratings were made on most of the subjects in the test situation. A measure of "confidence" was obtained by noting whether the dog would eat from a dish placed first on the stand, then beside it, and finally at some distance. Some dogs would not take food at all. A measure of "attraction" was obtained by rating the subject on his approach to the handler between trials. Each dog was called by the handler after he had finished recording each trial on the data sheet. Results of these two measures could not be converted into stanines, and subjects were merely classed as low, medium, and high with respect to "confidence" and "attraction." The object in recording these measures of emotional and social behavior was to see whether they had any consistent relationship with degree of trainability.

Finally, a special procedure was adopted during the program to test the effects of changing handlers. Each litter was assigned to two handlers, one man and one woman. Within each breed or hybrid group, male and female dogs were divided separately at random between the handlers. Among the pure breeds, the retest on Day 5 was conducted by the other handler; among the hybrids, the retest was administered by the original handler. This design permits evaluation of the efficacy of male and female trainers and the extent to which obedience training is generalized as a response to human beings as contrasted with specific individuals. One would predict that there would be some deterioration of performance on retest with another handler and possibly that the effect would be more extreme in breeds with the name of being one-man dogs. Among our subjects, the basenjis have such a reputation.

All the breeds showed marked improvement over the three days of preliminary training. Figure 9.3 demonstrates that the order of per-

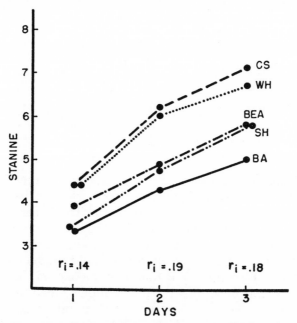

FIG. 9.3.—Level of training (stanine units) on successive days of obedience test.

formance does not shift; in fact, the spread of the means increases with training. The proportion of the total variance attributable to breeds, r_i, also increases, but not significantly. The same order of breed scoring was maintained on the test day and in the retest. Results of both tests are shown in Table 9.4. Differences between breeds were again highly significant, with intraclass correlations in the first test of .149, and on retest, .262.

The reliability of the individual scores on test day (Day 4) can be estimated by the correlation of .939 between the two halves of the test. The correlation between test and retest was .863. These figures indicate that a few days of training produce highly consistent behavior on the obedience test, even more so than in the leash-training test.

It is interesting to note that the average performance of dogs trained by men and those trained by women were almost exactly the

TABLE 9.4

MEAN STANINE SCORES ON OBEDIENCE FOR PURE BREEDS

BREED	SEX OF EXPERIMENTER		N	MEAN SCORE		DIFFERENCE BETWEEN DAY 5–DAY 4
	Day 4	Day 5		Day 4	Day 5	
Basenji...................	F	M	11	4.00	3.40	−0.60
	M	F	5	3.20	3.20	0.00
Beagle...................	F	M	10	4.90	4.25	−0.65
	M	F	11	5.50	4.45	−1.05
Cocker spaniel............	F	M	13	5.80	5.41	−0.39
	M	F	11	6.50	6.04	−0.46
Shetland sheep dog.........	F	M	9	6.50	5.50	−1.00
	M	F	10	5.05	4.70	−0.35
Wire-haired fox terrier......	F	M	5	6.60	6.20	−0.40
	M	F	8	6.19	5.44	−0.75
All...................	F	M	48	5.42	4.82	−0.60
	M	F	45	5.47	4.88	−0.59

same (5.47 and 5.42, respectively). The average change in score on retest was also practically identical (−0.60) whether the change was from male to female trainer or the opposite. The hybrids, who were re-tested by the same individual, fell less than half as much (−0.27) on retest. These results would have been predicted on the argument that any change in the testing situation would produce detrimental effects upon performance. We found no indication, however, that any one breed was more affected than another breed by the change in experimenters. The number per group is rather small, but the experiment does not support the notion that some of the tested breeds attach themselves only to one individual, while others shift their responses readily to another person.

It was somewhat surprising to find that performance on the obedience test as measured by the tendency to inhibit jumping was correlated neither with our measure of "attraction" nor by "confidence." Striking breed differences were found in these two ratings. Within any one breed, however, high or low confidence or attraction ratings were unrelated to obedience scores. It is perhaps odd that training was equally effective whether or not the subject was attracted to the trainer, and whether or not the subject was willing to approach the stand to receive food. The results suggest that the various parts of the dog's behavioral repertoire tend to be independent. His entire behavior is affected by genotype, but general factors of trainability and emotionality are not readily found.

To summarize the results of this test:

1. Breeds differ widely in the ease with which they can be trained to remain quiet on command.

2. All subjects learn under training, but the differences between breeds persist, at least over a one-week period.

3. The contribution of heredity to total variance does not change greatly over the period of training.

4. Changing the experimenter for a retest produces a moderate decrement in score, which is above that found when the retest is performed by the original experimenter. The effect is about the same in all breeds examined, but great individual variation is found.

Comparison of breeds.—All of these forced-training tests involve teaching the animal to be inactive. The puppy is taught to stay quietly on the scales by hand signals and restraint. When placed on a leash, he is being taught to restrain any activity except that of following the experimenter. Finally, on the obedience test, he is taught to be inactive on a platform at a distance from the trainer.

When the breeds are ranked on all three tests (Table 9.5), the

TABLE 9.5

Rank of Five Pure Breeds under Forced Training

Test	Controlled Activity ⟶ Overactive				
Scale activity......	CS	SH	BA	BEA	WH
Leash control......	CS	WH	BA	BEA	SH
Obedience.........	CS	WH	BEA	SH	BA

cocker spaniels emerge as the easiest to train in all three situations. Basenjis and beagles are consistently hard to train, but shelties and wire-haired fox terriers may do well in one situation and poorly in another.

We are forced to conclude that there is no simple general capacity involving restraint for accepting forced training. The trainability of cockers is related to their former selection for the ability to accept training to crouch. This response is easily elicited by hand signals, and these were used to some extent in all three tests. The shelties are easily taught restraint by directly touching their bodies but tend to be active when away from an experimenter. They do, however, crowd against the experimenter's legs during the leash-control test. Wire-haired terriers respond to hand contact with playful fighting but accept restraint training at a distance. On the leash they tend to pull ahead of the experimenter. Basenjis respond to most kinds of restraint with attempts to escape, while beagles give the appearance of being restless in any restraining situation. These observations are borne out by the low correlations between the three tests (Table

TABLE 9.6

CORRELATIONS BETWEEN TESTS OF FORCED TRAINING
(FROM C. L. BRACE)

Population	Number	Inactivity—Leash Control	Inactivity—Obedience	Leash Control—Obedience
Basenji........	22	−.04	.20	.17
Beagle........	24	.07	.12	−.32
Cocker........	23	.02	.24	+.24
Sheltie........	16	−.26	.34	−.38
Fox terrier.....	15	.03	−.26	−.66**
All pure breeds	100	.13	−.11	−.32**
F₂ hybrids.....	73	−.10	−.03	.04
All dogs...	294	.06	−.12*	−.17**

* Sig. at .05
** Sig. at .01

9.6). The only correlation of any importance is a slight negative correlation between leash fighting and learning to stay on the platform.

We therefore concluded that the basic trainability characteristics of the different breeds tend to be specific to particular test situations, and that they are based on a large variety of capacities. In some test situations, the possession of two different capacities will produce widely different results between breeds; in others it may produce almost identical performances.

REWARD TRAINING

In these tests the animal is given a reward as soon as he performs the desired act. The usual technique is to break the training down into a series of steps, first teaching the animal to do a simple first step, then adding more and more requirements in sequence until a complex series of acts has been learned. This stepwise training has long been successfully used to train circus animals and has been developed into an art with the techniques of operant conditioning devised by Skinner (Skinner, 1938; Breland and Breland, 1961). Food is the most common reward, but the dog is also notable for its responsiveness to words and attitudes of a human handler.

In the section which follows, we have summarized results of the habit formation, retrieving, and motor-skill tests. The motivation test described in chapter 10 as a preliminary to discrimination training also falls in this category.

Goal orientation (habit formation) test.—This test given when the puppies were 9 weeks of age, had two functions. We intended it as basic training for running to a goal (as a foundation for more complex tests), and we also attempted to measure the rapidity with which a puppy could change a habit once it was formed. The test itself consisted of showing the puppy a small bit of fish in a wooden box and then placing him in a cage with a wire gate in front as in the photograph (see plates). The experimenter left the room, raised the gate with a string running through a pulley, and recorded the behavior of the puppy. The puppy was given a total of four trials on two successive days, and then the position of the box was changed to a different corner of the room, a distance of about 6 feet. The puppy then ran six more trials over a period of three days.

As Figure 9.4 shows, differences between breeds were greater on

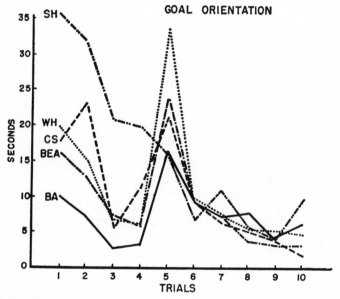

FIG. 9.4.—Mean time scores on goal-orientation test.

the first two trials than on the third and fourth. We selected the second trial for analysis, in the hopes that this would prove a measure of basic motivation for running to a goal after the puppies had a chance to learn the object of the test. Basenjis gave the best performance, and shelties were the poorest. As on many other tests, the behavior of the beagles showed a high degree of variability. Differences between breeds were significant, but the proportion of variance attributable to this cause was only 7 per cent (Table 14.3).

When the position of the goal was changed, large differences be-

tween the breeds reappeared, the basenjis again performing best and
fox terriers most poorly. Unlike other breeds, shelties continued to
improve, performing as well as the basenjis. All the breeds performed
very similarly on the subsequent trials, and there was no indication
of any differences in the ability to switch from one habit to another.

The superiority of the basenjis on the goal-orientation test may
reflect simply their superior running speed at this age, but more
probably it shows a greater tendency toward visual orientation.
There was no indication that differences between cockers and
basenjis showed any regular form of inheritance. The general conclusion is that all breeds performed in a very similar fashion on this
simple kind of reward training.

Training to retrieve.—Most dogs can be taught to retrieve a stick
and return it to their masters without any apparent reward other
than the execution of the act itself. Some dogs "beg" their masters to
throw objects for retrieval. In addition, many bird dog breeds are
regularly trained to retrieve wounded or killed birds and return them
to their masters. The cocker spaniel has in the past been selected for
its ability to be trained in this fashion and also for a "soft mouth,"
meaning that a cocker will pick up the bird without biting or crush-
ing it.

We originally did a few retrieving experiments with 9-week-old
cocker spaniel puppies, an age when such training usually succeeds
easily (Pfaffenberger, 1963). The training schedule became crowded,
however, and the test was postponed to a much later age (32 weeks).
We trained our animals by using a modification of the play-retrieving
procedure which most owners employ with their dogs. An unfamiliar
object which could be easily carried was left in the pen for a couple
of days in order that the dogs might familiarize themselves with it.
On the succeeding three days each animal was tested for retrieval of
a dumbell-like object (two thin blocks of wood connected by a heavy
wooden dowel) on a total of 9 trials.

The results were disappointing. Only 11 per cent of the purebred
dogs actually returned the dumbell and released it in at least one
trial, and none of the differences between breeds was significant
(Table 9.7). More detailed analysis gave a similar result. The act of
retrieving can be broken down into smaller behavior patterns for
purposes of analysis. The dog must follow the object thrown, pick it
up, carry it, return it, and, finally release it when the trainer attempts
to take it away. Very nearly 100 per cent of the dogs would follow the
object, and about 25 per cent would pick it up and carry it, but very
few would actually return it to the trainer. Again, the differences are

TABLE 9.7

PERCENTAGE OF ANIMALS SUCCESSFULLY TRAINED IN RETRIEVING TEST

Population	Number	Follow, Pick Up, and Carry (Per Cent)	Complete Pattern (Per Cent)
Basenji	34	35	6
Beagle	37	22	11
Cocker	41	22	15
Sheltie	25	32	8
Fox terrier	20	15	15
Total pure breeds	157	25	11
Total hybrids	196	27	14

not great enough to indicate that there are any important breed differences. Occasionally we observed puppies carrying other objects in their mouths about the pen—a spontaneous manifestation of part of the complete pattern. However, the rarity of such observations permits no conclusions regarding genetic or environmental causation.

We concluded that the ability to learn retrieving is rather widely distributed throughout dog breeds and that the special abilities of the cocker spaniel, if any, are confined to retrieving birds in the field rather than the play retrieving which was the object of the test. We were also impressed with the relative difficulty of training the older animals compared to the few which were trained at 9 weeks. There is possibly a critical period for learning to retrieve, that is, a time when the probability of executing the complete pattern is relatively high, so that reward training can be maximally effective.

The motor-skill test.—About one-third of the way through the experiment we observed basenjis jumping to the tops of their houses and surveying the world from these vantage points. Some members of this breed could scramble halfway up the fence of their enclosure and cling to a narrow shelf while peering through the wire mesh. None of the other pure breeds did this, and we therefore designed a test to measure degrees of climbing and jumping capacity.

The motor-skill test, as it was named, required the dogs to climb to the top of a pile of boxes, and later to cross a bridge leading from the pile to another high point in order to obtain a food reward. On the first day we stacked two wooden boxes in the pen and placed a dish containing a spoonful of fish upon it. A broad ramp led to the top, and the height of 2 feet was just enough so that a dog could not reach the food from the ground, and must either jump to the top or climb the ramp in order to eat. We first placed each subject in turn on the pile and allowed it to eat, then placed it back on the ground and measured the time required to reach the top again. The test was repeated on Day 2, and on Day 3 was modified by increasing the height to 3 feet. As the pile grew higher, the ramp, of course, grew steeper. On Day 4 the stacks of boxes were 4 and 5 feet high, at which point the test became most difficult. On Day 5 the stack was reduced to the 3-foot level, but the dogs now had to walk across a plank to reach the food 6 feet away on the top of the house. The plank was at first one foot wide, and on the second trial was replaced by one only 6 inches wide. Many dogs found this difficult, and some lost their footing and fell off.

If a dog did not reach the goal after two minutes, the experimenter took a leash and led him up the ramp so that he could eat. Thus, forced training was secondarily combined with reward training.

The whole test may also be considered the canine equivalent of an athletic event similar to a high jump. The early part of the test on the low boxes consisted of training the animals in motivation and in the technique of climbing and jumping, while the last two days were equivalent to a contest in which the height of the object to be climbed was gradually raised to one of considerable difficulty.

Failures, defined as failure to reach the food within two minutes, whatever the cause, were so many that the results of the test are best expressed in terms of percentages of success rather than time (Fig. 9.5). As expected, the basenjis did better than any other breed on the first trial of the test; but unexpectedly, the cocker spaniels did

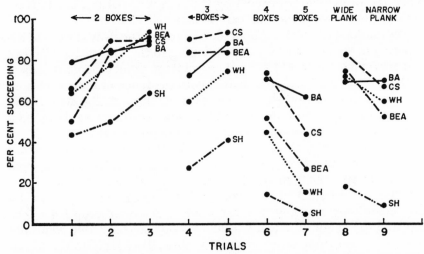

Fig. 9.5.—Proportion of animals succeeding on each trial of the motor-skill test.

equally well on the total test results (Table 9.8). The basenjis obviously had superior climbing ability, but the cockers were turning in an equally good performance. What actually happened was that the basenjis reacted suspiciously to the strange objects placed in their pen, and tended to approach the boxes slowly and cautiously, finally jumping or climbing to the top. They also showed very little improvement in preliminary training. On the other hand, the cockers showed no fear of the test equipment and usually backed off and ran rapidly at the ramp so that their momentum would carry them to the top. The result was that the cockers actually had a lower percentage of failures than the basenjis, except on the first trial and on the highest set of boxes. Beagles and fox terriers fell into second rank with about the same percentage of failures. Both breeds showed up badly on the 4- and 5-foot stacks. The Shetland sheep dogs were outstandingly poor all the way through the test, apparently being quite fearful of heights and being relatively little interested in the food reward under these conditions. They were approximately 95 per cent unsuc-

TABLE 9.8

OVER-ALL PERFORMANCE ON MOTOR-SKILL TEST

Breed	Number	Per Cent Success
Basenji.........	34	75.8
Beagle.........	32	66.7
Cocker.........	32	76.7
Sheltie.........	22	30.3
Fox terrier......	20	63.3

cessful on the most difficult parts of the test. The over-all results can be evaluated by the chi-square analysis. Significant breed differences were obtained on all trials at 5 feet ($P = .001$).

The most interesting part of these results is the way in which the animals organized their abilities to meet the actual situation. The basenjis with their superior height and jumping ability were obviously superior in climbing ability. Yet the cockers, by organizing their own capacities, were able to compete successfully except under the most extreme conditions, and even then the differences were not great. On the most extreme test, with the boxes stacked to 5 feet, 62 per cent of the basenjis succeeded against 44 per cent of the cockers, a difference of only 18 per cent. The test is not, therefore, a favorable one for genetic analysis through the crossbreeding experiment. Nevertheless, the performance of the hybrids was of considerable interest. The F_1 hybrids differed considerably from each other and tended to resemble their mothers. The basenji backcrosses were very similar to basenjis, and the F_2's somewhat intermediate between the two breeds. The backcrosses to the cockers made by far the best records in the preliminary training phase of the experiment, but made the worst record on the most difficult trial. We may conclude that there are two genetic tendencies operating here. One is the capacity of the cocker to respond quickly to the type of motivational training used in this experiment, and the other is the climbing and jumping ability of the basenji. These two interact in complex ways.

SUMMARY

The results of all training tests indicate that the ability to respond successfully to training is highly specific, with the result that the expression of any one ability is highly dependent upon the type of training used. Furthermore, it appears to be extremely difficult to devise a test which measures only one simple capacity. Even when the puppies were being trained to be quiet on the scales, the cocker spaniels responded with their tendency to sit, while the fox terriers employed another capacity, that of reacting aggressively. The leash-control test brought out both the characteristic resistance to restraint of basenjis and the sheep dog tendency to crowd next to a person, or to a sheep. In short, in any particular situation, a puppy organizes the capacities which it happens to possess, usually selecting the combination which gives the best results. This was especially evident in the motor-skill test, in which cocker spaniels were able to compensate for a lack of jumping ability by superior motivation and effort.

Where it has been possible to measure the components of behavior involved in training, it is obvious that emotional reactions have a highly important effect. Such reactions appear not only in the special situations devised to test them, as described in chapter 8, but they also either facilitate or inhibit training of almost any sort.

The importance of breed differences varied from test to test, ranging from very large differences in the leash-control tests to none in retrieving, although in the latter it might have been possible to demonstrate differences by selecting a more favorable age for training.

Training has various effects upon the expression of breed differences. In the quieting test, training actually increased the magnitude of differences. In most tests, the magnitude of differences was decreased, but the *proportion* of variance attributable to breed might nevertheless remain constant, as it did in the leash-training and obedience tests. Demanding a higher degree of performance after preliminary training may increase the magnitude of differences, as it did in the motor-skill test.

Finally, each test of training brought out individual conclusions of great interest. The obedience test, for example, showed that dogs of all breeds tend to generalize their training from one handler to another, giving no support to the "one man dog" theory. Whether such dogs actually exist in other breeds independent of differential training is an open question. The leash-training test showed breed differences throughout the period of training, but the form in which these differences were expressed varied as training progressed. In short, the effect of heredity upon trainability is highly complex, both because of the number of specific basic abilities involved and because of the complicated interaction between them made possible by behavioral adaptation.

CHAPTER 10

THE DEVELOPMENT AND DIFFERENTIATION
OF PROBLEM-SOLVING BEHAVIOR

A newborn puppy cannot survive without the constant care of its mother or of a conscientious human caretaker acting as a substitute mother. Until the puppy emerges from the nest box at about 3 weeks of age, life offers few challenges and we see little evidence of problem solving. This does not mean that the puppy learns nothing during these 3 weeks. There is some evidence that simple discriminations are acquired, and we know that maturation of the sensory and motor systems is occurring during this time. The point is that under ordinary conditions problem-solving behavior is not conspicuous and is hard to demonstrate under experimental conditions, perhaps because critical tests are difficult in an animal with restricted response capacities.

At 3 weeks of age a large number of behavioral changes occur. Of particular importance is the appearance of the capacity to form stable conditioned reflexes with rapidity comparable to that of an adult. However, the retina is still undeveloped, and the puppy is unable to perceive differences in depth until 4 weeks of age. Changes in the alpha rhythm of the EEG indicate that full visual capacity is not reached until 7 to 8 weeks. Moreover, motor abilities are still undeveloped, so that a puppy is clumsy and tires easily. Consequently, 6 weeks of age is very close to the earliest practical time for administering a problem-solving test. For the next few weeks the puppy is remarkably adaptable and responsive to almost any sort of learning experience which his developing motor capacities and short attention span will permit.

By 4 months the puppy has developed into an independent creature who can find his own food, negotiate complicated pathways, and in general cope with ordinary environmental challenges. The puppy at this age is, of course, far too young to have acquired the specialized abilities for which dogs are famous, abilities such as sheepherding, guiding the blind, or performing in circuses. He can, however, be easily trained to obey simple commands and can solve complicated problems provided that no highly developed motor skills or great endurance are required.

Even the simplest training may be considered as a form of problem-solving behavior. From a multitude of possible responses, the puppy must learn to select those bringing the largest or the quickest reward and to avoid those which result in pain or discomfort. Several training procedures were considered in the previous chapter. Here emphasis is placed upon more complicated tests involving spatial relationships and visual cues. Attention will be given also to the barrier test, manipulation test, and the maze test, which were carried out with younger subjects in an effort to learn more about the development of problem-solving ability.

DEVELOPMENT OF PROBLEM-SOLVING BEHAVIOR

Several difficulties beset the investigator who attempts to study the development of learning in the young puppy. The newborn subject has few simple responses, and one can demonstrate learning only by an increase or decrease in the frequency of one of these few patterns. In older puppies, the capacities for response are enormously greater and it is easier to show whether or not a particular response occurs as a result of learning.

It is difficult to note the exact time of the appearance of any particular learning capacity. Complex tasks cannot be mastered suddenly even by experienced subjects, and attaining a criterion level requires many trials, often spaced over a considerable period of time. Should we establish our criterion of maturation as the day on which the puppy first meets our statistical requirement for learning, or as the day on which training is begun? If it takes a puppy 2 weeks to demonstrate that he has learned a particular task, we cannot determine the age at which the ability was acquired more precisely than sometime within the 2-week period. Obviously, greater accuracy is possible when testing is restricted to simple problems potentially solvable within a few trials conducted in one day. Under such conditions, maturational changes occurring during the period of learning can be neglected. Such restrictions, however, eliminate the most interesting maturational problems.

Since our experiment was designed primarily to detect differences in problem-solving behavior attributable to heredity, we do not have data to demonstrate rigorously the age at which various kinds of learning become possible. Studies on monkeys by Mason and Harlow (1958a) and by Harlow (1958) with this objective used the same procedures beginning at different ages on genetically similar subjects. Several studies from our laboratory have been concerned with learning in newborn puppies (Fuller, Easler, and Banks, 1950; Cornwell and Fuller, 1960). In order to evaluate hereditary factors, however, it was more practical to give each test at a particular age. To an extent, our series of tests represent steps of increasing difficulty corresponding to the increased maturity of our subjects. It would have been interesting, had our resources been unlimited, to have investigated both the genetic and age variables, since some of the breed differences found may have been simply manifestations of different rates of development and would have disappeared if our tests had been carried out at a later age. Such an interaction between rate of development and heredity is probably more important in young animals than in older ones. We must, however, remember that there is a possibility of some transfer of effect from one situation to another. Conceivably, puppies who failed in their first tests because of immaturity also did poorly in later tests because of lack of motivation. Thus, far-reaching effects might follow from such seemingly trivial factors as the exact scheduling of the tests.

BARRIER OR DETOUR TEST

The barrier test which was given at 6 weeks of age was the first performance test for the puppies. It was also their first experience outside their own rearing room, aside from the confinement that they experienced in small cages in the hall while tests were being set up inside. Consequently we had an opportunity to experiment with a totally unsophisticated puppy, and we attempted to develop a test which would be suitable for animals of this age and at the same time show us whether or not there was some native problem-solving ability which did not depend on previous learning.

The test was basically the same as the detour problem devised by Koehler (1927) for a variety of animals, and details of preliminary training have already been described in chapter 1. The first real problem was presented on Day 3. On this day the puppy found a barrier between him and the dish of food. This barrier was a wooden frame, 6 feet long and 3 feet high, covered with poultry wire, and made opaque with heavy paper except for a slit one foot wide immediately in front of the puppy. Thus the puppy could see the food and the experimenter, but it had to move away from and out of sight of the food in order to reach this reward. The ends of the barrier were clearly marked by supports which extended out on the puppy's side of the barrier.

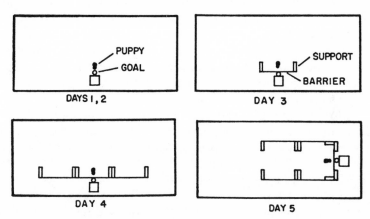

FIG. 10.1.—Problems in the barrier test. This test was performed in an arena 12 × 27 feet in size. The barrier was opaque, except for a wire-covered opening one foot wide immediately in front of the goal. In order to reach the goal, the puppy had to move away from it.

A puppy could make a perfect score by running directly around the end of the barrier to the food dish. If it stopped or reversed its direction an error was counted. In addition, the total time required to reach the food and the number of vocalizations that the puppy made in each minute of the test were recorded. If the puppy did not get to the food within 10 minutes, the experimenter called the puppy around to the food dish and allowed it to eat.

On the fourth day of the test the barrier was extended on either end so that it was three times as long as on the first day. On the fifth day the barriers were converted into a U shape, with one side of the U parallel to the original side of the barrier so that the goal was now in a different location. Three trials were given on each day of the test, and all puppies thus had an opportunity to learn one situation before proceeding to the next.

It was extremely interesting to watch the puppies work out solutions to these simple problems. A few animals immediately ran around the end of the barrier, without mistakes. Others ran first to one end, took a look, then returned to the other end, took a second look, and then ran to the food, usually back around the end where they first started. Still others moved only a short distance from the food, which was of course inaccessible behind the barrier. Such puppies soon began to run back and forth and to paw vigorously but ineffectually on the wire, vocalizing as they did so. Sometimes this would go on for several minutes. Then the puppy would stop, look around calmly, and run around the end to the food dish. Vocalization was clearly a sign of frustration, and no puppies solved the problem while vocalizing.

Is insight learning developed without previous experience? An adult dog would immediately solve such problems by taking one look and running around the barrier to the food. As shown in Table 10.1, only a very few puppies were able to do this, some 8 out of the 203 purebred puppies tested. If the capacity for solving a detour problem can in fact be developed without experience, this must have occurred in only a very small proportion of puppies of any breed. On the other hand, if the capacity is developed by experience, there should be a much higher percentage of success on the next day when a similar but longer barrier is presented. Four times as many puppies made perfect scores on the first trial of the second test as they did on the first, and large breed differences began to appear. Basenjis improved a great deal more than any other breed, and beagles and wire-haired terriers were second. The cocker spaniels and shelties showed no improvement. These differences seem to result from different reactions to frustrating situations. Basenjis at this age are usually highly active and well developed physically, while the puppies of most other breeds are still fat and clumsy. In many cases the cocker spaniels which failed would simply lie down and go to sleep until the test period was over. The basenjis remained active and were thus more likely to solve the problem by chance.

Most of the puppies were not able to solve the third problem (the U-shaped barrier) without errors. Even the number of successful basenjis was considerably reduced. This indicates that puppies of this age have limited capacities for generalization. In one of the preliminary experiments, the straight barrier was simply rotated 180° in the same room, and the puppies reacted to this as if it were an entirely new situation. However, the differentiation between the breeds remains, with basenjis most successful, fox terriers and beagles next, and cockers and shelties least successful.

These conclusions concerning the lack of generalization are borne out by the correlations between the first trials on each test. When all breeds and hybrids are considered, performances on the long and

TABLE 10.1

NUMBER OF ANIMALS WITH EXCELLENT PERFORMANCE ON THE DETOUR
TEST—FIRST TRIALS ONLY

BREED	NUMBER TESTED	NUMBER OF ERRORS			TOTAL	PER CENT
		0	1	2		
Test 1—Short Barrier						
Basenji.........	44	2	4	2	8	18
Beagle.........	49	0	1	5	6	12
Cocker.........	57	2	3	5	10	18
Sheltie.........	26	1	0	1	2	8
Fox terrier......	27	3	1	1	5	18
Total......	203	8	9	14	31	15
Test 2—Long Barrier						
Basenji.........	44	18	4	10	32	73
Beagle.........	49	9	4	3	16	33
Cocker.........	57	5	1	3	9	16
Sheltie.........	26	0	0	1	1	4
Fox terrier......	27	3	3	4	10	37
Total......	203	35	12	21	68	34
Test 3—U-shaped Barrier						
Basenji.........	44	13	2	2	17	39
Beagle.........	49	2	1	3	6	12
Cocker.........	57	1	1	0	2	4
Sheltie.........	26	0	0	0	0	0
Fox terrier......	27	3	1	1	5	19
Total......	203	19	5	6	30	15

short barriers are correlated at .16, but the correlation between performances on the long barrier and the U-shaped barrier is only .07.
Correlations within the pure breeds and hybrids are highly variable.
The behavior of the puppies indicated that easily learning Problem 1
might actually be a handicap in dealing with Problem 2, because the
puppy might turn too soon when running around the long barrier and
thus become confused, whereas an animal which had made several
mistakes would have explored the apparatus more thoroughly and
thus recognize changes more quickly.

MANIPULATION TEST

As we worked with the basenjis we observed that they were considerably more skillful with their paws than the other breeds, sometimes using them in an almost catlike fashion. The manipulation test

was designed to measure this difference objectively. The puppies had been previously trained to run to a wooden box containing a food dish (see goal-orientation test, chap. 9). During the following week they were confronted with the problem of removing the dish from the box, which was now covered.

On the first problem, the top of the box was covered with hardware cloth so that the puppies could see and smell the food, but could reach it only by pulling or nosing the dish out through the open side of the box. On the first trial, the dish protruded enough so that the puppy could easily get the food by sticking its nose into the top of the dish and forcing it outward. Very nearly 100 per cent of the animals were able to do this. This preliminary training was repeated on each day of the test so that the puppies would be rewarded by success and food at least once per day.

On the second day, the puppy went through the first trial as before, but on the second trial found the dish farther back in the box, making it more difficult to pull out. On the third day, the dish was still farther back but was attached to a small wooden dowel and a string, so that the dish could be pulled out with the string. The animal could also paw it out directly, but this was quite difficult to do. The test on the fourth day was similar to that on the third, except that the dish was still farther back in the box, making it almost impossible to pull out the dish except with the string. Finally, on the fifth day the whole problem was changed. This time the box had no bottom, and its one open side was against the wall, so that the puppies had to pull the box off the dish rather than pulling the dish out of the box.

There were a very large number of failures, so that the test is best analyzed in terms of proportions of successes rather than in those of time required to solve the problems.

A glance at Figure 10.2 indicates that the performance of the breeds was generally consistent on all parts of the test. Basenjis always did the best and cocker spaniels the worst. Beagles, wire-haired terriers, and shelties were intermediate, their relative ranks fluctuating somewhat. The chief effect of the different forms of the problem apparatus was that the presence of the string made the solution easier for all breeds.

One of the most interesting observations was the effect of repeated failure upon the behavior of any particular animal. On the first failure it would persistently try a variety of ways to get at the food, scratching, nosing, and even dragging the heavy box around. On the second trial it would make a much briefer attempt, and by the third or fourth the puppy would usually go up to the box, take one look, and then return to the gate to wait out the remainder of the two minutes allotted to the trial. Thus the animal very quickly formed habits which effectively prevented any future success. This points up the importance of success in enhancing motivation. The results are, of course, consistent with the known effects of reinforcement on the maintenance or extinction of conditioned responses.

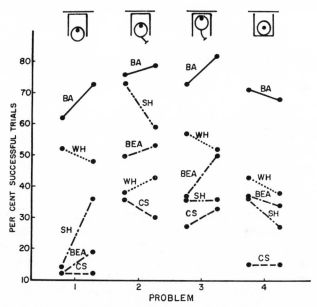

FIG. 10.2. Results of manipulation test on successive days. Note that basenjis are consistently the best, cockers the worst.

The significance of this phenomenon in relation to the effect of heredity upon behavior is that it demonstrates the importance of maintaining high motivation in order to bring out the maximum genetic capacities of the individual. If an animal is subjected to a problem even slightly beyond its powers at the outset, it may simply learn to inhibit approach and manipulative behavior which has not been reinforced. A slightly less difficult problem at the outset of training could make the same animal into a highly persistent individual. We observed that a puppy which had consistently succeeded in this test for several trials would respond to a difficult problem with an almost desperate effort, and would still be attacking the box at the end of the 2-minute period.

Because there were so many failures, we continued the test in an attempt to find out whether dogs could learn by following a human example. The animals which failed were divided into two groups, experimentals and controls. With the experimentals the handler attempted to show the puppy the solution by scratching on the box with his finger and moving it. Many puppies became very excited, and sometimes joined in the scratching. However, there was little evidence that the puppies learned to perform the task independently. Some puppies could probably be taught in this way with considerable time and patience, but the method does not work efficiently.

A MAZE TEST FOR DOGS

By the age of 13 weeks puppies have a large repertory of responses and can learn spatial relationships with considerable speed. At this age they were introduced to a 2-week testing period on a 6-unit maze. The pattern of right and left turns in this maze could be readily reversed. Half the puppies from each litter were tested with the arrangement RLLRRR and the other half on LRRLLL (Fig. 10.3), the puppies being alternated so that any scent cues from the previous animal would be confusing.

The maze was constructed of wooden frames covered with poultry mesh in an attempt to simulate a pathway through a dense thicket. Previous experiments with enclosed mazes had not worked well because of the puppies' apparent distress when confined within opaque barriers.

On the Monday closest to its 13-week birthday, the litter was brought to a holding cage outside the testing room. Each puppy in turn was placed in the final section of the maze and allowed to emerge and eat a small portion of food while the experimenter stood by. The procedure was repeated on Tuesday in order to habituate the subjects to the test situation and to build up their motivation to reach the goal by repeated reinforcement. Judging from their behavior, the puppies were motivated not only by the food but also by escape from a restricting space and by the opportunity to come in contact with the experimenter. A second person, the observer, sat on a raised platform behind a screen and recorded the subject's responses.

Maze runs commenced on Wednesday. The experimenter placed the puppy in the initial section, then walked rapidly around the outside of the maze on the side with the most blind alleys. Experimenters were directed not to attract the attention of the subjects or to provide any cues to the correct path at the successive choice points.

Records were made of the time in seconds required to run the maze and of the number of errors, defined either as reversals or complete stops. Each error was recorded on a plan of the maze as

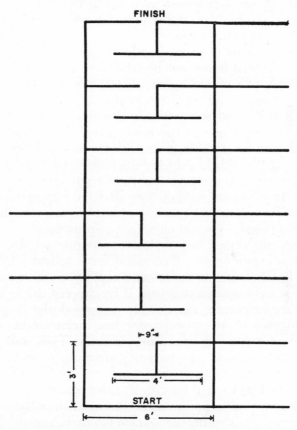

FIG. 10.3.—Floor plan of the maze test, left-hand pattern. External projections resulted from the use of 6-foot sections in erecting the maze and had no effect on the performance of the pups.

a consecutive number, so that a line drawn between them would give a complete record of the animal's movements through the maze. The average running time and error scores (excluding Day 1) were taken as the principal measures of individual performance. However, these composite scores gave an inadequate picture of performance on the maze. For example, an animal which learned the maze gradually and slowly would receive approximately the same average error score as one which made a large number of errors at first and then made a quick reduction and ran consistently thereafter. Also, on the basis of other performance tests, we expected that the principal genetic differences would show up in the ways in which the animals solved the problem rather than the end result. We therefore scored the tests in many ways.

As it turned out six of these, in addition to the average time and error scores, showed important breed differences. These were: (1) minimum time; (2) minimum errors; (3) range of time (probably reflecting minimum time); (4) range of errors (related to minimum errors); (5) the speed score on the first day; and finally (6) the habit score, which reflected the tendency of the animal to form a stereotyped pattern of running through the maze during the latter part of the test. In addition, during the first two days of preliminary training, the observer gave the subjects a confidence rating based on body posture and tail carriage just after the puppies found themselves in the maze. This gave an opportunity to measure the possible effects of emotional responses on maze performance.

When first introduced into the maze, a puppy usually attempted to follow the experimenter, then stood for a moment by the wire mesh, and finally began to run around rapidly inside the apparatus, apparently searching for a way out. If he was not immediately successful, he might become emotionally distressed and begin to yelp. As with puppies on the first barrier test, behavior became quite stereotyped at this point, the puppy running back and forth and pawing at the barriers. Fox terriers sometimes tried to bite their way out. Finally the puppy would work its way through the maze, eat the fish, and be carried back to its litter mates.

On subsequent days, performance improved rapidly. A few subjects after 4 or 5 trials were able to solve the maze visually by observing the path ahead through the transparent walls. But careful, slow, errorless progression through the apparatus never persisted, because these animals, like others which never developed a visual solution at all, soon adopted a stereotyped habit of alternating right and left turns, thus making two errors on each trial. A few individuals formed the habit of always turning to one side even though this might involve entering 4 blind alleys per trial.

Average errors and times for the five pure breeds are shown in Figures 10.4 and 10.5. The two families of curves are very similar, and differences between breeds are manifest, particularly during

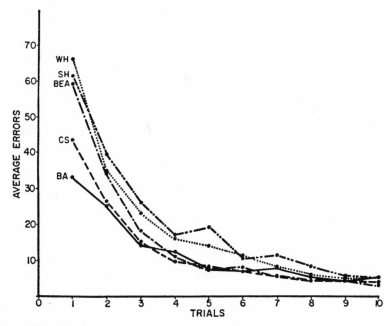

FIG. 10.4.—Learning curves: errors in maze test.

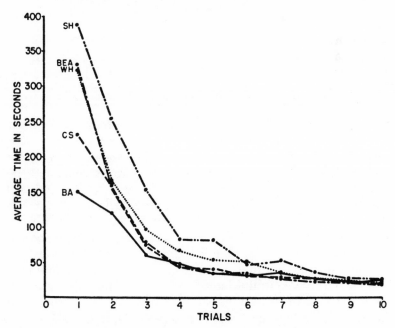

FIG. 10.5.—Learning curves: times of solution of maze test.

the first days. Elliot and Scott (1963) performed a complex analysis of covariance on this experiment, taking into account the possible effects of background conditions such as conditions of testing and rearing. In this detailed analysis of the data, the variance associated with breeds, matings, litters, and individuals was calculated for each score. In addition, the proportions of variance related to environmental, error, and background factors were calculated. These included room temperature, age of mother, order of litter, litter size, sex of litter mates, sex, body size, and emotional factors (measured by the confidence rating).

TABLE 10.2

PERCENTAGE OF TOTAL VARIANCE OF MAZE SCORES RELATED TO BREED, MATING, LITTER, AND BACKGROUND VARIABLES

	Minimum Time	Minimum Errors	Range Time	Range Errors	Habit	Speed Score
Sum of squares...........	545	325	507	498	517	458
Breeds:						
Total.................	19	21	16	20	9	38
Residual..............	9	7	3	7	9	10
Habit*...............	6	4	4	2	...	4
Background...........	4	10	9	11	0	24
Matings................	12	13	15	10	11	11
Litters.................	4	6	11	12	8	9
Individual:						
Habit*...............	12	7	4	3	...	6
Background...........	8	9	3	3	6	7
Sum of explained variance	55	56	49	48	34	71
Remaining unexplained individual variance.....	45	44	51	52	66	29

* Portions of breed and individual variance related to habit score.

When the background variance was subtracted, there were no significant breed differences in the average error and time scores, confirming our original impression that these are inadequate measures of performance. Significant breed differences remained, however, in the minimum time and minimum error scores and in the speed score on the first day—the remaining variance due to breed ranging from 11 to 15 per cent of the total. Significant differences also remained in the range of errors and range of time scores, as well as the habit score, forming 7 to 9 per cent of the total (Table 10.2). In contrast, the total variance due to breeds, including that associated with background factors, was in the neighborhood of 20 per cent for most of these 6 measures of performance, although that of the habit formation score was only 9 per cent, due to the fact that almost all the differences were caused by one breed, the beagle.

The highest proportion of explained variance, 71 per cent, was obtained with the speed score, which is based on the first day's performance. The other variables (compounded from several days' scores) stand at approximately 50 per cent. The difference is accounted for by the large effect of background factors (24 per cent of the total variance) associated with breed on the speed score. We can conclude that background factors associated with breed, such as litter size, order of litter, age of mother, waist circumference, confidence, and temperature, have their maximum effect on the first day's performance. This suggests that such factors might have an important effect on success motivation in tests in which initial failures were permitted, such as the manipulative test.

In most of the measures, there is about 50 per cent of unexplained individual variance, presumably caused by individual genetic factors and unmeasured environmental factors, such as previous experience and random distractions. To take the minimum time score as an example, about one-fifth of the total variance (19 per cent) is associated with breed differences. Of this, about one-fifth is caused by miscellaneous background variables peculiar to the breeds—sex, size of litter, and order of litter. Differences between matings within a breed, which are also presumably genetic in origin, account for 12 per cent of the total variance. Excluding the variance associated with the habit score, the variance which can be definitely assigned to genetic factors is 21 per cent, plus an unknown amount of individual genetic variance within the breeds.

From these data one can characterize the breeds in the following ways, remembering that considerable overlap exists. Fox terriers and Shetland sheep dogs make more errors and consequently show longer times than the other three breeds. On the four measures summarized in Figure 10.6, beagles appear generally superior and Shetland sheep dogs inferior, with the other three intermediate. The success of beagles is attributed to their avoidance of stereotyped habits and characteristic rapid random investigation of their surroundings. Although they often result in poor initial performance, these forms of behavior contribute to rapid solution of the maze. The Shetland sheep dogs appeared timid and hesitant in the maze and developed strong stereotyped habits. The other notable feature of the data is the good performance of basenjis on the first day. Basenjis appear more observant and less excited than the other breeds, as if making more use of visual cues.

Thus heredity seems to affect performance in this test largely by determining emotional reactions affecting confidence and the choice

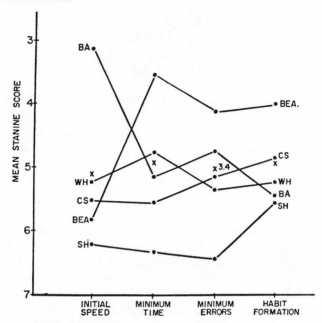

Fig. 10.6.—Relative performance of different breeds in the maze. Best performers at top of scale. X's denote means of all breeds.

of behavior patterns. The actual organizing processes in the central nervous system seem to be little affected, except in the case of the tendency in all breeds except the beagle to form stereotyped habits rapidly.

CUE RESPONSE AND DELAYED RESPONSE

Each of our subjects, beginning at 22 weeks of age, was given a series of three related tests in a T-maze in order to discover what effect genetic differences might have on this form of problem solving. An important part of solving any problem is the association of particular cues with appropriate responses, or learning to discriminate. Learning to go on cue to one arm or the other of a T-maze to receive a food reward is representative of this basic class of problems.

The apparatus is shown in Figure 10.7. Subjects were restrained in the starting box, which had a mesh door and permitted a clear view of the runway and the discriminative cues. These were 12 × 30-inch hinged panels, each painted black with a 3-inch white stripe down the middle and mounted at the entrances to the escape corridor on each side. By pulling cords, the experimenter could move one panel or the other through an arc of about 30°. Movement, rather than level of illumination or pattern, was chosen as the discriminative stimulus because dogs attend readily to motion, and it was hoped that learning would be more rapid than with conventional stationary stimuli. The moving panel also made some noise which served to attract the attention of the subjects.

Fig. 10.7.—T-maze used for cue-response, discrimination, and delayed-response tests.

A barrier extending down the middle of the runway forced subjects to choose a side before they could see which of the corridors was open. The barrier was hinged so that a portion could be swung to either side, thus blocking one of the exit corridors and forcing the animal to go to the opposite side. Equal numbers of forced trials to both sides were used as a training procedure for three days before discrimination testing was begun. Forced trials were interposed under specified rules into the discrimination-learning series when spatial preferences seemed to be interfering with learning. In ordinary test trials, retracing and correction of errors was permitted.

After emerging from the apparatus, dogs could run to the vicinity of the starting box and receive food. Some dogs, including some which learned readily, seldom ate in the test situation. Escape from the maze enclosure and handling by the experimenter were apparently sufficient in some instances to motivate acquisition of a discriminated response.

Preliminary training (motivation test).—The three days of preliminary trials served a dual purpose, habituation to the apparatus preparatory to discrimination training and a test of motivational differences between breeds and hybrids. On each of three consecutive days, the subjects were given 10 forced trials in the order of RLRRLRLLRL. The moving cue was presented on each trial; time was taken from the opening of the starting gate to emergence from the apparatus. The results of this test were used in chapter 7 to illustrate the method of computing stanine scores. For certain purposes, we have also chosen to present results as speed scores obtained by multiplying by 1,000 the reciprocal of the total time in seconds for 10 trials.

The average stanine scores for the three days are set forth for each breed in Figure 10.8. All breeds show day-by-day improvement, and the five pure breeds form two definite clusters: beagles, fox terriers, and spaniels run faster than basenjis and Shetland sheep dogs.

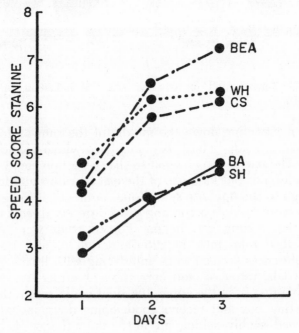

Fig. 10.8.—Speed scores in cue-response test.

Cue response (discrimination).—On the day following comple-
tion of the motivation test, testing for a discriminated response was
begun. Each subject was given 16 trials per day with the barrier
in a central position to allow running to either side. Two series of
trials were given: the first terminated either when a criterion of
learning was attained or after 128 trials; the second, given only to
subjects which met the criterion, consisted of 50 additional trials.
In order to equalize insofar as possible the levels of performance
at the end of Series I, three independently sufficient criteria were
set up for its termination. A run of consecutive correct responses,
computed according to the tables of Grant (1947), was the usual
basis for a "pass." In this system, the length of the criterion-run
increases as total trials accumulate. A second criterion was 14 correct
out of the 16 trials on one day. Finally, at the end of the 128 trials
(the upper limit), a total of 79 or more correct responses over the
series was sufficient to indicate better than chance performance at
the 1 per cent level of significance. The 50 additional trials given
to subjects which met the criterion measured consistency of response and provided some overlearning as preparation for the delayed-response test which followed.

Performance on the cue-response test can be measured in several
ways. In Table 10.3 are data on the attainment of the criterion by
the five pure breeds.

TABLE 10.3

ATTAINMENT OF CRITERION IN CUE-RESPONSE TEST

BREED	NUMBER OF DOGS	CRITERION MET BEFORE TRIAL 64		CRITERION MET AFTER TRIAL 64		CRITERION NOT MET	
		Number	Per Cent	Number	Per Cent	Number	Per Cent
Basenji.......	23	12	52	5	22	6	26
Beagle........	24	18	75	4	17	2	8
Spaniel.......	29	17	58	8	28	4	14
Sheltie........	19	4	21	6	32	9	47
Terrier.......	20	14	70	6	30	0	0
Total.....	115	65	57	29	25	21	18

The value of the chi square for comparing animals meeting the
criterion before trial 64 with all others is 22.686 with 8 degrees of
freedom and $P < .01$. Because of the small expected numbers in
some cells, comparisons between breeds are somewhat risky; but it
is safe to conclude that our strain of Shetland sheep dogs was not
as good in discrimination as either wire-haired terriers or beagles.

One difficulty with the technique of training to a criterion is that some subjects did not learn the problem during the limited number of trials which could be allowed for any one procedure. In retrospect, we believe that many of the failures would have met the criterion if training had been continued for two additional days (160 trials total), but the rules could not be changed in the middle of the experiment.

The total number of correct responses is another measure of performance which has certain advantages because every individual, including failures, receives a determinate score. Under the rules of procedure adopted, it was possible for a subject to complete the test in four days by meeting the criterion on the first day and receiving 50 confirmatory trials on the following three days. Since some subjects were tested only for the minimum of 64 trials, we have calculated the results in terms of correct responses out of 64. A few animals acquired the discriminatory response during the preliminary training with forced runs and obtained 60 or more correct responses over the four days. Others, particularly among the basenjis and Shetland sheep dogs, were still performing at a chance level after four days. Only stanine scores of 7 or better indicate significant learning, but scores in the middle range may represent animals on the way to meet the criterion.

Mean stanine scores for correct responses over the first four days of cue-response training were: wire-haired terriers 5.95; beagles, 5.75; cocker spaniels, 4.58; Shetland sheep dogs, 4.42; and basenjis, 4.39. None deviate far from the mean, and the overall differences between breeds were not significant. Among spaniels, sheep dogs, and beagles, however, individual litters deviated significantly from the breed means. Whether this was caused by common experiential factors or by characteristics of a particular mating is not known.

Delayed response.—Subjects which met the criterion for cue response were tested in the same apparatus on delayed response. For this procedure, the appropriate panel was moved, a curtain was then dropped over the door of the restraining cage, and the door raised following a specified interval of time, after all panel motion had ceased. The subject was, by this procedure, forced to respond in the absence of the discriminative stimulus. The delayed-response test, originated by Hunter (1913), has been widely used to assess the results of brain lesions, particularly in the frontal regions, and we were interested in possible detection of hereditary differences in brain function.

Obviously, there are many difficulties in a genetic experiment in applying any test which depends upon prior training, and especially when subjects are eliminated by lack of success in preliminary training. No real measure of delayed response could be obtained from subjects which did not meet the learning criteria of the cue-response test. Yet we cannot conclude that these non-learners were deficient in attributes measured by delayed response, for the group of animals receiving delayed-response training was a biased sample of our total group. Hence, it is impossible, on the basis of our experiment, to come to definite conclusions regarding the distribution of delayed-response ability among dog breeds. However, the data are still of interest because of the great amount of individual variation shown.

TABLE 10.4

DELAYED-RESPONSE PERFORMANCE OF FIVE PURE BREEDS

BREED	DISCRIMINATION		FAILED DELAY TEST		PASSED 1-SEC. DELAY TEST		PASSES 5-SEC. OR BETTER ON DELAY TEST		MAXIMUM DELAY, SEC.
	Number Trained	Met Criterion	Number	Per Cent	Number	Per Cent	Number	Per Cent	
Basenji........	23	17	5	29.4	7	41.2	5	29.4	15
Beagle.........	24	22	5	22.7	8	36.4	9	40.9	15
Cocker spaniel..	29	25	8	32.0	3	12.0	14	56.0	240
Shetland sheep dog.........	19	10	4	40.0	6	60.0	0	.0	1
Wire-haired fox terrier.......	20	20	5	25.0	6	30.0	9	45.0	60
All breeds..	115	94	27	28.7	30	31.9	37	39.4	240

The delayed-response test was designed as a "power-test" to probe the upper limit of each dog's ability. Thus, the measure adopted was the maximum length of delay in seconds permitting a correct response in the T-maze under the procedure described. Delay intervals were 1, 5, 10, 15, 30, 60, 120, and 240 seconds. A maximum of 40 trials was allowed to meet the criterion at 1-second delay; 30 trials was the limit for longer delays. Once a level was failed, testing was discontinued. A summary of the results of this test is shown in Table 10.4. It is apparent that the range in any one breed is great. Among the cocker spaniels, one individual passed the test at a delay of 240 seconds, an interval which placed a strain on the memory of the experimenters, but 32 per cent were confused by the briefest possible delay. The Shetland sheep dogs had the poorest record, but so few were tested that it is not legitimate to consider them as an adequate sample of the breed. In fact, no over-all differences between the breeds can be demonstrated by the chi-square method. Once the original hurdle of learning the discrimination is surmounted, the probability of success on delay is about equal for all breeds. It is not equal for all dogs. Over 28 per cent fail to meet criterion on the shortest delay, and about 32 per cent are able to meet only this requirement, failing on longer trials. About 40 per cent can delay 5 seconds or more, while an occasional "genius" reaches 60, or even 240 seconds. Part of the success is due to the holding of a fixed attitude during the delay. The animal aims itself when the stimulus is presented and pulls the trigger later. This is not the whole story, however. Some subjects with success at long delays actually circled in the restraining cage during the delay, and many moved their heads from side to side or pawed at the starting gate. The point of this experiment is that a power-test can uncover greater individual differences within a species than does a test of learning which is within the ability of all subjects. Other experiments in our laboratory have shown that even prolonged practice does not appreciably increase the maximum delayed-response time of a dog. Under stable testing conditions, an animal's limit fluctuates within a relatively narrow range, perhaps between 15 and 20 seconds or between 20 and 30 seconds. But this test with its demonstrated ability to detect individual variation failed to demonstrate breed differences. Thus the hereditary determination of delayed-response ability is neither proved nor disproved. A selection experiment to determine the heritability of the character might yield important results.

THE TRAILING TEST

This test was developed in an effort to study variations in the tracking behavior that is so characteristic of dogs. The difficulties of standardizing an outdoor test at different times of the year and under varying weather conditions were so great that we developed a completely artificial indoor test.

There is no doubt that breeds differ in their capacities to find prey under natural conditions. Preliminary experiments with beagles in an outdoor situation indicated that if an animal had found an object in a particular place, on the second trial it went directly to the original place and started searching from there. We set up our experimental apparatus to take advantage of this fact.

The next problem was to set up some sort of artificial trail which could be exactly duplicated in test after test. We took some 1-inch boards, 10 feet long and 2 inches wide, and attached aluminum plates to them at 1-foot intervals. We then laid a trail along the metal plates by placing a drop of fish juice in the center of each with the end of a ⅜-inch woolen dowel, beginning at the point where the dog was to be started. A spoonful of fish in a small Syracuse watch glass was placed at the far end of the board.

The next step was to train the animals to associate the artificial trail with finding a reward. On the first day an experimenter led the dog that was to be tested over the entire loft in which the test was to be given and let him eat at the future starting point of the trail. The idea here was to let the animal become reacquainted with the room, which was the same one in which it had been previously tested on the detour and maze tests.

On the second trial, a trail 20 feet long was laid out, starting from the point where the animal had been fed the day before. We found that many dogs started off in the wrong direction and, in order to make sure that they observed the trail, we surrounded the starting point with a U-shaped fence which forced them to move in the right direction. Nevertheless, many animals failed to find food within the time limit. If they did not succeed, they were placed on a leash, lead to the fish, and allowed to eat. This procedure was repeated once on the first day. On the following day, 2 trials were given without the U-shaped fence. On these 4 trials the dog had the opportunity to associate the scented board with the reward of fish at its end. Many animals would run along the board with their noses directly over the drops of fish juice.

The next day we began tests on track selection. Two branches were added on the end of the trail, a scented one which led to the fish and another, unscented, which led to an empty dish. If the dog had associated the sight of the trail with the food reward, he might as easily take one branch as the other. Most of the dogs ran to the fork of the V and began searching from there. The next part of the test was more difficult. A second V was added at the last point where the fish had been found, but this time the branch on the opposite side from the first held the reward.

On the final day a simple Y-shaped apparatus was set up in a different part of the room and the dogs given trials with the reward first on one side and then on the other, to test whether the dogs could generalize to a new location. In all, there were 3 trials with a single choice point and one with a double choice.

The whole test was set up so that the only reliable cue to the location of the reward was the scent trail. However, the dogs could and did use a variety of methods for solving the problem. They could, for example, learn that the reward was always at the end of one of the trails and go first to the ends and check these. A second type of visual solution consisted of looking around the room as they ran over it, paying no attention to the trail. The beagles usually employed still a third method, circling rapidly around the room until they apparently detected an airborne scent. Then they dropped their noses to the ground and investigated the floor inch by inch until they located the fish.

The results indicated that the problem as presented by us was learned by few dogs during the brief period of training. As Table 10.5 shows, only 7 dogs out of 143 solved the last problem perfectly

TABLE 10.5

ANIMALS SHOWING PERFECT PERFORMANCE ON THE FINAL TRIAL
OF THE TRAILING TEST

Breed	Number	Perfect Scores on Last Trial	Number of Animals Having No Failures
Basenji..........	34	1	5
Beagle...........	32	2	6
Cocker..........	32	2	11
Sheltie..........	25	2	3
Fox terrier.......	20	0	2
Total........	143	7	27

by following directly along the fish-scented trail. These animals were distributed throughout all the breeds except the fox terriers. Failure to follow the trail closely, however, must not be confused with failure to solve the problem by one of the alternative methods; 27 subjects had no failures on any trials.

One surprising result was that the cockers and shelties did as well as the beagles on the test (see Table 10.6). Terriers and basenjis

TABLE 10.6

RELATIVE RANKS OF BREEDS ON PERFORMANCE ON MOST
DIFFICULT TRIAL OF TRAILING TEST (TRIAL 7)

Average Time (sec.), Successful Animals		Per Cent of Successes	
1. Beagle............	51.4	1. Sheltie..............	56
2. Cocker...........	53.9	2. Cocker.............	53
3. Sheltie...........	54.6	3. Beagle..............	50
4. Basenji..........	74.5	4. Fox terrier.........	30
5. Fox terrier........	77.8	5. Basenji.............	23

did less well, and it soon became apparent that the poor performance of the basenjis was caused chiefly by timidity.

As the experiment went on, we observed that some animals were apparently failing the test because of fear of the artificial trail itself. They refused to cross the arms of the Y in order to get to the food; and when led to the goal, they vigorously resisted being forced to cross the board track. In the most extreme cases the dogs never crossed the arms of the trail in 4 consecutive trials. We used this behavior as a measure of timidity.

As shown in Table 10.7 there are many more cases of fear reac-

TABLE 10.7

DISTRIBUTION OF FEAR REACTIONS IN TRAILING TEST

BREED	NUMBER OF ANIMALS	NUMBER OF TESTS IN WHICH TRAIL NOT CROSSED					PER CENT OF ANIMALS SHOWING FEAR IN 2 OR MORE TESTS
		4	3	2	1	0	
Basenji............	34	2	3	2	7	20	20
Beagle............	32	0	0	0	2	30	0
Cocker............	32	0	0	0	4	28	0
Sheltie............	25	1	0	1	2	19	8
Fox terrier........	20	0	0	0	2	18	0
Total.........	143	3	3	3	17	115	6

tion in basenjis than in cocker spaniels. If the basenjis which failed because of fear reaction are excluded from the summary, very little difference remains between the two breeds. Therefore, the fear reaction to the apparatus accounts for most of the difference between the scores of the cocker spaniels and basenjis.

Fear of apparatus is not the only factor producing breed differences in performance, since the fox terriers, which make few timidity responses, were also poor in the trailing test. However, results with other tests in which strange apparatus was used indicate that fear was frequently a complication. Even after extensive habituation, some subjects avoided apparatus and were excessively distracted by sights and sounds in the test space. This matter is discussed further in chapter 14.

THE SPATIAL-ORIENTATION TEST

The series of problem-solving tests was climaxed by a procedure which we named the spatial-orientation test. Fundamentally, the apparatus consisted of an elevated goal table on which food was placed. Access to the food was possible on any one trial by only one set of three ramps leading up from the ground. The construction was such that an immediate visual solution was possible if the subjects utilized visual cues. It is often stated that some breeds are guided predominantly by sight, others by olfaction; and the apparatus design was intended to detect such differences. In order to minimize disturbance produced by unfamiliar objects and places, the equipment was set up in each home pen one day prior to training.

An objective of the test, in addition to clarification of hypothesized differences in utilization of visual cues, was the relationship between motivational factors and success in reducing errors. Measures of running speed during the test and of the persistence of approach behavior following complete blocking off of the food were included for this purpose.

Test design.—The arrangement of the equipment is shown in Figure 10.9. This apparatus, constructed of heavy wood, was set

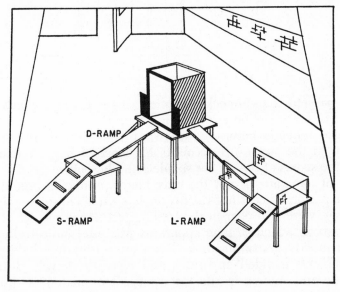

F<small>IG</small>. 10.9.—Apparatus used in the spatial-orientation test.

up for each litter in its regular living space. The goal cage was a wooden frame covered on three sides by heavy brown canvas over a wire mesh. In a later model, plywood sides were used. Wings extended laterally on both sides of the entrance and prevented dogs from edging around the table into the cage. They also served to emphasize the position of the open side. The cage was rotated on the goal table so that one pathway was open at a time. The goal table was 36 inches high and 30 inches square on top. The intermediate tables were 28 inches high. Ramps were 5 feet long and 20 inches wide. The free span of the 18-inch bridges was about 5 feet. Just previous to testing, all members of a litter were removed to small retaining cages outside the living quarters. Animals were introduced into the test area one at a time through the door shown in Figure 10.9. The apparatus, with the exception of the three-sided cage, was placed with the subjects at least 24 hours before training was begun in order that they might become accustomed to it. It remained in the pen during the entire period of testing so that all subjects had an opportunity for thorough exploration.

Experimentation was divided into three parts: preliminary training to meet a criterion (a minimum of 2 days up to a maximum of 12); testing on a set of standard problems (6 days); and persistence testing (1 day).

Preliminary training.—Preliminary training was carried out in two stages. In the first stage, the dog was released into the pen by an assistant, while the experimenter stood beside the apparatus holding a dish containing a teaspoonful of cooked herring pieces. The food was placed on the goal table with the cage in place, and guidance and assistance were given as needed to induce the dog to walk up a ramp and obtain food. Training was given in rotation on all ramps until, on 3 consecutive trials, the subject walked up and received food within 1 minute without any actual handling by the experimenter. From 3 to 20 trials were required to attain this criterion.

In the second training stage, the apparatus was set up with only one ramp in place, and the goal cage was turned so that the open side corresponded with this ramp. Food was placed in the goal cage, and the animal released into the pen. The experimenter remained outside the pen, partially concealed by a wooden barrier. The procedure was the same as that in testing except that there was no choice of ramps to ascend. Ramps and goal-cage position were rotated after each trial until 3 consecutive successful trials were finished. A successful trial was defined as reaching the goal within 2 minutes. On unsuccessful trials the experimenter entered and led the subject to the goal.

Test and scoring.—Two different arrangements of the apparatus were employed. During training and the first three days of testing, the tables were set up for Problem 1 as shown in Figure 10.9. On Day 4 through Day 6 the apparatus was rotated 180°. This was called Problem 2.

The three pathways to the goal were designated as (D) direct ramp; (S) straight ramp and bridge; and (L) L-shaped path of ramp, table, and bridge. The unit of performance taken for analysis was a set of 3 consecutive trials, each with a different correct path. Two sets (6 trials) were given each day for 6 days. Sets 1 through 6 on Problem 1 occupied days 1 through 3. Sets 7 through 12 on Problem 2 were given on Day 4 to Day 6.

Records of behavior included a count of errors and time in seconds to a correct solution. Errors were defined as (1) climbing an incorrect ramp, (2) retreating from the correct pathway, (3) circling, running back and forth, and other movements not part of a direct progression from ramp to ramp, (4) leaping directly at the goal table or the elevated bridges, and (5) failing to leap directly down from the elevated portion of an incorrect pathway. At first an attempt was made to record each type of error separately on a diagram, but it was found that the various types were positively correlated and only sums are treated in this paper. Repetitions of Type-1 errors within a single trial were, however, recorded separately, since this appeared to represent a particularly marked persistence of an incorrect response.

The time between the animal's entrance into the experimental field (defined by a line drawn 6 feet from the ramp nearest to the entrance) and the reaching of the food was the solution time. The logarithm of this time was the time score for each trial. The average of these time scores gives the logarithm of the geometrical mean of the actual times in seconds. A limit of 5 minutes was set for each trial, but the trial was discontinued at any time when a subject for a period of 1 minute either remained more than 6 feet from the apparatus or stayed in one place with no attempt to solve the problem.

Persistence test.—On Day 7 the subjects were tested for one trial on Problem 2 with the open side of the goal cage turned away from the three approaches and covered with a wire screen to prevent any dog from leaping directly onto the table. A dish containing food was placed inside the cage. A count was made of the number of attempts to get at the food. For this purpose any entry onto a ramp or standing up to any part of the apparatus was considered as an attempt. The test was continued for 5 minutes or until 1 minute passed without any attempts to reach the goal.

Results of spatial-orientation test.—In Table 10.8 are listed mean error scores, adjusted time scores, and persistence scores, all expressed in stanines. The scales have been oriented so that a high E score indicates good performance (few errors); a high T score

TABLE 10.8

COMPARISONS BETWEEN PURE BREEDS ON SPATIAL-ORIENTATION SCORES

Breed	Number Tested	Number Failed*	Number of Litters	E_1	E_2	T'_1	T'_2	P
Basenji.............	24	4	5	3.88	3.56	3.16	3.80	3.20
Beagle.............	25	0	6	5.48	5.60	6.50	6.16	6.76
Cocker spaniel.....	27	2	6	5.81	5.11	5.52	5.33	4.44
Sheltie............	11	8	4	3.27	3.45	5.09	5.50	5.25
Fox terrier........	18	2	5	3.66	3.06	3.54	3.11	4.88
All breeds.....	105	16	26	4.80	4.34	4.83	4.82	4.86

* Not included in number tested.

indicates a speedy solution of the problem and hence fast running; and a high P score many attempts to reach the goal on the insolvable problem. The adjusted time scores (T') take into account the additional running time attributable to extra distance produced by errors. The mean regression of time on errors was computed and the adjustment made according to the following equation: $T' = T - 0.0713\,N$, where T' is the adjusted time score; T is the time score in log seconds; and N is the number of errors.

Before considering the differences between breeds in detail some of the general features of the test will be considered. Among the 121 purebred dogs which were tested, 16 failed preliminary training and were eliminated. Disqualifications were especially prevalent among the Shetland sheep dogs, where only 11 out of 19 could be tested. This breed also was the poorest performer in terms of errors.

The raw error scores for the five pure breeds are shown in Figure 10.10, which also depicts results for the cocker spaniels (most successful) and Shetland sheep dogs (least successful). The connected points for Set 1 through Set 6 form a typical learning curve; errors rose when the apparatus was rotated and fell again after practice. The most striking curve is that of the Shetland sheep dogs on Set 7 through Set 12. Rotation of the apparatus produced a very great decrement in performance which remained throughout the test. Some sheep dogs, however, performed well. Figure 10.11 shows the individual error scores of the best and poorest subject among them. The extreme individuals among the spaniels are also shown. The range of performance was great. SH-2953 and CS-2724 solved the problems with a minimum of retracing and random activity; CS-1913 showed a learning curve very much like the group average; SH-1945 did not improve systematically, though it scored well on Trial Set 4.

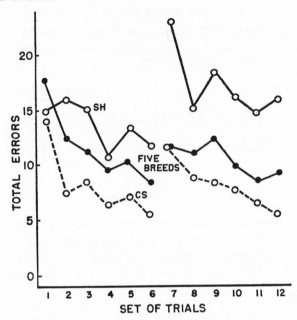

FIG. 10.10.—Raw error scores for the combined five pure breeds (*solid circles*) on the spatial-orientation test compared with scores for shelties and cocker spaniels (*open circles*).

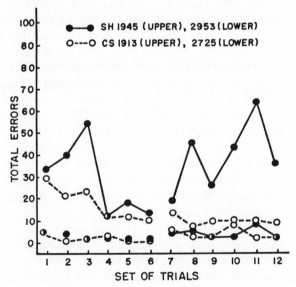

FIG. 10.11.—Individual error scores, spatial-orientation test.

Despite individual variability, differences between breeds in error scores for both series of tests were found to be highly significant (Table 10.9). Evidence was also found for significant litter effects within breeds. Since the test was carried on outdoors under varying weather conditions, it is possible that a common test environment was largely responsible.

Differences between breeds on adjusted time and on persistence were also significant, although litter effects were not found. It is evident that the breeds with the highest scores on persistence (P) are not necessarily those with the fewest errors or fastest times. Product-moment correlation coefficients between persistence and performance on Problem 2 are shown below:

$$E_2P, r = .040 \text{ (degrees of freedom, 88)},$$
$$T_2'P, r = .259 \text{ (degrees of freedom, 88)}.$$

The second of these correlations reaches the 5 per cent significance level. It indicates a slight tendency for the faster dogs to be more persistent in accord with the hypothesis that both measures were closely dependent upon motivation.

TABLE 10.9

Analysis of Variance of Spatial-Orientation Measures

Measure	Breed			Litters*			Within Litters	
	Degrees of Freedom	Mean Square	F	Degrees of Freedom	Mean Square	F	Degrees of Freedom	Mean Square
Errors$_1$......	4	63.02	12.30‡	21	5.12	2.33‡	79	2.20
Errors$_2$......	4	27.30	4.91‡	21	5.55	2.09†	80	2.65
Adj. Time$_1$....	4	42.17	15.44‡	20	2.73	1.31	81	2.07
Adj. Time$_2$....	4	34.82	8.84‡	21	3.94	1.70	81	2.31
Persistence...	4	22.97	4.55†	17	5.04	1.18	70	4.24

* Variations in the number of litters are attributable to accidental loss of data and to variations in the number of subjects given the persistence test.
† $P < .05$
‡ $P < .01$

It is, of course, impossible to estimate directly the relation of speed per se to errors, since the adjusted time scores were computed by a process which involved subtraction of that portion of the time score which could be related to error scores, and the non-adjusted time scores are correlated with errors because each error lengthens the distance run. The scores used in the statistical analysis, however, were transformations of the actual time in seconds made to bring the data into a more normal distribution. It can be shown that the expression for adjusting time scores, $T' = T - bE$, is equivalent to adjusting the actual times by the expression $t' = t/k^E$, where t is time in seconds. The correction term is an exponential function of E rather than a linear function. Slow dogs do tend to make more errors, and the correction term should be correspondingly expanded for high-error dogs in order to adjust their time scores to an error-free basis.

Even after transformation and adjustment[1] an association remained between high errors and slow times which is apparent in Table 10.8. Correlation coefficients were calculated between E_1 and T'_1, E_2 and T'_2 after normalization of the scores and separation of the within breeds and between breeds sums of squares and cross products as follows:

Sample	Degrees of Freedom	$E_1 T'_1$	$E_2 T'_2$
Total sample	102	.333	.349
Between breeds	4	.610	.798
Within breeds	98	.273	.079

There is a strong tendency for the slower running breeds to make more errors, but within breeds the correlation is low or negligible. The over-all correlation appears to be largely spurious.

Effect of practice on genetic effects.—Repeated trials on the spatial-orientation apparatus resulted in increased correlation of error scores obtained in consecutive sets of trials. Figure 10.12 shows the product-moment correlations (inter-trial) between consecutive sets for the purebred dogs as a group. The reliability of this test did not reach the level of our best tests as the maximum correlation was about .60. On the same figure are depicted the intraclass correlations computed trial by trial. Although lower in magnitude, all intraclass correlations except those for sets 1, 3, and 6 are significant at better than the .01 level. During the first part of testing, the intraclass correlations fluctuated, but they were relatively steady during the second portion of the test period. It will be remembered that an intraclass correlation measures the relative contribution of breed differences to total variance. One must conclude that training does not decrease, and possibly stabilizes, the contribution of heredity to individual variation. The point is important, for it is sometimes assumed that change in a behavior pattern following practice is evi-

---

[1] The question may be raised concerning the validity of correcting time scores by the described method. A scatter plot or errors against log time showed good fit to a linear relation as judged by eye. The calculated value of the regression coefficient was about ten times its standard error. However, no claim is made that the transformation is more than a statistical convenience, and a better mathematical model might yield better results.

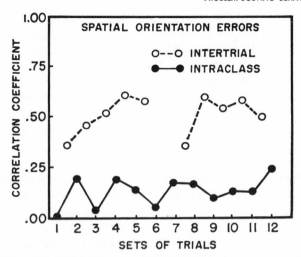

FIG. 10.12.—Reliability (inter-trial correlation) improves as the test progresses, but breed differences (intraclass correlation) remain constant in the spatial-orientation test.

dence against its heritability. The argument follows the line that demonstration of modifiability by training makes it unnecessary to account for any variance from heredity. The logic of this argument is clearly false and experimental data contradict it.

CONCLUSIONS

Interpretation of many of these problem-solving tests is rendered difficult by the fact that not all subjects met preliminary criteria for testing. These failures were not evenly distributed among the five breeds, hence the actual samples were biased. The effect was probably to decrease calculated values of genetic effects, since the failures came predominantly from breeds which had low average scores.

Since correlations of measures of learning and of motivation, particularly in the spatial-orientation test, were not high, it does not seem likely that all variations in problem solving were secondary effects of fearfulness in the test situation. Yet in conducting the tests, the experimenters often observed avoiding behavior in poor-scoring subjects. The poor performance as a group of basenjis on cue response and Shetland sheep dogs on spatial orientation appeared to be largely a function of fearfulness. This explanation does not account for poor scores of terriers on spatial orientation.

Certainly no evidence was found for a general factor of intelligence which would produce good performance on all tests. On this point we agree with Searle (1949), who compared maze-bright and maze-dull rats on a variety of tests. The genetic transmission of problem-solving ability will be discussed in chapter 12, but it can be stated now that genetic control appears to be largely a matter of non-allelic interactions with relatively small additive genetic effects. Such a finding is not unexpected when dealing with so complex a phenotype.

Each of the different problem-solving tests produced interesting ideas and results. The first barrier or detour test given at 6 weeks of age indicated that the capacity for insight or solving a problem without trial-and-error is developed by experience rather than appearing spontaneously. Since young puppies at this age have very little capacity for wide generalization, their insights are limited to problems very similar to their previous experiences.

The manipulation test emphasized the importance of success and failure in increasing or reducing motivation. Once an animal has learned to fail it is extremely difficult to get him motivated enough to make the effort to solve even a simple problem.

In the maze test we made a detailed analysis of the various effects of environmental and other background factors. The results suggest that these random, uncontrolled factors are indeed important in determining problem-solving behavior, as psychologists have long suspected. However, there still remained in this test a hard core of variance attributable to breed differences and thus to heredity.

The cue-response and delayed-response tests were unusual in that they showed no breed differences, but did indicate wide individual differences. These were tests which most clearly depended on pure intelligence. The results indicate that in this area individual differences within a breed may be more important than differences between population means.

The trailing test emphasized the importance of emotional factors in determining performance. The basenjis showed a fear of strange objects which accounted for almost all the breed differences shown in this test. The same reaction affected their behavior to a lesser extent in any of the tests involving strange apparatus.

Finally, the spatial-orientation test showed that practice need not reduce genetic variation. While the total variance might be reduced as the animals become more expert, the proportion of variance attributable to breeds remained constant.

With the exception of the cue-response and delayed-response

tests, all the tests showed clear-cut breed differences, many of them quite important. This finding raises the problem of whether there are general differences in intelligence between the breeds. Can we take the sum of these tests as a sort of canine IQ and determine the relative capacities of each breed?

In the first place, a human intelligence test includes dozens and sometimes a hundred or more problems which the student solves within an hour or so. This is possible because the students have been subjected to years of previous training in similar problems, and it is meaningful because their educational and environmental backgrounds have been reasonably similar. Comparing such a test with our problem-solving experiments with dogs, it is fair to say that each of the canine tests is equivalent to a single question on an IQ test. Furthermore, the tests are done at different times and different ages. It is impractical to give dogs the same variety and intensity of training to which human pupils are subjected. For all these reasons, the canine results are much less accurate than those obtained by a human IQ test.

Nevertheless, it is interesting to put all the results together and look at the over-all picture (see Table 10.10). In the first place, there is no breed which comes out uniformly with the highest rank. Even on individual tests, a breed may be first in one part and last in another. In the maze test, for example, beagles achieved the best minimum scores but performed badly on the first trial.

In general, the four hunting breeds (beagles, basenjis, terriers, and cockers) performed best on the tests. This is probably because most of the tests were deliberately designed to test independent capacities motivated by food rewards; and it is noteworthy that the beagle, which is normally used for hunting without direction, shows the best over-all performance in terms of number of first ranks. By contrast, the Shetland sheep dogs, whose ancestors have been selected for their ability to perform complex tasks under close direction from their human masters, performed rather badly. Indeed, in many of the tests, the shelties gave the subjective impression of waiting around for someone to tell them what to do. Furthermore, while all the hunting breeds are strongly motivated by food, sheep dogs in general have been selected away from this trait. The difference between shelties and other breeds in this respect is strongly brought out in the simple task of running to a food reward (see Fig. 9.4).

Obviously, a better answer to the question of differences in problem solving ability needs to be found. Such experiments would in-

TABLE 10.10

RANKS OF THE DIFFERENT BREEDS IN VARIOUS PROBLEM-SOLVING TESTS

TEST	BREED				
	Basenji	Beagle	Cocker	Sheltie	Fox Terrier
Detour test:					
Short barrier...................	2	4	2	5	2
Long barrier...................	1	3	4	5	2
U-barrier.....................	1	3	4	5	2
Manipulation test:					
Moving dish..................	1	4	5	3	2
String pulling (1)..............	1	3	5	2	4
String pulling (2)..............	1	3	5	4	2
Uncovering dish...............	1	3	5	4	2
Maze test:					
Speed........................	1	4	3	5	2
Minimum errors...............	2	1	3	5	4
Lack of habit formation.........	4	1	2	5	3
T-maze and delayed-response tests:					
Speed........................	4	1	3	5	2
Discrimination................	4	1	3	5	2
5-second delay................	4	3	1	5	2
Trailing test:					
Success, most difficult trial......	5	3	2	1	4
Speed, most difficult trial.......	4	1	2	3	5
Spatial-orientation test:					
Few errors....................	3	1.5	1.5	4.5	4.5
Speed........................	4.5	1	2	3	4.5
Persistence...................	5	1	4	2	3
Total first and second ranks...	9	8	7	3	10

clude more of the breeds which are thought to be unusually intelligent, such as poodles and border collies, and the testing program should include training for complex tasks similar to those for which these breeds are usually employed. On the basis of the information we now have, we can conclude that all breeds show about the same average level of performance in problem solving, provided they can be adequately motivated, provided physical differences and handicaps do not affect the tests, and provided interfering emotional reactions such as fear can be eliminated. In short, all the breeds appear quite similar in pure intelligence. On the other hand, we have evidence from the delayed-response test that there are enormous individual differences within breeds for developing certain capacities. Whether or not these are inherited can only be determined by selection and crossbreeding experiments within breeds.

PART III

INHERITANCE
OF DIFFERENTIAL CAPACITIES
AMONG HYBRIDS

THE INHERITANCE OF BEHAVIOR PATTERNS: SINGLE-FACTOR EXPLANATIONS

As we pointed out in chapter 3, the behavior of dogs has been principally modified in connection with agonistic and investigative behavior. The latter is difficult to measure under laboratory conditions, since most of the investigative behavior of dogs has to do with hunting and can only be seen adequately under field conditions where it is almost impossible to keep random environmental factors from interfering with an experiment. Furthermore, many of these patterns of hunting are developed only after extensive training.

On the other hand, the patterns of agonistic behavior are much easier to study, since they appear over and over again under laboratory and kennel conditions. The two breeds which were chosen for the cross showed wide differences in this respect. Cocker spaniels have been selected for non-aggressiveness. Occasionally a "mean" cocker will appear in the breed, but this is considered a serious fault, and such animals are not used for breeding. Shyness or timidity is also considered a fault in this breed, as it is in most others, so that cocker spaniels as a group are relatively non-aggressive and confident in their relations to people.

The basenji breed is by contrast highly aggressive, although not so much so as the wire-haired terriers. When first discovered, basenjis were sometimes called "Congo terriers." Also in contrast to the cockers, young basenjis reared under our conditions showed a great deal of timidity at 5 weeks of age: running away, yelping, snapping when cornered, and generally acting like wild wolf cubs. It is probable that in the African jungle villages such wariness has considerable survival value. In spite of this early timidity, basenji puppies tame down very rapidly with handling and human contact.

GENETIC ANALYSIS OF BEHAVIORAL TRAITS

Our original prediction was that all behavior traits would prove to be affected by many genetic factors; i.e., by polygenic systems, and we therefore designed our experiment so that the results could be analyzed according to the relative amounts of variation produced by genetic factors in the hybrids and pure breeds (Chapters 12, 14). As the results came in, however, we found that some differences in behavior patterns seemed to be inherited as if they were produced by one or two genes, so that the results of the cross looked like those from a simple Mendelian experiment and very similar to those obtained from measuring a simple physical trait in the same cross. For example, all basenjis are short haired and all cockers are long haired. When the two breeds are crossed, the puppies in the F_1 generation, whether from cocker or basenji mothers, all have short hair, indicating that short hair is dominant over long. If this result is produced by a single gene, we would expect that the backcross to the basenjis would all have short hair, that approximately one-half of the backcross to cockers would have long hair, and that one-fourth of the F_2's would also have long hair. As shown in chapter 13, this was actually the case.

In many experiments with the inheritance of physical traits, early experimenters deliberately chose characteristics in which there was no overlap between the parent strains, simply for convenience and ease of analysis. However, we were unable to find any single behavior trait in which there was no overlap between the breeds. As we have pointed out before, selection has emphasized or de-emphasized certain behavior traits but has not created anything new. In many cases selection has simply modified the amount of stimulation required to bring out a particular behavior trait. Basenjis can be stimulated to fight with relative ease and cockers only with difficulty, but it is quite difficult to find an amount of stimulation which will cause 100 per cent of basenjis to become aggressive and leave all cockers unaffected. Consequently there is always some overlap between the breeds. This means that when we analyze the inheritance of breed differences in behavior we are always forced to work with differences ranging from 50 to 90 per cent and to allow for considerable variability within each breed. We can nevertheless compare the results with models of simple Mendelian inheritance in various ways, and this chapter describes the outcome of surveying the data on the assumption that some of the data may be explained as a result of single-factor inheritance and discontinuous variation rather than multiple-factor (polygenic) inheritance and continuous variation.

METHODS OF GENETIC ANALYSIS

The chief purpose of experiments with hybridization is to discover whether or not the pattern of inheritance follows Mendelian theory. Here we have a choice of analysis by two methods: quantitative inheritance or qualitative inheritance. Most of the behavioral data looks like cases of quantitative inheritance, since behavior is usually measured on continuous scales of frequency, latency, or speed. On the other hand, there are certain cases in which behavior is either present or absent and apparently consistent with qualitative inheritance. From a mathematical viewpoint these two types of data can be described as being distributed either in multiple categories or alternate categories. The latter is often referred to as a dichotomous distribution, in contrast to a continuous distribution. Whichever distribution is present, the data must give indications of being explainable by a reasonably simple Mendelian hypothesis in order to make possible any further analysis along these lines.

Quantitative inheritance: analysis of variance.—Castle & Wright (Castle, 1921) developed a formula for estimating the number of genes involved in a particular cross, basing their calculations on two major assumptions. The first of these is that the genes have simple additive effects. Such an assumption is often a reasonable one, as it is in the case of the number of times which a dog jerks on the leash during training. If one gene causes a certain number of jerks, then a second gene might well add a few more. If the second gene should instead cause a percentage increase in the number of jerks, transforming the data into a stanine scale will cause this effect to appear additive.

The second assumption is that each gene has an effect equal to that of every other gene. This is a pure assumption, and there is no way of verifying it unless the genes can actually be identified. This means that the result of such analysis always gives an estimate, and a minimum estimate at that. The formula therefore can be used most conservatively to distinguish between single-factor and multiple-factor inheritance.

Because of these assumptions the method has certain limitations. It will not handle cases of complex interaction between genes since the analysis of such an interaction requires that the presence or absence of each gene be recognizable in some way. Also, it will not apply to certain special patterns of inheritance like that seen in playful aggressiveness, where the F_1's and F_2's show the same amount of variance and overlap both parent strains (the "Tryon effect").

TABLE 11.1

MODIFICATIONS OF THE CASTLE-WRIGHT FORMULA
FOR THE ANALYSIS OF QUANTITATIVE INHERITANCE

	Backcrosses	F_2
No dominance	$\dfrac{\Delta^2}{16(\sigma^2_{BX} - \sigma^2_{F_1})}$	$\dfrac{\Delta^2}{8(\sigma^2_{F_2} - \sigma^2_{F_1})}$
Dominance	$\dfrac{\Delta^2}{4(\sigma^2_{BX} - \sigma^2_{F_1})}$	$\dfrac{\Delta^2}{\frac{16}{3}(\sigma^2_{F_2} - \sigma^2_{F_1})}$

The formulas for analyzing backcross and F_2 generations are given in Table 11.1 In this notation Δ is the difference between the means of the parent generations, and σ^2 is the variance from the mean of each population. Mather's (1949) formula for K is a more general form of these mathematical statements, but its use is still subject to the same limitations.

As applied to our particular experiment, certain modifications need to be made. One is a correction for differences caused by maternal effects and/or the selection of different parents in the reciprocal crosses. This can be done by subtracting the difference between the means of the F_1's from that of the parents, leaving only the true breed differences. In addition, the variance caused by the same effects can be removed by subtracting the variance due to the difference between the means of reciprocal crosses in the F_1 and F_2 generations. The most accurate estimate of the number of genes involved is that based on the corrected variances of the F_1 and F_2 populations, since we have the largest numbers in these two groups. Working formulas for these corrections are given below:

$$\Delta = (\bar{x}_{P_1} - \bar{x}_{P_2}) - (\bar{x}_{BCSF_1} - \bar{x}_{CSBF_1})$$

$$\sigma^2_{F_1} = \frac{n_1\sigma^2_{BCSF_1} + n_2\sigma^2_{CSBF_1}}{n}$$

$$\sigma^2_{F_2} = \frac{n_1\sigma^2_{BCSF_2} + n_2\sigma^2_{CSBF_2}}{n}$$

Qualitative or Simple Mendelian Inheritance.—The following methods can be applied only where there is some evidence of a real discontinuity in distribution such that the data can be legitimately divided into two categories. They apply particularly well to cases in which measurements are based on the presence or absence of a behavior pattern, and animals can be classified as either showing the pattern or not showing it. Since transforming the data into the stanine distribution will inevitably obscure the points of discontinuity, the best evidence for the existence of two categories comes from the raw data. Here we are faced with a new difficulty; nowhere in our data was there a case in which there was no overlap between the two parent strains. We are therefore faced with the problem of modifying the methods usually applied to the analysis of qualitative inheritance.

In a dichotomous distribution, the mean is always the same as the proportion of individuals in one class or the other; i.e., in a population composed of 0.5 animals showing a trait and 0.5 not showing it, the mean is 0.5. It is always the mean, or proportion, which is used in the conventional analysis of data falling into two categories with no overlap between the parent strains.

We can extend analysis by comparison of means to populations in which overlap exists, using the following formula based on a comparison between the means of backcross (BX) and parent (P) strains,

$$\frac{\Delta_{BX}}{\Delta_P}, \quad \text{or} \quad \frac{\bar{x}_{BX_1} - \bar{x}_{BX_2}}{\bar{x}_{P_1} - \bar{x}_{P_2}}.$$

Table 11.2 gives the expected ratios for various numbers of genes. The calculated figures are the same whether or not dominance is actually involved, since dividing the data into two categories in effect assumes dominance. If only one factor is involved, the figure is the same whether or not heterozygosis is present in the parent strains (Scott, 1954). Therefore this method provides a test of a single-factor theory which can be rapidly applied to any data. One can also use the formula to calculate theoretical ratios in the segregating generations, basing estimates on the empirical difference between the parents. Naturally, the validity of the results will depend on the size and representativeness of the samples measured. A similar formula can be applied to the differences between the F_1 and F_2 means, with the limitation that the expected differences are not as great. However, in our experiment, these are the largest populations and deserve attention.

Like the formulas for quantitative inheritance, these must also be corrected for differences due to maternal effects and accidental selection of different strains in reciprocal crosses. This can be done

TABLE 11.2

THEORETICAL RATIOS FOR THE DIFFERENCE BETWEEN BACKCROSSES,
AND THE DIFFERENCE BETWEEN F_1'S AND F_2'S, COMPARED TO THE
DIFFERENCE BETWEEN THE MEANS OF PARENTAL STRAINS

Number of Genes	$\frac{\Delta_{BX}}{\Delta_P}$	$\frac{\Delta F_1 - F_2}{\Delta_P}$
Populations with a Threshold Cutting Off a Terminal Class		
1........	.5	.25
2........	.25	.125
3........	.125	.0625
4........	.0625	.03125
Populations with a Central Threshold (Tryon Distribution)		
1........	.5	
2........	.75	
3........	.875	
4........	.9375	

by subtracting the difference between the F_1 populations from both that of the parental and backcross populations. In the comparison of F_1 and F_2 populations, this correction can be eliminated on the assumption that both are equally affected. The calculation is made by simply subtracting the mean of the combined F_2's from that of the combined F_1 populations.

This general method will also apply to the Tryon distribution, in which the F_1 is as variable as the F_2 and there appears to be a central threshold. There is no difference between the means of the F_1 and F_2, but that between the two backcrosses gives an estimate of the number of genes involved (see Table 11.2).

The limitations of the method are quite similar to those of analysis of variance applied to quantitative inheritance. The basic assumptions are those of equal and additive effects of genes, at a threshold in this case instead of over an extended scale. Like the quantitative method, this one will not apply if gene effects are non-additive or extremely unequal. The formula has the advantage, particularly useful with our data, of not being affected by heterozygosity in the one-factor case (Scott, 1954). In short, like the Castle and Wright method, it chiefly provides a test of single-factor versus multiple-factor inheritance, and the number of genes beyond one is only a minimum estimate.

THE INHERITANCE OF WILDNESS AND TAMENESS

The relative wildness of the basenjis is expressed in two obvious traits. One is in avoidance and vocalization in reaction to handling as young puppies, and the other is the tendency to struggle against restraint, which appears in many different situations, but is particularly marked in leash training.

Both cocker spaniels and basenjis, however, are readily tameable as young puppies. The general trait of tameability is a basic one, because it must have been involved in the original domestication of the dog. If domestication occurred but once, it is likely that there was an early selection for whatever genes for tameness were present in wild wolves at that time, and that these genes would thenceforward be present in all dog breeds. If domestication occurred more than once, or if domestication occurred from more than one species, we would expect that different genes might be involved. If this were true, there is a possibility that an F_1 cross might be much wilder than either parent, or, if both factors for tameness were dominant, that excessively wild individuals would appear in the F_2 or other hybrid generations.

Inheritance of avoidance and vocalization.—We may think of the handling test as a series of mild to strong stimulations which may cause a young puppy to be fearful. The strongest stimulus is that of walking rapidly toward the puppy, and the weakest is walking away or kneeling with an outstretched hand. The test score reflects the number of times which such stimulation pushes the puppy over a threshold into the fearful behavior of running away or yelping with fright. Some puppies become fearful only under the strongest stimulation, while others react with fear to anything which the handler does and thereby receive much higher scores.

TABLE 11.3

Scores for Avoidance and Fearful Vocalization at 5 Weeks of Age in Response to a Human Handler

Score	Basenji Matings 2 and 4	Backcross to Basenji	F_1	F_2	Backcross to Cocker	Cocker Matings 4 and 7
0	0	1	1	3	4	16
1–6	7	6	12	25	10	6
7–12	6	14	20	20	8	3
13–18	3	6	10	15	8	1
19–24	8	10	4	4	...	...
25–30	2	4	4	4	...	...
31–36	1	1	1	1	...	...
37–42	...	...	...	1	...	...
43–48	...	...	...	...	...	...
49–51	...	...	...	1	...	...
Total	27	42	52	74	30	26
Percentage with low scores (0–6)	26	17	25	39	47	85

As can be seen in Table 11.3, about 62 per cent of the cockers show no fearful reactions at all, whereas all basenjis show at least some fearful behavior, with a great deal of variability in the amount produced. The F_1 generation is very similar to the basenjis, with only one animal showing no fearful reactions, and we can therefore conclude that the differences are produced by one or more dominant genes. On this basis we can construct the hypothetical results of a single dominant gene. The backcross to the dominant basenjis should be like the basenjis, whereas the backcross to the recessive cockers should be intermediate. Finally, the F_2's should be halfway between the basenji and cocker backcross.

The actual observed percentages fit this hypothesis fairly well, and better than they fit that of two dominant genes. On the basis of this evidence, we can say that a single dominant gene causing wildness in the basenji accounts for the observed results reasonably well.

Consequently, the contrasting gene for tameness in the cockers must be a recessive one. This fits our general ideas about mutations, since, in all animal and plant species, most of the wild-type genes for structure and color are dominant rather than recessive. Also, the F_1 hybrids do not show excessive wildness, indicating that both cockers and basenjis have the same general genes for tameness. This is evidence in favor of a remote common ancestry for both breeds, although it cannot, of course, exclude the possibility of separate domestications which acquired the same genes for tameness from the wild wolf population.

Inheritance of the tendency to fight and jerk the leash.—In the routine of our "School for Dogs," the puppies were carried everywhere until they were 16 weeks of age and were ready to be moved outdoors. At this time both distance and their increasing size made it necessary to teach them to follow on a leash, as described in chapter 9. Beagles and cocker spaniels usually went along readily and peacefully, but basenjis often pulled back strongly on the leash, attempted to bite it in two, or jumped into the air, shaking their heads like a trout on the end of a fishing line. One of the scores obtained from this behavior was the number of times that a particular puppy fought the leash. The final results showed big differences between cocker spaniels and basenjis (Table 11.4). Furthermore, there were large differences between the two F_1 populations, the hybrids tending to act like their mothers. This difference was carried on into the second generation to a lesser degree. It looks as if there is an important effect of either the maternal environment or differences between the parents of the reciprocal crosses. The former seems at

TABLE 11.4

Tendency to Bite and Jerk Leash (in Stanines)

Breed or Hybrid	Number	Mean	Variance
Basenji................	34	7.38	1.86
CSB F₁................	28	5.50	0.68
Backcross to basenji.......	43	5.74	1.02
CSB F₂................	41	5.10	1.82
Cocker spaniel...........	31	3.87	1.60
BCS F₁................	24	4.96	0.78
Backcross to cocker.......	28	3.96	1.93
BCS F₂................	32	4.63	1.69

first glance most likely because of the strong resemblances to their mother's behavior in both generations, but this is not correct, as will be shown later.

Nevertheless, there is still evidence that biological heredity has an effect, because the hybrids are not exactly like their mothers, and there is increasing variability in the backcross and F_2 generations, as we would expect from Mendelian segregation.

Inspecting the data we can form the hypothesis that the genetic factor or factors involved do not show dominance and that we are thus dealing with the simplest form of Mendelian inheritance. When the data are analyzed by the quantitative method, estimates of the number of genes involved can be obtained from the backcross and F_2 generations. These agree fairly well with each other, for we derive the figure 1.07 from the F_2 data and two estimates of 1.60 and 0.48 from the separate backcross populations. Averaging the backcross estimates gives 1.04, and we can conclude that the inheritance of the difference in behavior between cockers and basenjis can be explained by one gene with no dominance, complicated by large differences between reciprocal crosses (see also chap. 12).

INHERITANCE OF OTHER PATTERNS OF AGONISTIC BEHAVIOR

Playful aggressiveness.—In the handling tests given at 13 and 15 weeks of age, a characteristic reaction of the puppies was to rush toward the handler, leaping against him or up at his hands, and nipping playfully. When patted, the puppy usually turned around and pawed or wrestled with the hand, chewing on it gently. It is obvious that a puppy of this age reacts to an outstretched human hand in much the same way as it reacts to the playful approach of another puppy. When two dogs interact in this way they rear up, throw their forepaws around each other's necks, and attempt to wrestle each other to the ground, biting and chewing on each other's necks and ears.

As we have seen in an earlier chapter, cocker spaniels have been selected against violent aggressive behavior in two ways. One is by selection for crouching in response to an upraised hand, and the other is for the highly restrained biting or "soft mouth" useful in retrieving. As we might expect, cocker spaniels respond to handling in a much more submissive and less violent way and consequently receive lower scores for playful fighting than do basenjis.

When we graph the scores of the cockers, basenjis, and hybrids, we find a pattern of inheritance which is distinctly different from any of those we have so far examined (Fig. 11.1). Most of the basenjis re-

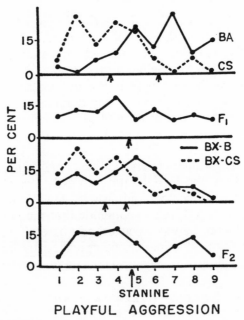

PLAYFUL AGGRESSION

Fig. 11.1.—Distribution of stanine scores for playful fighting at 13–15 weeks of age. Note the highly variable F_1, overlapping the complete range of both parent breeds. Arrows indicate means.

ceive high scores for aggressiveness, and most of the cockers receive low scores. Unexpectedly, the F_1 hybrids include both high and low scores and are highly variable, and the F_2's are very similar to the F_1's. The backcrosses to the basenjis have a large percentage of aggressive animals, and the backcrosses to the cockers have a low percentage.

From the viewpoint of genetic theory, we would expect that both pure strains would show little genetic variability. Likewise, the first generation hybrids should all have the same heredity, each animal receiving the same genes from either parent. The F_2's, on the other hand, should be genetically variable, because the genes have been segregated or sorted out into the parental and hybrid types. In this actual case the parent breeds are variable, the F_1's are highly variable, and the F_2's no more variable than the F_1's. How can we reconcile this with the theory of Mendelian genetics?

The simplest explanation is that we are dealing with a threshold of response to stimulation. The puppies are either stimulated to playful fighting or not. If they have a low threshold, as in the basenjis, they react with playful biting and chewing to the slightest human stimulus, but if they have a high threshold like the cockers, they rarely reach this threshold and do little biting or chewing. The F_1's have an intermediate threshold; so that this may or may not be reached during the test. Consequently the F_1's may show either a large amount of playful fighting or almost none, just as in the two parental types. The second generation hybrids should contain all genetic types; but the intermediate ones, or heterozygotes, should respond in the same way as the F_1's, so that the F_2 generation will show variability exactly like the F_1. On the other hand, either of the backcross generations should have fewer of the intermediate or heterozygous types than the F_1's and consequently be more like the parent generations.

The only way to estimate the number of genes involved is to calculate the expected and observed differences between the two backcross generations. For example, if only one gene is involved, 50 per cent of the backcross should be heterozygous, and according to our theory half of these heterozygous animals would fall into one type and half in the other. Where the F_1's are half like one parent population and half like the other, and only one gene is involved, the backcross should be three-fourths like one parent and one-fourth like the other. If two genes were involved, only one-fourth of the backcross would be heterozygous, which would mean that the backcross would be seven-eights like one parent and one-eighth like the other (Table 11.2).

The theoretical results must be adjusted to allow for the fact that there is some overlap between the two pure breeds. The observed result is closer to that expected for two factors, in which the back-crosses are quite far apart, than it is for one factor. We can conclude that the inheritance of this particular trait can be explained by the hypothesis of two genes, each of which lowers the threshold of stimulation by approximately the same amount (Fig. 11.2). This, of

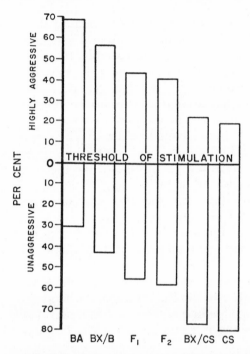

INHERITANCE OF PLAYFUL AGGRESSIVENESS
IN PUPPIES

Fig. 11.2.—The occurrence of playful fighting transformed to a scale of threshold of stimulation. This graph is based on raw scores, with a threshold arbitrarily located at the point of maximum separation. Scores for the pure breeds are based only on those animals closely related to the hybrids. Note that F_1 and F_2 are almost identical and intermediate between the two parent strains.

course, does not eliminate the possibility that there is a more complex type of inheritance involved. Our results indicate only that the results can be explained on the basis of two genes acting in the manner described above.

This unusual but basically simple sort of inheritance turns up fairly frequently in behavioral data. It was probably overlooked in the early classical studies of Mendelian genetics for several reasons.

One was the fact that the early geneticists were interested primarily in the mode of chromosomal transmission rather than the nature of any particular trait, and so they usually disregarded any inherited trait which did not give clear-cut differences between strains. Another is that threshold phenomena are much more frequent and more obvious in traits of behavior than in traits of form and color, although the genetics of color can also be analyzed in terms of thresholds. A mouse may be either completely colorless (as in an albino) or all black, and we can say that there is a threshold for the biochemical process of pigment formation which is reached only when a particular gene is present. However, it is difficult or impossible for an environmental factor to affect this threshold in such a way that a mouse could be shifted from one color to another by an environmental accident. A better example is the inheritance of the number of sacral vertebrae in the mouse, where certain pure genetic strains have variable numbers of such vertebrae, presumably produced by random environmental factors acting early in embryonic development. The concept of a threshold raised or lowered by gene action and environmental factors is therefore basic in genetics.

In order to have the type of inheritance shown in this example, with a high degree of variability in both F_1's and F_2's, there must not only be a threshold but one which is easily affected by environmental factors, conditions which are always present in behavior. If the F_1 hybrids fall close to the threshold, the above pattern of inheritance will automatically appear.

The frequency of this kind of inheritance is therefore dependent not only on thresholds but on the fact that these thresholds are easily affected by environmental factors.

Barklessness.—One of the striking characteristics of the basenji breed is the fact that these dogs rarely bark. We can only speculate as to why this trait was developed, since it was already present in the breed when it was brought out of Africa. It is possible that barking, which is an alarm signal given by dogs whenever a strange animal or person approaches their home territories, is not conducive to survival in the African forests. Leopards are reputedly fond of dog meat, and it may be that the dog which barks simply attracts attention to himself and comes to an untimely end. Although travelers have described basenjis as being very noisy in their native African habitat, especially at night, none of the sounds produced are like barks, being variously described as "crowing," "yowling," and "howling." This suggests that the basenjis may have developed sounds with unusual acoustic qualities. The barking of most dogs, as analyzed on the sonograph, consists of a succession of short, sharp, monotonous sounds, which are very easy to localize. That is, barking conveys accurate information as to the location of the barking animal. On the other hand, sounds which vary in pitch, loudness, and duration are much more difficult to localize with respect to direction and distance, as anyone who has had experience with the vocalizations of coyotes will recognize. While distinctly different from coyotes, the basenji sounds have similar qualities of variability and may serve the same adaptive and protective function. In any case, basenjis bark very little compared to other dog breeds and, whenever different breeds of dogs live together, the basenjis' relative silence is extremely noticeable. As a stranger walks by the dog runs at our laboratory, a chorus of barks arises from a group of cocker spaniels and from a nearby group of Shetland sheep dogs. In a pen between, a litter of basenjis look up without opening their mouths.

Darwin thought that wolves do not ordinarily bark and that when they do it is because they have learned the habit from dogs. All modern observers of wolves under any conditions, whether in zoos or in the remote wilderness, agree that they bark, although not as much as many dog breeds. The barklessness of basenjis is therefore not a primitive ancestral trait but rather a new and unusual characteristic, produced by some sort of selection.

Our best data on the relative barking capacities of the different breeds comes from the dominance test in which we took each pair of litter mates and let them compete for a bone during a period of 10 minutes. While they were doing this we recorded the vocalizations of each animal, including barks, whines, and growls. We have such data from 5, 11, and 15 weeks of age. As can be seen from Figure 11.3, the maximum amount of barking occurs at 11 weeks in all breeds, with the possible exception of the Shetland sheep dogs, in which the amount of barking seems to be still rising at 15 weeks. The graph also shows that the cocker spaniels at 11 weeks do the most barking of any breed and the basenjis do the least.

Obviously, basenjis (or at least the strain which we have) are not completely barkless. When sufficiently excited, they will bark. Table 11.5 shows that basenjis barked during 20 per cent of the opportunities given them during the dominance test, whereas the cocker spaniels barked during 68 per cent. The basenjis usually gave only one or two low "woofs" when they did bark, the average number being about two. At 11 weeks of age, the largest number of barks given by any basenji during the dominance test was 20, and the next highest number was 12. More than this, the sound which the basenjis

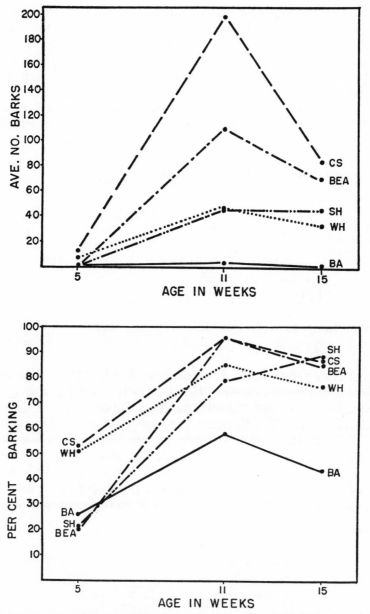

FIG. 11.3.—Occurrence of barking during the dominance tests at different ages. *Upper:* average number of barks. *Lower:* per centage of animals barking. (Note that these two characteristics are not directly related.)

make has a different quality from that in other breeds. Thus there are three different aspects of what looks offhand to be a simple behavior trait. One is the threshold of stimulation—very high in the basenji and very low in the cocker spaniel. A second trait is the tendency to bark only a small number of times rather than to become excited and bark continuously as do many cockers. The maximum number of barks recorded for a cocker in a 10-minute period was 907, or more than 90 a minute. Still a third difference is the tone quality, which we

TABLE 11.5

PER CENT OF ANIMALS BARKING PER OPPORTUNITY
COMPARED WITH EXPECTED PERCENTAGE

Breed or Hybrid	Observed	Expected 1 Factor Dominant	Expected 2 Factor Dominant
Basenji..................	19.6	...	...
Cocker..................	68.2	...	...
F_1.....................	60.1	68.2	68.2
F_2.....................	55.5	56.0	65.2
Backcross to cocker.......	65.1	68.2	68.2
Backcross to basenji.......	50.0	43.9	56.0

did not attempt to analyze, as the equipment necessary for producing a sound spectrogram was not available when we started the experiment.

In analyzing the effects of heredity, we first find that there are no important differences between the sexes with regard to this type of vocalization. Analyzing the inheritance of the two traits is therefore reasonably simple. The F_1 generation shows almost as many animals barking as do the cocker spaniels (Table 11.5) but not as many animals barking to excess (Table 11.6; Fig. 11.4). There is no great difference between puppies born of basenji mothers and those born of cockers, so that we can rule out the possibility that either sex-linked inheritance or learning by example from the mother has any great importance.

TABLE 11.6

PER CENT OF ANIMALS BARKING 19 OR FEWER TIMES AT 11 WEEKS

Breed or Hybrid	Number Tested	Per Cent
Basenji....................	42	95.2
Cocker....................	49	18.4
F_1........................	39	51.3
F_2........................	66	53.0
Backcross to cocker........	16	18.8
Backcross to basenji........	35	62.9

This looks as if the trait of being easily stimulated to bark is inherited as a dominant one. If we now compare the percentage of animals which barked during each opportunity with the percentage calculated by assuming inheritance through a single dominant gene, we find a reasonably close agreement, and in fact a better agreement than with a two-factor theory (Table 11.5). This does not mean that

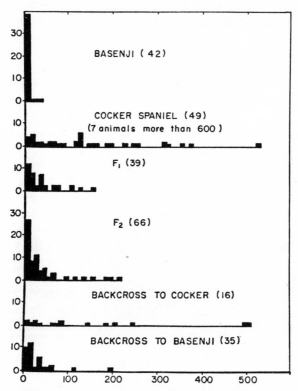

FIG. 11.4.—Distribution of barking to excess at 11 weeks of age in the pure breeds and hybrids. The figure shows the number of animals producing a given number of barks. Note that almost all basenjis barked less than 10 times.

the possibility of a larger number of genes is completely ruled out. A test to check this would require a much more complicated experiment and larger numbers of animals. The most that we can say is that the results could be explained in this fashion.

If we now look at the inheritance of the tendency to bark to excess (Fig. 11.4), we find a different situation. The F_1's are now intermediate between the two parent strains and the F_2's very much like the F_1's. This looks like a case where the hybrids between two strains have a genetic constitution close to a threshold. Consequently, each animal may be pushed over the threshold or held below it by a few chance environmental factors. The results from the backcross agree with this hypothesis, since the backcross to the basenjis is more like the basenjis and the backcross to the cockers is more like the cockers. Again, the results could be explained as the result of a single gene.

In summary, the apparently simple trait of barklessness turns out to involve at least three characteristics, of which two were analyzed. These two involve different genetic mechanisms; and, while they affect each other, they do not combine in any simple fashion. Obviously, a dog will not bark to excess if it does not bark at all, and the appearance of the former trait is thus conditional on the presence of the latter. Consequently, the effects of the two traits together cannot be represented by simple addition. Furthermore, these traits can best be described in terms of probabilities rather than absolutes. Given a particular sort of opportunity, such as the competition over a bone, there is a 20 per cent chance that a basenji will bark at 11 weeks of age and a 68 per cent chance that a cocker will bark. The inheritance of this trait, like so many others, illustrates the extreme complexity of behavior which can result from the action of a small number of genetic factors acting in different ways.

The results also raise an interesting question regarding the physiological mechanism behind the trait of excessive barking. Some animals, given the stimulus of another puppy in a competitive situation, will bark a few times and then stop. Others continue to bark, becoming more and more excited and taking on an almost hysterical tone. It looks as if they may be stimulating themselves by their own barking. If so, this would account for the "either-or" behavior in barking—either a few barks or a great many. Excessive barking is not only interesting as a genetic trait but also as it gives hints of the nature of emotional stimulation.

INHERITANCE OF DIFFERENCES IN SEXUAL BEHAVIOR

All the members of the genus *Canis* have similar patterns of sexual behavior, as we have shown in a previous chapter. One of the peculiarities of their breeding cycle is the fact that the period of estrus or receptivity in the female is preceded by bleeding. In wolves and coyotes the bleeding may begin quite early in the year, in January or February, and there are some reports of wolves in which this began in December. Actual mating of wolves usually takes place in late January or February and is not repeated until the following year.

In contrast, the females of most breeds of domestic dogs have estrus cycles at any season of the year and approximately 6 months apart. The cycle is regulated by hormones. After pregnancy and also after a cycle in which no eggs are fertilized, the corpus luteum of the ovary continues to secrete progesterone for several months, suppressing any further cycles as long as it functions. The length of the cycle is therefore regulated by the length of time during which the corpus luteum persists.

The basenji is different from other dog breeds in that females come into estrus at a particular time of year, close to the autumnal equinox, and show only one cycle per year. The inheritance of this trait is particularly interesting because we can be positive that our parent stocks are genetically pure in this respect. We have no records of cockers showing seasonal cycles and no records of basenjis showing 6-month cycles.

However, when we look at individual records in detail, we see that there is some variability in both stocks. The modal month for the basenji cycle is September, and the vast majority of basenji heat cycles begin during the last two weeks of September and the first week in October. Moreover, a certain number of cycles begin in late August and a few as late as November. This means, of course, that the interval between cycles is not exactly 12 months. The same dog may come into estrus early one year and late the next, so that we have records of intervals as short as 10 months and as long as 14, with the great majority falling between 11 and 13.

When we look at cockers, we see that the 6-month cycle is only an approximation. Some individuals have quite regular cycles, coming into heat at approximately the same times each year, as shown in Figure 11.5. Others are more variable. The interval may run anywhere between 3 and 11 months, with the mode at approximately 6 months. This means that the shortest cycles of the basenjis can overlap some of the exceptionally long cycles of the cockers. There is no indication, however, that the cockers show any remnant of the annual spring cycle of their wolf ancestors, or that basenjis ever run two cycles 6 months apart.

In order to study the inheritance of this trait, we retained all the female hybrids until they had gone through at least two cycles. A few of the F_1 females which were parents of the F_2's were saved until they were several years old, and we were thus able to get records on many repeated cycles for these individuals.

The actual records were results of weekly examinations, at which times we looked for the onset of bleeding and noted it down, if present. The dates are therefore only accurate within a week, since bleeding might have begun as much as 6 days earlier. There are certain possibilities of error in that a slight amount of bleeding could be overlooked. In a few cases this happened and the animals became pregnant. In such cases we estimated the onset of the estrus cycle by counting back 63 days and adding 1 week.

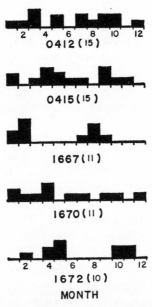

Fig. 11.5.—Month in which bleeding began in five related cocker spaniels: mother (0415), double grandmother (0412), and 3 daughters. Numbers in parentheses indicate numbers of cycles observed in each animal. Note that some animals were highly variable, but some (1667 and 1672) ran fairly regular cycles at 6-month intervals.

Our results showed that the F_1 females were different from both parent stocks. Many of them had two cycles in the latter half of the year in the period of declining light. A female might come into heat in July and again in September if she was not mated and made pregnant. Females might also come into heat in the spring, and in those animals which were kept several years, our records show two peaks, in February and August. Since they might skip the spring period, however, the F_1's were capable of having cycles quite far apart as well as quite close together. In short, the F_1 hybrids are more variable than either of the parent strains, although from the standpoint of genetic theory they should all be alike. The explanation is obvious. The F_1's have inherited the basenji tendency to respond to declining light, but they have also inherited the cocker tendency to run cycles every 6 months. The interaction of both these traits produces a high degree of variability.

When we look at the other hybrids, we see that the F_2 is quite

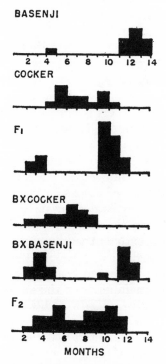

FIG. 11.6.—Intervals between first two estrus periods (in months). Note indications of segregation in the backcross to the basenji and wide variation in the F_1. F_1's also showed cycles of intermediate length in later periods.

similar in variability to the F_1, so that a comparison of these two generations gives us no clues to the inheritance of the trait. The backcrosses, however, appear to show Mendelian segregation. About one-half the basenji backcrosses show long cycles like a basenji and about one-half show shorter cycles which are typical of the F_1 hybrids. The backcrosses to the cockers, on the other hand, are quite similar to cockers. We can then formulate a hypothesis that the basenji type of cycle is inherited as a single-factor recessive, with the heterozygotes showing the peculiar type of interaction described above. Since very few of the F_1's have cycles as long as 11 months, we can tentatively classify all individuals having cycles of 11 months or longer as basenji types and then test our hypothesis in the backcross and F_2 generations. In the backcross to the basenjis the theoretical ratio matches the observed ratio almost perfectly. In the F_2 generation there are not as many animals with long cycles as we might expect, which brings up the possibility that there could be two factors instead of one. However, the two-factor hypothesis fits the backcross quite poorly (the probability of greater deviation from the expected ratio is only two in one hundred). By contrast, the one-factor theory fits the F_2 considerably better, with a probability of .08 (Scott, Fuller, and King, 1959).

We may therefore conclude that the most probable genetic explanation of the observed facts is one-factor inheritance. The possibility of two-factor inheritance is not definitely excluded however, since we did not do the crucial experiment for a one-factor hypothesis, i.e., repeated backcrosses to the recessive.

There are several important conclusions that can be drawn from these data. The first is that we have a clear-cut case of a highly variable F_1 which cannot be explained by possible heterozygosis of the parents. The explanation therefore lies in the physiology of gene action rather than in Mendelian segregation. Second, we have some clear indications as to what this physiology could be, i.e., one member of a gene pair from the basenji causing a responsiveness to light changes and the other member from the cocker spaniel producing a tendency to run 6-month cycles in addition to the first effect, with both members of the pair interacting with each other to produce the final effect. It should also be noted that this is quite a different kind of explanation from the threshold hypothesis, which explains a similar variability of the F_1 scores on playful aggressiveness. Finally, this case shows how the simplest sort of Mendelian inheritance, that involved in one pair of genes, can produce a highly complex and variable result on behavior by acting upon physiological processes which affect behavior more directly. As we have seen in all the above cases, such complexity seems to be characteristic of genetic effects upon behavior. Superimposed upon the almost infinite variability possible from Mendelian mechanisms, we have a still greater variability made possible through the physiological effects of genes. Of course, this makes it highly unlikely that any two individuals in a genetically variable population will ever behave exactly alike, even if they are placed in identical environments. It also makes the exact genetic analysis of any sort of highly complex behavior, such as success in problem-solving situations, almost impossible by present methods.

EFFECTS OF THE MATERNAL ENVIRONMENT

In almost every test that we did there was some difference between the reciprocal crosses in the F_1's, and these differences were perpetuated in the F_2 in many cases. There are three different ways in which such differences could be brought about.

Sex-linked inheritance.—If a trait is produced by a sex-linked recessive gene, this trait should appear in all the males and none of the females born from recessive mothers mated to dominant fathers. In the reciprocal cross, the F_1's from recessive fathers should all be like their dominant mothers. If sex-linked inheritance is present, it can be easily recognized by the fact that there is an important difference between males and females in one of the F_1 populations but not in that from the reciprocal cross.

No indication of this sort was found in any of the 40 variables selected for detailed study. The lack of any case of sex-linked inheritance in behavior characteristics in a sample of this size should not surprise us, since there are 39 pairs of chromosomes in the dog and only one of these includes the sex chromosomes. Assuming that there are equal numbers of genes on all chromosomes, a very large number of genes would have to be studied to find even one case of sex-linked inheritance. The large number of sex-linked genes which have been found in man is probably related to the fact that this type of inheritance is very easy to recognize in family pedigrees compared with the pattern of inheritance produced by a recessive gene from another chromosome.

Accidental selection of different strains.—There is every indication that a great deal of individual and strain variability exists within the pure breeds of dogs, and in selecting the 4 pairs of parents for each cross we may have accidentally included two different sorts of animals. With such accidents, we would theoretically expect that the resulting difference between the two F_1 crosses might either be in the same direction as the difference between the parent strains or in the opposite direction, depending purely on chance. The latter seems to have been the case with regard to physical size. By chance, we included an unusually large basenji male and one or two large cocker spaniel females among the parents in the BCS cross. The result was that the offspring born to these animals were, on the average, larger than those born to the basenji mothers. Thus, the difference between the F_1's is in the opposite direction from that of the parents. We would theoretically expect that such a difference would be perpetuated into the F_2 generation, but this is not true (Table 11.8). That the F_2's are alike is accounted for by the fact that the excessively large F_1 animals were not chosen as parents.

If chance had worked in the opposite direction, we might have selected parents for the F_1 generation whose genes would produce a difference in the same direction as that between the original pure strains. In this case, accidental selection as a cause of the difference could not easily be distinguished from the maternal environment.

Effects of maternal environment.—If a mother influences the behavior of her offspring, either by some prenatal environmental effect or by a postnatal effect involving learning from the mother, we would expect that the effect would be usually to make the offspring more like the mother. As pointed out above, there is no way of distinguishing this effect from that of strain differences in any individual case, but we can make a test of the over-all experiment. If selection of strain differences were the important factor, we would expect that in half of the tests the differences between the F_1's would be in the same direction as that between their parents and in the other half differences would be in the opposite direction. As it turned out, 12 out of 40 tests showed unlike differences, while 28 out of 40 showed like differences. Thus, we have evidence that some sort of maternal effects have played an important part in behavior.

If such effects do exist, they should be continued into the F_2 generation in an undiluted form. Examining the actual data (Table 11.7), we find that there is one case in which the differences in the F_1 and F_2 generations are very nearly as large as those between the parents. This is the attraction score on the handling test at 13 and 15 weeks of age, which is chiefly based on the number of times which a dog follows or comes toward an individual. There would be ample opportunity for puppies to learn such behavior from their mothers, as experimenters and caretakers entered their pens week after week.

Another case with a less obvious explanation is the vocalization and time scores of 6-week-old puppies when confronted with a barrier between them and their food. In these correlated scores the differences are in the same direction in both F_1 and F_2 and seem to be little reduced in the latter generation. Other tests in which there is a consistent maternal effect persisting to the F_2 generation, are the trailing test, the time score in the spatial-orientation test, and the speed score in the delayed-response test. In the first two, the differ-

TABLE 11.7

TESTS INVOLVING PERSISTENT DIFFERENCES BETWEEN MATERNAL LINES
(STANINE SCORES)

Test	Basenji — Cocker	CSBF₁ — BCSF₁	CSBF₂ — BCSF₂
Attraction	−1.60	−1.39	−1.57
Noises, barrier test	−2.10	−.27	−.45
Trailing	1.85	1.18	.28
Spatial orientation, time	−1.51	−.65	−.18
Motivation, speed	−1.28	−.20	−.37
Reactivity, investigation	2.43	1.07	.37
Reactivity, escape	2.58	1.06	.53
Heart rate, adult	1.95	1.01	.95
Heart rate, bell	−1.14	−.80	−.34
Leash control	3.41	.54	.47

ence between the F_2's is considerably reduced, indicating the loss of the effect and probable complications by genetic factors.

Several scores on the reactivity tests give indications of important maternal effects, particularly the scores on investigative behavior and attempts at active escape. Likewise, the adult heart rate score shows a very important effect, as does one of the other heart rate scores. Finally, the leash-control test, involving the tendency to bite and jerk a leash, shows a large and consistent effect which is very little reduced in the second generation.

In conclusion, three of the tests, the attraction, heart rate, and leash control, give evidence of unusually important and persistent effects running through the maternal line. There is an obvious explanation of how these could be produced in the attraction score, and even the heart rate may represent a learned emotional reaction to handling. The effect of leash control is more difficult to explain in such terms and, as indicated above, may actually be an accidental result of strain selection (see chap. 12).

RESULTS OF ANALYSIS OF OTHER TESTS

Performance tests.—Some of the performance tests which we devised were so difficult that a large number of puppies failed to solve them. This automatically forced the data into a two-category distribution of failures and successes, and this in turn suggested the possibility of analysis by the methods of qualitative inheritance.

In the barrier test given at 6 weeks of age, basenjis show much more ability to solve the problem without mistakes than do cocker spaniels. Likewise, in the manipulation tests, the basenjis show comparatively few failures and cocker spaniels show a large number.

However, there is no simple pattern of Mendelian inheritance apparent in the hybrid populations. The reasons for this are fairly obvious in the manipulation tests. Some animals start off well on the simplest tests and fail as the tests get harder, presumably because of a lack of capacity. Other animals do poorly at the outset and get better as they go along. These two very different types of reactions to the situation probably involve quite different hereditary mechanisms, and they certainly lead to complex results.

Most of the other performance tests were so designed that all animals would succeed and obtain some sort of score on a continuous scale. These results presumably could be analyzed by the methods of quantitative inheritance, but here again none of the tests gives evidence of any simple pattern of Mendelian inheritance such as is necessary for this type of analysis. There are still other methods which can be used to analyze such data, namely the calculation of the amounts of variance due to breed differences, and the results are presented in another chapter.

Physical differences.—The size measurements are theoretically well adapted to analysis by the methods of quantitative inheritance. However, even this part of the experiment is complicated, first by the fact of strain differences referred to above and second by hybrid vigor. Both the F_2's and F_1's are on the average very nearly as big as basenjis, the larger of the two parents. The increased body size of the F_1 can be attributed to hybrid vigor, and the F_2's are in turn affected by this same hybrid vigor in that their F_1 mothers had unusually abundant milk supplies and gave them excellent care. Nevertheless, the backcross and F_2 generations do show greater variation than the F_1's, as might be expected from Mendelian segregation (Table 11.8). Calculations based upon this give approximate esti-

TABLE 11.8

DISTRIBUTION OF ADULT WEIGHT MEASUREMENTS IN BASENJIS,
COCKERS, AND THEIR HYBRIDS (STANINE SCORE 5 = 9 KG.)

| BREED OR HYBRID | NUMBER | STANINES | |
		Mean	Variance
Basenji	34	6.18	0.63
CSB F_1	28	5.71	1.42
Backcross to basenji	43	5.88	1.88
CSB F_2	41	6.15	3.53
Cocker spaniel	31	4.81	1.53
BCS F_1	24	6.66	1.88
Backcross to cocker	28	4.79	2.76
BCS F_2	32	6.09	2.82

mates of one gene accounting for the difference between basenjis and cocker spaniels.

The heart rate of young puppies also shows a continuous distribution and evidence of increased variation in the segregating populations (Table 11.9). The general pattern of inheritance looks like one of no dominance except that the backcross generations are almost exactly like their maternal parents. Analysis of the F_2 data gives a figure of less than one gene; but it is possible that if the variance were adjusted in various ways, the result would be more clear-cut. At any rate, the evidence indicates that not more than the one gene is involved.

TABLE 11.9

DISTRIBUTION OF AVERAGE HEART RATES, 11–16 WEEKS, CONVERTED TO STANINES

Breed or Hybrid	Number	Mean	Variance
Basenji	34	6.47	1.49
CSB F_1	28	5.89	1.41
Backcross to basenji	43	6.81	3.13
CSB F_2	41	5.46	2.14
Cocker spaniel	31	5.16	1.06
BCS F_1	28	6.00	1.42
Backcross to cocker	28	5.21	2.65
BCS F_2	32	6.16	2.01

THE DEVELOPMENT AND DIFFERENTIATION OF BEHAVIOR

As we saw in our study of the development of the traits of barking and barklessness, the maximum differences in that test situation appeared at 11 weeks of age. At 5 weeks the basenjis and cockers were too much alike to make a genetic analysis possible, and by 15 weeks both breeds had begun to bark so little that genetic analysis was again impossible. The following examples give an even more clear-cut case of the development of differences in behavior.

Inheritance of the tendency to be quiet while weighed.—Once per week each puppy was placed on a platform scale for one minute and given mild forced training in being quiet, as described in chapter 9. Figure 11.7 shows the development of the behavior of being completely inactive for 1 minute. By 11 weeks of age approximately 70 per cent of the cocker spaniels keep quiet, and this proportion climbs slowly thereafter. By the same age the percentage of basenjis reaches approximately 20 per cent and stays at the same level. We can there-fore say that the behavior of the two breeds has become *differ-entiated,* although some overlap between individuals still remains.

We can now analyze the inheritance of the trait at the time when maximum differences appear, using methods for qualitative inherit-ance. In order to give a somewhat more accurate estimate, we can average the scores at 14, 15, and 16 weeks of age. The accuracy can also be improved by including only those cockers and basenjis which are most closely related to the parents in the cross; it turns out that they are somewhat more different from each other than are the two

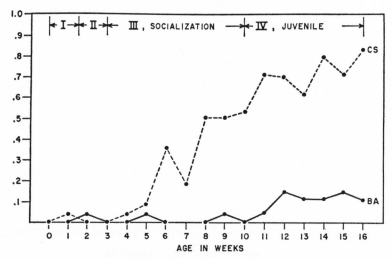

Fɪɢ. 11.7.—Changes with age of proportion of animals rated as completely quiet for the one-minute period while weighed, in the cocker and basenji breeds.

general populations of each breed. Analyzing this result, we get figures which approximate those which we would expect from two-factor inheritance, with the cocker spaniel tendency being recessive. The figures depart widely from what might be expected if a single factor was involved.

The inheritance of postural responses.—As part of the same test, the initial posture of the puppy as it was placed on the scale was recorded. There are four possible behavior patterns: lying flat, crouching with the legs bent and head down, sitting, and standing.

In the cockers, behavior on the scale is related to the tendency of the breed to lie flat when threatened, which goes back to their medieval use in setting birds for the net. As "setters" they were sup-posed to lie flat upon the ground. However, the top of the scale was only 10.5 × 12 inches square, not an ideal place for lying flat, especially as the animals grew older, so that the favorite position of the cocker spaniel puppies was usually sitting.

As Figure 11.8 shows, puppies of both breeds initially respond by

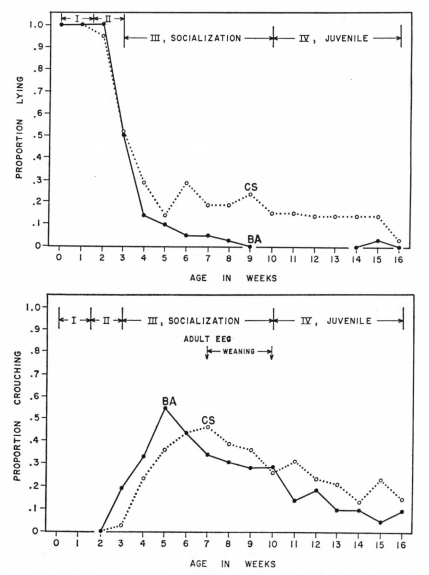

FIG. 11.8.—The development and differentiation of postural resonses in the cocker and basenji breeds. The majority of cockers eventually sit while weighed, but the majority of basenjis stand.

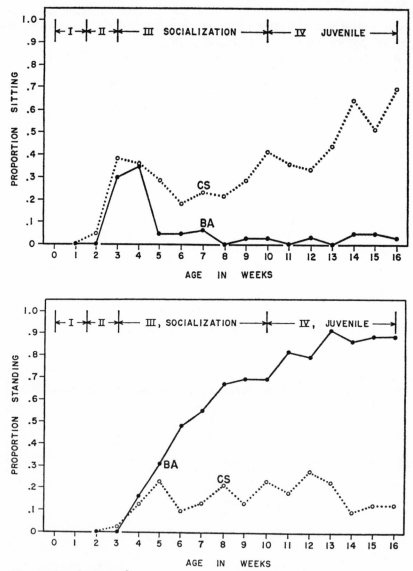

Fig. 11.8.—Continued.

lying flat on the scale and continue to do this through the neonatal period. During the transition period they begin to adopt other types of posture. At 3 weeks of age both basenjis and cockers have a tendency to crouch instead of to lie down, but some of them are beginning to sit as well. By 5 weeks of age the most common posture is crouching, while the frequency of sitting has declined. Crouching reaches its peak at 5 weeks in basenjis but not until 7 weeks in cockers and thereafter declines in both breeds with very little difference between them. As crouching disappears it is replaced by standing in the basenjis and by sitting in the cockers, so that at 16 weeks of age about 70 per cent of the cockers are sitting and nearly 90 per cent of the basenjis are standing. The curve for sitting in the cockers is still rising, indicating that the behavior has not been fully differentiated in this breed. However, in both breeds the percentage of animals standing is pretty well stabilized.

Analyzing the ratios based on scores at 14-16 weeks, when the tendency to stand is most completely differentiated between the two breeds, we find that the ratios fit what might be expected if there were two genes involved, those for the cocker spaniel being dominant. The fit is not as good as it was in the trait of being quiet, but this may result from the fact that body size has some effect on the posture on the scales. In both cockers and basenjis, there are consistently more females standing than males, probably because their smaller size makes it easier to do so. Likewise, the F_2 animals sometimes became unusually large due to the excellent nutrition provided by their F_1 mothers, and fewer of the F_2's took the standing posture than might have been expected from a simple genetic theory.

One interesting result of the experiment is that the behavior of the cocker spaniels, which looks like a simple behavior pattern of lying quietly on the scale, is actually more complex. Being quiet is inherited as a recessive trait, the F_1's being active like basenjis. The tendency to adopt some other posture than standing, on the other hand, is inherited as a dominant trait, with the F_1's being like their cocker parents. We are obviously dealing with two different traits, and this is backed up by the fact that there is no correlation between them in the F_2 puppies, in which both traits can appear. Since our results indicate that at least two genes are involved in each one of these traits, the genetic mechanism is quite complex. The chance of getting an animal genetically identical to one of the original pure breeds in the F_2 generation is $(0.25)^4$, or 1 in 256. Since we only had 74 animals in this generation, it is quite likely that none of them were exactly like the original pure breed in respect to this trait.

The other interesting result of this experiment is the almost diagrammatic illustration of the process of differentiation of behavior resulting from maturation and training. The graphs show all puppies starting out very much the same and gradually becoming more and more different until finally they show the behavior which we ordinarily think of as being characteristic of the breed.

SUMMARY

These analyses of the inheritance of simple behavior traits show that inheritance can be explained in many cases on the basis of one or two genes. This is somewhat surprising, as we would expect that anything as complex as behavior would be affected by many genes. We should remember, however, that these particular differences in behavior were produced by artificial selection and that the effect of a gene on a character as variable as behavior could probably not be recognized unless it was quite a large effect. A breeder choosing between two dogs usually cannot distinguish between behavioral differences produced by heredity and those produced by training and other environmental experiences. Unless a gene produces a relatively major result, it is likely to pass unnoticed and be lost. Minor modifying factors seem to have their chief importance in contributing to individual and strain differences within a breed.

We should also remember that the behavior pattern is, in one sense, a natural unit of behavior. All selection, whether artificial or natural, is done on the basis of effect, and the modification of behavior patterns having specific functions is consequently the most direct way in which selection can affect behavior. Of all the measures of difference in behavior, those involving simple behavior patterns are most likely to have direct and reasonably simple genetic explanations.

While the possibility of single-factor inheritance is thus a reasonable one, this conclusion cannot be final in the absence of a critical test for single-factor inheritance, namely, a second backcross. Confirmation will therefore have to await further experimentation. Assuming that the single-factor explanation is a valid one, we should point out that it works best when applied to cases in which simple behavior patterns are either present or absent. This is consistent with the theory of a threshold of response to stimulation modified by gene action.

On the other hand, the theory of a threshold modified by a single gene, or by the additive action of several genes, does not apply to any case of a threshold of successful adaptation to a complex problem-solving situation. Here the underlying processes of gene action appear to be much more intricate. There is evidence in the case of the inheritance of the annual breeding cycle that the action of even a small number of genes can produce extraordinarily complex effects through interaction on a physiological level. Beyond this, the combination of two simply inherited patterns of behavior multiplies the number of possible genetic combinations for the whole. Even a small number of genes can thus produce an enormous amount of variability in a population.

While both simple qualitative and quantitative inheritance are sometimes found in relation to simple patterns of behavior, the combination of these patterns into complex adaptive patterns through learning and problem solving inevitably results in an extremely complex genetic situation. The multiple-factor theory of genetic effects of behavior is correct in the sense that each behavior pattern appears to be affected by a different set of genes. On the other hand, extremely complex effects can be explained by a small number of genes, so that the genetic mechanisms are probably based on a finite, and possibly relatively small, number of mutations.

We also find that behavior has to be developed: that it only appears at particular times in development and under particular environmental conditions. The inheritance of a simple physical trait like long and short hair seems to contradict this finding, but the contradiction is only an apparent one. Even in this case we find that the trait undergoes development because puppies at birth have hair which is much the same length, and before birth there is a time when the fetus has no hair at all, short or long. This emphasizes the important fact that genetic characteristics or phenotypes are not in a strict sense inherited, but always *developed* under the influence of genetic and environmental factors. The big difference between behavior and the anatomical characteristics usually studied by geneticists is that the great bulk of behavioral development occurs after birth rather than before and that behavior continues to develop and differentiate throughout life.

In this chapter we have indicated two difficulties in dealing with the data. One of these is that most of the mathematical methods used so far assume that the parent populations are pure breeding, although we have every reason to believe that the dog breeds are genetically variable in many respects. The other difficulty is that of applying quantitative methods to cases involving complex interaction, whether on the genetic, physiological, or behavioral levels. In the next chapter we shall attempt to deal with these difficulties by the use of quantitative methods which do not depend upon the assumption of homozygosity, and we will also see if these specialized methods apply to the inheritance of complex patterns of behavior where it is highly likely that many genetic factors are involved, and hence that some of them are heterozygous.

CHAPTER 12

BEHAVIOR IN HYBRIDS: COMPLEX BEHAVIOR

In previous chapters we have demonstrated that breeds of dogs differ significantly in their scores on a large number of behavioral tests. On occasion, the relative importance of the breed differences has been expressed as an intraclass correlation, defined as the proportion of total variance ascribed to breeds. Correlations of .2 to .5 were common; 20 out of 34 variables analyzed in Table 14.3 fall within this range. On some measures the degree of overlap between the breeds was considerable, but examples were also found of almost complete separation of scores in some breeds. In this chapter we shall attempt to extend this sort of analysis to hybrids, and particularly to cases of complex behavior in which no simple pattern of inheritance is apparent.

The general biological significance of estimates of genetic effects derived from breed comparisons is limited by the peculiarities of the mating system by which breeds are formed and maintained. A pure breed is far from being a homogeneous group. The cocker spaniels, beagles, and Shetland sheep dogs showed segregation at coat color loci, the beagles for hair length, and all breeds showed heterogeneity in blood type (Cohen and Fuller, 1953). Thus, to use between-breed variance as a measure of the effects of heredity and within-breed variance as a measure of environmental effects is simplification which could lead to error. If some of the behavioral variation between members of the same breed is genetic, the intraclass correlation tends to underestimate the importance of heredity. On the other hand, fixation within a breed of special combinations of genes could produce peculiarities of behavior. Such combinations, maintained by restricting matings within a breed, would be genetic in the broad sense. In a random mating population, these combinations would be broken up because of the independent assortment of chromosomes and the peculiarities of behavior associated with them would be rare. If this argument holds, breed differences might provide an exaggerated estimate of the importance of heredity for behavioral variation in a population with a more open breeding system.

METHODS

Breeding plan—genetic implications.—In order to extend our information on the transmission of behavioral characters through genes, crosses were made between basenjis and cocker spaniels. These two breeds were chosen for the hybridization study because they showed marked phenotypic differences on many tests and were also about as remotely related as any two breeds could be. It will be recalled that four of our five pure breeds originated in the British Isles and that the basenjis is of African descent. Historical evidence indicates that there has been no genetic interchange between the basenjis and the cocker spaniels for at least several centuries and possibly for thousands of years. We were also aware of conditions which made these two breeds less than ideal for a genetic experiment. Preferably the stocks employed as parents for a cross should be pure-breeding in a strict genetic sense, at least for the loci of interest. It is unlikely that this condition was met in our breeds except possibly for loci controlling barklessness and an annual breeding cycle, but it is also true that on many phenotypic measures the degree of overlap between the basenjis and cockers was very small. Another difficulty arose from the need to breed from a small number of parents because of kennel space limitations. In a highly inbred stock one parent is genetically equivalent to any other of the same stock, but in a variable population the choice of parents introduces a sampling error. Sampling error was possibly responsible for some of the differences found between reciprocal F_1 crosses, though an alternative explanation involving maternal effects has not been excluded.

As described in chapter 1, the basenji-cocker spaniel cross was essentially an ordinary Mendelian cross with certain modifications made necessary by the nature of our animals. Reciprocal basenji × cocker matings were made to produce two F_1's, each of which was bred *inter se* and brother to sister to produce an F_2 population. Ideally, backcrosses to the parental strains would have been made in reciprocal fashion, but limitations of space and time led us to breed in only one direction, F_1 males to a basenji or cocker dam. This cross enabled us to compare offspring of purebred dams by F_1 and purebred sires.

Since the puppies had no physical or social contact with their sires, any difference between the CSB F_1 and the CSB $\times$ B backcross or between the BCS F_1 and the BCS $\times$ CS backcross (see Fig. 1.1) can be attributed to the genetic contributions of the sires. The plan also allowed for two comparisons between backcross and F_2 groups with common sires (F_1) but different dams (F_1 or purebred). We found that the F_1 females provided better care than either type of purebred, this being reflected in lower infant mortality. Furthermore, offspring were associated with their dams for 10 weeks, offering ample opportunity for learning and imitation. Thus, the finding of any difference between F_2's and the backcrosses from the same sire can reasonably be ascribed in part to non-genetic transmission from dams to offspring.

Analysis by groups.—The data from the hybrids were treated in two ways—by group and by mating. The general procedure is available in various sources such as Wright (1952) and Falconer (1960). In the analysis by group, we provisionally assumed that basenjis and cocker spaniels were pure breeding for the major loci producing behavioral differences between these strains. Basenji, cocker spaniel, and F_1 hybrid groups were considered genetically homogeneous and all sires or dams within a group as genetically equivalent. Furthermore, we predicted that the heterogeneous F_2 and backcross groups would be more variable than the homogeneous parental and F_1 groups since genetic variation between members of these groups would be superimposed upon environmental variation. These expectations were not often realized, and a consideration of the outcome of specific matings without respect to the group classification of the parents was frequently more enlightening. Nevertheless, on many measures in which the spaniels and basenjis differed sharply, we found that the means of the hybrid groups were arranged in an orderly fashion corresponding to the genetic contribution of the original pure breeds. Had the original stocks been more nearly isogenous, the results would have probably been even more conclusive.

Analysis by matings.—In this type of analysis, it is not necessary to assume that the parent stocks were homozygous. On the contrary, we can treat the data as though the parents came from a mixed population. Two methods were used in the analysis of outcomes of specific matings. Regressions between the mean ($\bar{o}$) of offspring from a mating and the phenotypic scores of their sire (s), dam (d), and midparent or average of sire and dam (mp) were computed. A regression equation of the form $X_{\bar{o}} = a + bX_p$ shows the relationship between the average of the offspring of a mating $(X_{\bar{o}})$ and some score of the parents (X_p). The constant a allows for the fact that the offspring may average higher or lower than the parents. The constant b, regression coefficient, has genetic significance (Falconer, 1960). For example, the regression coefficient $b_{\bar{o}.mp}$ is a direct estimate of heritability (h^2); $b_{\bar{o}.d}$ and $b_{\bar{o}.s}$ are estimates of $h^2/2$. Since the only connection between the sires and their offspring is the contribution of genes, $2b_{\bar{o}.s}$ is ordinarily the most dependable estimate of h^2. In some circumstances, however, we found that our sires were too uniform in behavior to provide a good estimate; for some tests scores of all parents were not available and parent-offspring correlation were unobtainable.

Analysis of variance.—A second method of utilizing data from specific matings was by analysis of variance, which enabled us to evaluate the genetic contribution of individual sires and dams, the importance of environmental factors operating on different litters of the same parentage, and to some degree the variation between the genetic contributions of parents classified in the same group. The details of the procedure and interpretation are described later in this chapter in connection with the fighting scores in leash training.

We shall deal with the scores of hybrids on a group of tests administered between 4 and 12 months of age. All of these tests have been described in previous chapters and provide a wide sample of kinds of tests and genetic results. Some of them were initiated somewhat later than others in the program and hence have more limited numbers in the parent strains. In these cases the use of the last described method was particularly applicable. Analyses are reported on the following scores: leash-control fighting, leash-control vocalization, motivation speed, obedience composite score, spatial-orientation errors, and total reactivity score. Additional measures from these tests are reported in less detail. The order of discussion corresponds to that in which the tests were administered.

TRAINING TO LEAD

Analysis by genetic groups.—The leash-training procedure and the results with the pure breeds have been described in chapter 9. The basenjis and cocker spaniels differed significantly, particularly on the category of demerit we called "fighting the leash." As training proceeded the average total demerits fell precipitously in all breeds until by 10 days most subjects were walking well on the leash, though basenjis were still charged with more demerits than cockers.

When one uses the total demerit score over 10 days as a measure of performance, all the hybrid groups are intermediate to the parental stocks and there is no significant differentiation between them (Table 12.1). The variances (V) of the F_2 and backcross generations are somewhat higher than those of the parents and the F_1's. Esti-

TABLE 12.1

LEASH-TRAINING DEMERITS (STANINES) OF HYBRIDS

Cross	Number	Mean Score	Pooled Within-Litter Variance
BA × BA............	34	6.35	1.82
CSB × BA...........	44	5.72	2.99
CS × BA............	29	5.67	2.82
CSB × CSB........	42	5.39	4.72
BCS × BCS........	32	5.06	2.68
BA × CS............	24	5.33	1.33
BCS × CS..........	31	5.27	3.65
CS × CS............	41	4.10	1.90

mating the heritability (h^2) of the total demerit score by the formula

$$h^2 = \frac{V_{F_2} - V_{F_1}}{V_{F_2}}$$

and using pooled estimates from the two F_1's, one obtains

$$h^2 = \frac{3.82 - 2.13}{3.82} = 0.44.$$

This value should be considered an approximation, however, since the assumptions of homogeneity involved in the group method have not been fully met. Furthermore, the total demerit score is obtained by adding together the occurrences of very different forms of behavior which may have quite different degrees of hereditary determination.

A more fruitful attack is to consider the performance of the hybrids and the parental stocks with respect to the individual measurements which made up the demerit score. These figures are presented in Figure 12.1. Comparison of the figures immediately demonstrates that the role of heredity in determining the type of demerits differs greatly for the tests. The parent strains differed considerably in fighting (F) and position (P) demerits, and the hybrids fell at intermediate positions. If the behavior were inherited in a simple additive manner, the hybrid means would fall on the line drawn between the parental means. (see Bruell, 1962). There is indication of a maternal effect, since animals resemble their mothers more than their fathers. Neglecting the sex chromosomes, the genetic contribution of the sexes is equal, but the sires had no contact with their offspring while the dams were with them for 10 weeks.

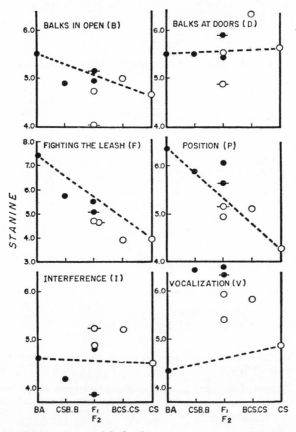

FIG. 12.1.—Mean scores of hybrids on six measures of leash training. Solid circles denote groups with basenji mothers or grandmothers. A line drawn through a circle indicates an F_2 group. The dashed lines between the parental means would pass through the hybrid means if the characters were inherited in a simple additive fashion.

On two measures, fighting the leash (F) and position (P), the parental stocks are distinct and the hybrids intermediate. In both scores the hybrids resemble their dams or granddams more than their sires or grandsires. On three measures, balks in the open (B), balks at doors (D), and interference (I), the parental scores are similar and the hybrids vary non-systematically. On the sixth measure, vocalization (V), all hybrid groups score well above either parental stock. We get a much clearer picture of the phenotypic differences and possible genetic mechanisms from this detailed analysis than from the overall performance score.

The leash-fighting scores were discussed in chapter 11 where a strong maternal effect was pointed out. The within-litter variances of the parental and hybrid groups are set forth in Table 12.2. Using the variances of the F_1 and F_2, one can estimate heritability as 0.52. This is a somewhat lower estimate than that based on the two pure breeds alone, 0.77 (see chap. 14) but is of the same order of magnitude. It is higher than that obtained for the total demerit score, reinforcing the conclusion that specific scores are more meaningful than complex combined scores.

TABLE 12.2

WITHIN-LITTER VARIANCE OF THE FIGHTING SCORE
OF LEASH TRAINING

Population	Number	Mean	Mean Square within Litters
Basenji............	34	7.38	1.54
Cocker spaniel......	41	3.94	0.97
F_1 (pooled).........	52	5.11	0.71
CSB × BA.........	44	5.75	0.85
BCS × CS.........	30	3.93	2.30
F_2 (pooled)........	73	4.89	1.50

Unfortunately the backcross variances give a much worse fit to a simple model than do the total variances analyzed in the previous chapter; the BCS-CS value is unexplainably high and the CSB-B value is too low. Dominance of the basenji genes could account for this finding in part, but such an interpretation is not supported by the means, which support the hypothesis of partial dominance of cocker spaniel genes. Similar inconsistencies in the variances of the hybrids repeatedly turned up in our analysis, possibly dependent upon the scales used for phenotypic measurement, and possibly reflecting accidents of random sampling. The ordering of means for the fighting and position scores is consistent with genetic determination of the scores, but the within-group phenotypic variances do not increase proportionately with genetic heterogeneity. We can seek the clues to the sources of within-group variation in a comparison of repeated matings.

Variation between litters of the same mating.—If environmental factors (such as those associated with specific testers and month of testing) were important, we would expect repeated litters from a single mating to vary widely from each other, since each litter would have had a different history. On the other hand, if heredity had played an important role, successive litters of a mating should be relatively similar. Maternal effects could also be consistent from litter to litter so that this kind of environmental effect cannot be excluded by the finding of consistency of performance between repeated litters from the same parentage.

Only a part of our sample (F_2's and backcrosses) was suitable for the litters-within-mating analyses, since each of these groups contained several duplicated matings. The results obtained with the fighting scores are presented in Table 12.3. Different litters of a par-

TABLE 12.3

ANALYSIS OF VARIANCE—FIGHTING SCORES OF HYBRIDS
BY MATINGS AND LITTERS

Source of Variation	Degrees of Freedom	Mean Square	F Ratio	Probability
Matings................	10	8.54	5.40	.01
Litters in matings.......	14	1.93	1.22	N.S.
Within litters...........	99	1.58	...	...

ticular mating differ insignificantly among themselves, although differences between matings are highly significant. The evidence for control of fighting by genetics or by maternal influence is therefore strong.

Effect of sire and dam.—In order to distinguish between genetic influences transmitted equally by both parents and associative influences derived solely from the mother, a more detailed analysis was carried out on data from matings in which a single sire was bred to two different females or a single dam to two sires. The groups available and their means on the fighting score of the leash-control tests are set forth in Tables 12.4 and 12.5. We have here the equiva-

TABLE 12.4

LEASH-CONTROL FIGHTING SCORE—OFFSPRING OF F_1 SIRES BY DIFFERENT DAMS

SIRE		DAM		OFFSPRING		
Type	Mean Score	Type	Mean Score	Litters	Individuals	Mean Score
CSB........	5.0	BA	7.8	10	43	5.77
		CSB	5.2	7	36	5.08
BCS........	5.3	CS	3.0	7	30	3.93
		BCS	4.7	6	32	4.62

lent of four experiments with some subjects, the backcrosses, appearing in both the common-sire and the common-dam experiments.

Looking at Table 12.4, we see that the offspring of F_1 sires have fighting scores which are on the average intermediate to the parents.

TABLE 12.5

LEASH-CONTROL FIGHTING SCORE—OFFSPRING OF PUREBRED DAMS BY DIFFERENT SIRES

DAM		SIRE		OFFSPRING		
Type	Mean Score	Type	Mean Score	Litters	Individuals	Mean Score
BA.........	7.8	CS	4.8	5	23	5.52
		CSB	5.0	9	44	5.76
CS.........	3.0	BA	8.0	4	20	5.00
		BCS	5.3	7	30	3.93

All the variation in the offspring can be attributed to the difference between dams, because the males used had almost identical fighting scores.

In Table 12.5 we have an opportunity to see whether the male parents influenced the amout of fighting. It appears to make no difference as to whether basenji females are bred to CS or CSB males, but inspection of the scores show that the two groups of males have extremely similar fighting scores. Cocker spaniel females produce very different offspring when bred to B and BCS males; here the means of the sires differ by 2.7. We conclude that the transmission of fighting behavior from males is demonstrated and that the effect must be genetic.

A more elaborate analysis of variance of the data summarized in Tables 12.4 and 12.5 appears in Table 12.6. In each of the four analyses, the between-mating sum of squares was partitioned in two ways: first, according to individual sires or dams, for example, the offspring of the CSB males in Part A; second, comparing the significance of the difference between the means of the two classes of offspring produced, for example, in Part A the difference between CSB × B and CSB × CSB. In all four analyses the litters-within-mating contribution to variance was non-significant, which is good evidence that changes in testers, weather, and the like had little effect upon the scores. The within-mating mean square was thus used as the denominator in the computation of F ratios.

We shall now consider the two sets of matings of F_1 sires to parental and F_1 dams. None of the components in the BCS × BCS and BCS × CS experiment was significant, though the difference between the offspring of the two types of dams came close. At first

TABLE 12.6

ANALYSIS OF VARIANCE—FIGHTING SCORE IN LEASH TRAINING

SOURCE OF VARIATION	DEGREES OF FREEDOM	MEAN SQUARE	F
CSB sires with BA and CSB dams producing CSB × BA and CSB × CSB			
Sires....................	3	5.82	4.92**
Dams in sires.............	4	5.51	4.65**
Type of dam.............	1	9.17	7.74**
Between like matings......	6	5.06	4.27**
Within matings...........	71	1.18	
BCS sires with CS and BCS dams producing BCS × CS and BCS × BCS			
Sires....................	2	1.04	0.50
Dams in sires.............	3	2.99	1.43
Type of dam.............	1	7.40	3.55
Between like matings......	4	0.91	0.44
Within matings...........	56	2.08	
BA dams with CS and CSB sires producing CSB and CSB × BA			
Dams...................	3	2.30	3.06*
Sires in dams.............	4	1.37	1.82
Type of sire.............	1	0.79	1.05
Between like matings......	6	1.93	2.57
Within matings...........	59	0.75	
CS dams with BA and BCS sires producing BCS and BCS × CS			
Dams...................	2	0.64	0.42
Sires in dams.............	3	7.27	4.82**
Type of sire.............	1	13.65	9.04**
Between like matings......	4	2.36	1.56
Within matings...........	44	1.51	

* $P < .05$
** $P < .01$

glance the CSB × BA and CSB × CSB experiments look quite different. The highly significant mean square for differences between sires could indicate great heterogeneity in the genetic characteristics of the males selected for the experiment. However, in the analysis of the dams contributions, we find a highly significant effect due to breed of dam (basenji vs. CSB) and heterogeneity between matings of the same class. Inspection of the phenotypic scores of the parents (Table 12.4) supports the view that the variation between sires is accounted for by the dams to which they were mated. All four CSB sires had identical scores of 5.0, and their matings were non-informative with respect to the inheritance of fighting through the male parent. Fortunately, in the CS dam crosses, type of sire was the major genetic factor influencing the fighting scores. The formal analysis confirms the significance of the deductions based on inspection of group means.

Thus it is demonstrated that leash fighting can be inherited from either parent, and the failures to demonstrate differences among litters of the F_1 males were apparently the result of a curtailed range of phenotypic differentiation. The matings which compared unlike parents mated to common parents, either sires or dams, yielded clear evidence of heritability.

We are hesitant to estimate heritability of the fighting score from the data available since not all conditions required by the various methods have been met. It is interesting, however, to note the regression of offspring means on midparent values, $b_{\bar{o},\bar{p}} = 0.454 \pm 0.135$. This value was calculated from all the hybrid and purebred litters for which data on both parents was available. The regression of offspring means on dams was 0.233 ± 0.080, in good agreement with a heritability estimate of about 0.45. The regression of offspring on sires was only 0.110, but the estimate is less accurate because the accidents of sampling produced a group of sires with only half the variance of the dams. The values are lower than those estimated from the group differences.

The fighting scores have been discussed in some detail because they illustrate the problems of analyzing the inheritance of a relatively simple quantitative character of moderately high heritability.

Comparison of vocalization with leash fighting.—Vocalizing during leash training is of some general interest, since yelping is a conspicuous response and has been used as an index of distress in puppies (Fredericson, 1952). It is also of interest to compare vocalization scores, which showed such a different pattern of means, with leash fighting. All of the hybrid groups scored well above the means of both parental stocks, which were quite similar to each other. The combination of genes from the basenji and cocker spaniel lines thus favors the elicitation of vocal responses by the mildly stressful process of leash training. It should be noted that vocalization on the leash is of a different sort from the barking analyzed in the last chapter.

A detailed analysis of the vocalization scores was made in the same manner as that used for the fighting scores. With vocalization, however, the offspring-midparent correlation was zero. The data were then arranged by matings which involved common sires or dams. It is clear from Tables 12.7 and 12.8 that the scores of the offspring have no systematic relationships to the mean scores of the parental

TABLE 12.7

LEASH-CONTROL VOCALIZATION SCORE—OFFSPRING OF SIRES BY DIFFERENT DAMS

SIRE		DAM		OFFSPRING		
Type	Mean Score	Type	Mean Score	Litters	Individuals	Mean Score
CSB........	6.8	BA	3.8	8	43	6.4
		CSB	7.0	7	36	6.5
BCS........	7.0	CS	4.3	7	30	5.8
		BCS	5.3	6	32	5.4

TABLE 12.8

LEASH-CONTROL VOCALIZATION SCORE—OFFSPRING OF DAMS BY DIFFERENT SIRES

DAM		SIRE		OFFSPRING		
Type	Mean Score	Type	Mean Score	Litters	Individuals	Mean Score
BA.........	3.8	CS	5.5	5	23	7.0
		CSB	6.8	9	44	6.4
CS.........	4.3	BA	5.0	4	20	6.6
		BCS	7.0	7	30	5.8

groups. An analysis of variance uncovered no significant differences between the separate matings.

We have here a perplexing difference between the hybrids and the pure breeds. Differences in vocalization between the five breeds were highly significant ($P < .01$) and were one of the most useful scores for discriminating breed behavior patterns. Furthermore, it is well known that vocalization responds to genetic selection. Evidence is provided each time a hound bays on the trail or a toy breed barks shrilly at a stranger. Vocalization in the leash-training test, however, occurs when two conditions are met in the same subject—it must be emotionally aroused and vocalization must be readily emitted. By and large, our cocker spaniels were noisy animals, but they were relatively docile on the leash and earned average vocalization scores. The basenjis were highly excited by the leash but directed their energies into channels other than vocalization. The hybrids were aroused like the basenjis, and vocalization was a favored response, possibly because of their spaniel genes. However, the quantity of vocalization, once the necessary conditions were met, was independent of heredity, as demonstrated by the lack of a parent-offspring correlation.

MOTIVATION TEST

The motivation test, described in chapter 10, was simply a measure of running speed in a T-maze. On any one trial a barrier forced the dog to one or the other arm of the T. Five trials on the right and five on the left were given on each of three consecutive days. The total running time on the third day was our measure of motivation.

The mean scores of the hybrid groups and the parental stocks are given in Figure 12.2. The parental strains were significantly different, and basenjis selected as parents for the hybrids had much lower scores than the breed as a whole. The most interesting feature of the data is the elevation of the hybrid scores as compared with the parents, a situation similar to that of vocalization during leash training. If we can exclude any environmental effect which favored the hybrids (which obviously were tested at a different time than their parents), it appears that hybrids ran faster, perhaps due to physical superiority based upon genetic heterosis. We postulate a dual mechanism—first, cocker spaniel genes as compared with basenji genes increase attraction to food and to human beings. (Alternatively one could interpret the data in terms of increased timidity produced by basenji genes, which is perhaps saying the same thing.) Secondly, running speed depends not only upon the degree of attraction for the reinforcement provided at the end of the runway,

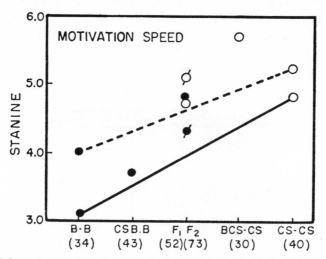

Fig. 12.2.—Mean motivation speed scores of hybrids. Points identified as in Fig. 12.1. The means of the actual parents of the F_1 hybrids are connected by a solid line; the means of the total breed sample by a dashed line.

but upon physical vigor as well. Figure 12.2 can be interpreted as illustrating an effect of physical vigor superimposed upon a genetic system controlling the relative strength of positive and negative reinforcement in the test situation.

We made additional analyses of the separate hybrid matings in an effort to estimate the quantitative contribution of heredity to the running-speed measure. The midparent-offspring correlation was .34 ± .20. We might take this as an estimate of heritability, but the error of estimate is very high. Analysis of variance of the groups in which one sire was mated to two dams, or one dam to two sires, showed considerable consistency. In no instance were there significant differences between litters of the same mating, indicating that rearing and testing conditions were well controlled over the course of the observations. On the other hand, individual matings differed significantly at the .01 level in each of the four sets of data and intraclass correlations of .15, .41, .24, and .24 were computed. All these computations agree in suggesting that something between 20 and 30 per cent of the variation in running speed is heritable. This compares with an estimate of 16 per cent based on the parent breeds alone (see Table 14.3).

DISCRIMINATION OR CUE RESPONSE

The motivation test on the T-maze was used as pre-training for discrimination in addition to being a measure in its own right. Subjects which did not reach the criterion within 128 trials (8 days with 16 trials) were classified as failures. In Figure 12.3 we present the performance of the basenji-cocker spaniel hybrids with respect to

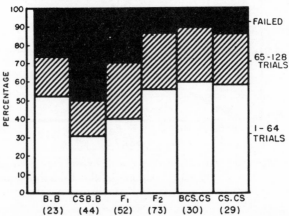

DISCRIMINATION CRITERION ATTAINMENT

FIG. 12.3.—Attainment by hybrids of criterion in the cue-response (discrimination) test. Failure is defined as not meeting criterion within 128 trials.

attainment of criterion within 128 trials. By using a simple pass-fail dichotomy, we found the data yield a χ^2 of 25.532 ($P < .001$). The ranking of the six groups is orderly except for the purebred basenjis which seem to be better than they should be considering that the basenji backcrosses were so slow in learning. When we consider the scores of the specific parents, we find a possible explanation. Unfortunately, only four of the seven basenji parents passed through the testing program after the discrimination procedure was standardized, and only three of the four met the criterion. It is almost certain that the basenjis selected as breeders were an inferior sample of the breed (for this test). Thus we feel safe in concluding from the ordering of group means that heredity makes a substantial contribution to successful discrimination learning. The conclusion is substantiated by an analysis of variance in those hybrid matings which produced two or more litters. Rather than use the attainment of a criterion as a measure of success, which would have forced the exclusion of failures, we used the number of correct responses in the first 64 trials. Table 12.9 demonstrates that the variations between matings were significant ($P < .05$), and that litters of the

TABLE 12.9

DISCRIMINATION OR CUE RESPONSE—ANALYSIS OF VARIANCE
OF HYBRID PERFORMANCE

Source of Variation	Degrees of Freedom	Mean Square	F Ratio
Matings....................	12	12.03	2.20*
Litters within matings........	15	5.45	2.36**
Within litters...............	117	2.30	

* $P < .05$
** $P < .01$

same parentage could also differ to a considerable degree ($P < .01$). Even though conditions specific to litters are important, the effects of hereditary factors show through. We have not, however, attempted to estimate heritability of this measure because of our reservations concerning its precision.

DELAYED RESPONSE

Subjects meeting the discrimination criterion were tested on delayed response following a series of 50 post-criterion discrimination trials. The sample is obviously not representative since in some groups up to 50 per cent of the individuals never qualified for de-layed-response testing. Nevertheless, the findings are of some interest and have been illustrated in Figure 12.4. The proportion of individuals meeting the criterion at delays of 5 seconds or more fluctuates unsystematically, but the proportions meeting the criterion at a minimum delay (1 second) appear fairly similar, an impression which is confirmed by a non-significant value (χ^2 of 8.263, 5 degrees of freedom, $P < .10$).

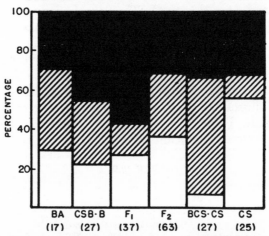

DELAYED RESPONSE PERFORMANCE OF HYBRIDS

FIG. 12.4.—Attainment by hybrids of criterion in delayed response. Failure is defined as not meeting criterion at 1-second delay within 40 trials. Black portion of columns indicates percentage of failures (defined as not meeting criterion at 1-second delay within 40 trials; barred area denotes percentage passing at 1-second delay; white portion shows the percentage attaining criterion at 5-seconds or longer delay.

These data provide no evidence for genetic determinants of short-term memory, but the variations between individuals were striking. The maximum delays at which criterion was met by any one subject were: basenji, 15 seconds; CSB × B, 5 seconds; F_1, 30 seconds; F_2, 30 seconds; BCS × CS, 15 seconds; and cocker, 240 seconds. Individual variation may have a hereditary basis which could be investigated by selecting both for good and poor delayed-response performance. One could retain for the low-line those individuals which learned the cue discrimination adequately, but failed when delay was introduced. Only by such a procedure can the higher order ability be disentangled from simple discrimination learning.

OBEDIENCE TEST

The obedience test was interpreted as a measure of ease of inhibitory training on the basis of the comparison between the five pure breeds. Since the cocker spaniels and basenjis differed sharply on the test, one might anticipate that the hybrids would be intermediate. The expectation is verified in Figure 12.5, which shows the

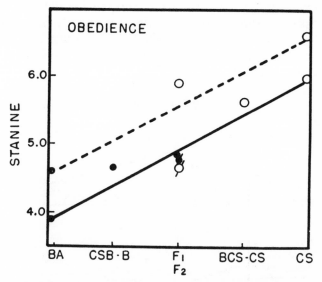

Fig. 12.5.—Mean scores of hybrids on obedience. High scores indicate greater inhibition of movement. Points identified as in Figure 12.1.

mean scores of parental and hybrid groups including the values for the basenjis and spaniels selected as parents of the hybrids. The data are not perfectly orderly, but the trend is clear. The more detailed analyses, unfortunately, are less informative than they were for the leash-training test. The offspring on midparent regression is only .16, not reliably greater than zero. Analysis of variance failed to uncover a consistent difference between matings, and successive litters of the same parentage were often as different as two matings. Despite the favorable mean difference between the breeds and the apparent orderliness of the group means, we ended with a rather unsatisfactory outcome of the genetic analysis. Intuitively the test seemed to be as good a candidate for high heritability as the leash-fighting score, but intuition proved a poor guide.

REACTIVITY TEST

As described in chapter 8, subjects in the reactivity test were restrained by loosely fitting loops while they were stimulated in sequence by the entrance of an experimenter, by the sound of a bell, and by four mild shocks applied to the hind leg. Measures included ratings of posture and various categories of behavior, and continuously recorded respiration, electrocardiograms, and electromyograms. Twenty-three measures were selected for detailed analyses, and the results on the five pure breeds have already been presented.

In addition to these part scores, the total reactivity score obtained by summing all of the overt-behavior ratings is useful as a general basis of comparison. A high total reactivity score indicates a gen-

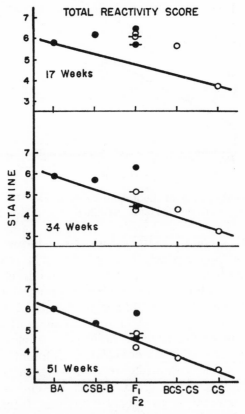

Fig. 12.6.—Mean total reactivity scores of hybrids at 17 weeks (*upper*), 34 weeks (*middle*), and 51 weeks (*lower*).

erally active animal. Basenjis were significantly higher on this measure (mean stanines between 5.5 and 6.0) than cocker spaniels (mean stanines between 3.5 and 4.0). The mean total scores of the hybrid groups are shown in Figure 12.6. One interesting feature of this figure is the change associated with maturity. At 17 weeks of age, all the hybrids are as reactive as the basenji; by 51 weeks the means approach the expectation based on additive inheritance. The CSB F_1's appear to be too high, but this may well be a sampling phenomenon. Unfortunately, the reactivity scoring system had not been standardized when some of the parents were subjects in the program, so that it is impossible to know whether the parents of the two F_1's were equal in reactivity scores.

The detailed analysis of group means on the separate measures at three ages did not add greatly to the information obtained from the total score comparison. On those subscores in which basenjis and spaniels were definitely unlike, the hybrids tended to be intermediate. In particular, the means of the backcrosses and F_2's usually fell close to a line drawn between the parent means as in Figure 12.6; the F_1 means varied more widely from the expectation based on additive inheritance and uniform environment. On some scores the parent stocks differed insignificantly and the patterns of the hybrid means were irregular. As illustrations we show in Figure 12.7 the means for four of the subscores from the 34-week test. The group means for the physiological measure (basal heart rate) fall close to the expectation based on additive effects of genes. The actual means in beats per minute were: basenji, 138; basenji backcross, 134; F_1, 117; F_2, 114; spaniel backcross, 109; and cocker spaniel, 102. Similar results for the heart rate of young puppies were reported in the previous chapter.

The data for body position did not fall as neatly in line with expectations, but all hybrid means were intermediate between the spaniels, whose low score indicates the prevalence of crouching, and the basenjis which almost always stood erect.

The distribution of biting and tail-wagging scores were discontinuous so that stanine transformations were not feasible. Seventy-eight per cent of the basenjis bit at their restraining loops, but only 10 per cent of the cockers did. The percentages of biters in the hybrid groups form an orderly array, although the spaniel backcrosses are as non-aggressive as the purebred spaniels. The dichotomous classification for graphic purposes into biters and non-biters obscures the fact that the 7 to 10 per cent of cockers and cocker backcrosses that were rated as biters were only marginally so.

Biting is an illustration of behavior which does not appear until a threshold of stimulation is reached. Given a standard stimulus, populations differ in the proportion which show any response. Once a response is elicited, it can be expressed in quantitative terms. Such combinations of continuous and discontinuous variation are not convenient for genetic analysis. Formal analysis can be done with Wright's (1934) threshold model (see Fuller, Easler, and Smith, 1950, for an application of this model to seizure susceptibility in the

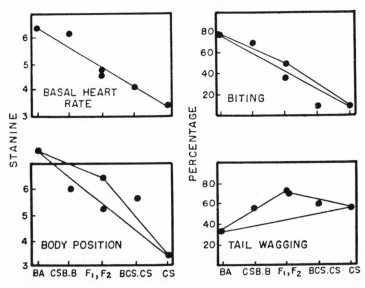

FIG. 12.7.—On the left: mean scores of hybrids on two reactivity-test measures. The two F_1 and F_2 populations were pooled. *On the right:* the proportion of each hybrid group scores as "high" on two reactivity test measures. Straight lines between parental means indicate location of hybrid means if inheritance were additive. Additional lines connect parental means with the F_1 to visualize "dominance deviation."

mouse), but it is not useful to carry out such procedures on small samples from non-inbred parent stocks. We can safely conclude that genotype affects the probability of eliciting biting under mild restraint; but, beyond asserting the probable applicability of a threshold model, we suggest no specific mode of inheritance.

The final part score illustrated in Figure 12.7 is for tail wagging and is distributed discontinuously like biting. The proportions of the two parent stocks surpassing the threshold of stimulation are not greatly different, and the hybrid group scores are likewise similar to each other, with means differing insignificantly from that of the cocker spaniels. The incidence of tail wagging changed markedly over the three tests. Figure 12.8 demonstrates that the general relationship between the genetic groups remains roughly the same, although the proportions showing the behavior are decreasing.

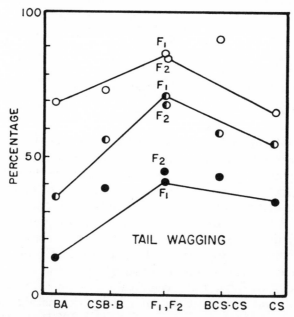

FIG. 12.8.—Proportion of each hybrid group rated as "high" in tail wagging at 17 weeks (*open circles*), 34 weeks (*half-filled circles*), and 51 weeks (*solid circles*).

Changes associated with aging appear to act uniformly over the variety of the genotypes tested. In general, the differences between the parent breeds are not as pronounced as earlier in development (see chap. 4).

It is not coincidence that most of the part scores in the hybrids follow the general trend of the total scores. The rating system was devised to yield high numerical scores for the dog which fought against restraint and which responded actively to each new set of stimuli. Low scores were assigned for inactivity and absence of physical evidence of arousal. The fact that the means for the total scores fall in an orderly sequence in the hybrids may signify that some common dimension of arousal vs. inhibition is manifested in a variety of ways. The precise manner in which arousal is expressed in the reactivity setting may not be as important as the level of activation attained.

TABLE 12.10

Analysis of Variance—Total Reactivity Scores of Hybrids

Source of Variation	Degrees of Freedom	Mean Square	F Ratio
Matings................	20	546.4	32.33**
Litters in matings.........	16	21.2	1.25
Within litters.............	143	16.9	...
Pooled error.............	159	17.35	...

** $P < .01$ $r_i = .736$

Further information on the contribution of heredity to the reactivity score was obtained by an analysis of the hybrid data by matings. The general form of this procedure has been described above. Reactivity scores for the three tests were summed for this analysis. Litters of a mating differed insignificantly from one another—evidence that the efforts to maintain constant rearing and testing conditions over the course of the experiment were successful. Variation attributable to matings was highly significant in each of the four sub-experiments into which we have grouped the data. The summary analysis of Table 12.10 was made without respect to genetic classification and includes data from 21 hybrid matings which produced 37 litters. The intraclass correlation of .736 is one of the highest found on any of the tests, much higher in fact than was found in the pure-breed comparisons. It should be emphasized that this high value comes from the classification of offspring by individual parentage and not by the breed classification of the parents. Although reactivity is heritable, different individuals from the parent strains pass on quite different genetic determinants to their offspring.

Had this been foreseen, we might have elected to disregard breed boundaries in planning matings and worked entirely with parent-offspring, sibling, and half-sibling correlations. The heterogeneity in phenotype of matings of the same type is illustrated for the reactivity score in Tables 12.11 and 12.12. Differences between matings of the same type were significant in these experiments: CSB ♂ × B ♀ or CSB ♀ ($P < .01$); BA ♀ × CS ♂ or CSB ♂ ($P < .01$); and CS ♀ × BA ♂ or BCS ♂ ($P < .05$). The number of different matings is too small to permit any conclusions on the relative con-

TABLE 12.11

Total Reactivity Ratings (Sum of 3 Tests): Means of Matings
of Purebred Dams by Purebred and F₁ Sires

IDENTIFICATION OF DAM	PUREBRED SIRE			F₁ SIRE			SIRES COMBINED
	Type	Number	Mean	Type	Number	Mean	Mean
BA-1515........	CS	11	16.9	CSB	5	17.0	16.9
BA-1517........	CS	5	20.8	CSB	11	18.7	19.4
BA-1757........	CS	2	20.0	CSB	8	12.2	13.8
BA-1759........	CS	4	18.0	CSB	18	17.7	17.8
Grand Mean			18.3			16.8	17.3
CS-1670........	BA	9	13.0	BCS	11	10.4	11.6
CS-1667........	BA	6	14.3	BCS	7	14.4	14.4
CS-1672........	BA	5	15.6	BCS	10	15.4	15.5
Grand Mean			14.0			13.2	13.6

tributions of dam and sire, but the data are consistent with equal contributions from both parents. The difference between reciprocal F₁'s could be interpreted as an example of maternal influence, but it is equally likely to be an example of sampling variation.

These data again confirm the conclusion that the breeds studied, though selected in part for behavioral characteristics, are still heterogeneous. Selection, of course, has succeeded in making breeds significantly different on the average, as we found in our breed comparisons.

TABLE 12.12

Matings of F₁ Sires by Purebred and F₁ Dams

IDENTIFICATION OF SIRE	PUREBRED DAM			F₁ DAM			DAMS COMBINED
	Type	Number	Mean	Type	Number	Mean	Mean
CSB 2036.......	BA	5	17.0	CSB	6	13.5	15.1
CSB 1993.......	BA	11	18.7	CSB	11	14.7	16.7
CSB 2506.......	BA	8	12.2	CSB	9	11.3	11.8
CSB 2507.......	BA	18	17.7	CSB	10	16.5	17.3
Grand Mean			16.8			13.9	15.6
BCS 1856.......	CS	11	10.4	BCS	6	14.0	11.7
BCS 2145.......	CS	7	14.4	BCS	17	15.9	15.4
BCS 2198.......	CS	10	15.4	BCS	9	14.7	15.0
Grand Mean			13.2			15.2	14.3

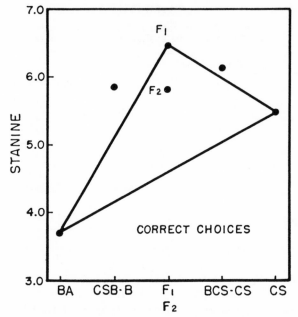

Fig. 12.9.—Mean correct choices scores of hybrids on the spatial-orientation test. Lines drawn as in Figure 12.7.

SPATIAL-ORIENTATION TEST

As with discrimination and delayed response it was necessary in the spatial-orientation test for subjects to meet a specified criterion during pre-training in order to obtain meaningful measures on the standard procedure. Since some dogs failed to meet the criterion, the comparisons between genetic groups were based upon selective sampling. The total number of subjects entering training for each group follows, with the number failing to meet the criterion for the test given in parentheses: basenji, 28(4); CSB × BA, 43(4); BA × CS, 24(0); CSB, 28(1); BCS × BCS, 32(1); CSB × CSB, 41(4); BCS × CS, 27(1); and cocker spaniel, 28(2).

The means in stanines of the groups for the three measures are shown in Figures 12.9 to 12.11. The cocker spaniels were definitely superior to basenjis in eliminating errors. All of the hybrid groups were superior on the average to the better parent. The persistence scores fell into a similar pattern except for the F_1's which were intermediate to the parents. All hybrid groups were similar to the cocker spaniels on the adjusted speed measure. Although this pattern of distribution of the means does not appear favorable for a quantitative analysis, we subjected the correct-choice scores to an

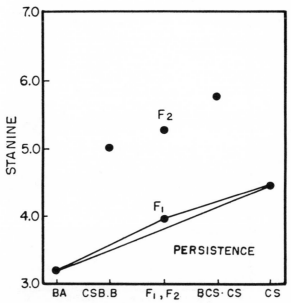

FIG. 12.10.—Mean persistence scores of hybrids on the spatial-orientation test. Lines drawn as in Figure 12.7.

analysis of variance to determine differences between matings and between repeated litters from the same mating. It was not surprising to find no significant effect of matings since the group means were so much alike. However, the differences between litters of the same mating were significant in three of the four separate analyses.

We conclude that something in the test situation or in the group life history exerted a powerful effect upon performance in the spatial-orientation test which overrode any inherited variations in learning ability. Since litter effects were not detected in other measures in which they were sought, it is probable that the effect in the spatial-orientation test arose from some factor peculiar to that test. The apparatus, because of its large size, was set up in the animals' home pens so that conditions of testing varied from litter to litter. The consequences of bad weather, disturbances from adjacent pens, etc. may be manifested in the analysis.

Despite the inappropriateness of the data for quantitative genetic analysis, we believe that the superiority of the hybrids over the basenji parents is a real genetic effect. The basenjis as a group were stereotyped in their responses, tending to repeat errors. They also were easily distracted by extraneous stimuli which increased their errors. When 25, 50, or 75 per cent of the genotype was of cocker

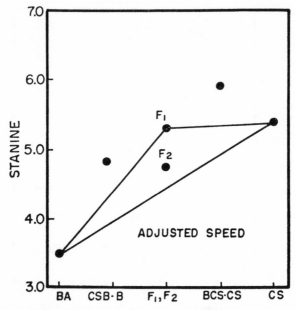

Fig. 12.11.—Mean adjusted speed scores of hybrids on the spatial-orientation test. Lines drawn as in Figure 12.7.

spaniel origin, the improvement was notable, although not correlated quantitatively with the proportion of spaniel genes. Again it seemed that when a threshold of timidity was surpassed, learning was reasonably adequate in all subjects. To be sure, individual differences were conspicuous and these may depend upon some heritable cognitive ability. The major group differences in solving the spatial-orientation problem, however, seem to be rooted in emotional reactivity to the test procedure (see discussion of "fear of apparatus factor," chapter 14).

In summary, methods of analysis which do not depend upon the assumption of genetic homozygosity, confirm the importance of hereditary effects upon the occurrence of simple patterns of behavior such as leash fighting and emotional reactivity. They do not, however, assist greatly in the analysis of genetic transmission of the complex adaptive behavior involved in problem solving. Important breed differences in problem solving exist, but the complexity of interaction between the basic capacities involved, both on the physiological and behavioral levels, is so great that these methods cannot cope with it.

Hereditary effects upon behavior related to ease of emotional arousal, specific emotional responses, and level of motivation were clearly demonstrated by the hybridization experiment. On the other hand, the capacities for various sorts of problem-solving behavior were not transmitted in a simple manner, and evidence for heritability was often weak or entirely lacking. It may be that learning ability is not heritable, and that the striking individual variations observed were reflections of differential experience. This seems unlikely, however, in view of the strong evidence of heritability, based on breed differences. We are reserving final judgment on this matter until selection experiments are performed using problem solving as the criterion. At present, we can definitely state that puppies resemble their parents more in "personality" than in intelligence, perhaps because primary gene products have more direct effects upon personality.

A GENERAL INTERPRETATION

The results of the hybridization experiment are at the same time illuminating and perplexing. The tendency of the F_1 and F_2 hybrids to receive scores intermediate to the parents, and for the backcrosses to be intermediate between the F_1 and the other parent was very strong. In contrast to tests cited in the last chapter, no evidence of segregation into distinct phenotypic classes was found in either the F_2's or the backcrosses, a result consistent with the interpretation that the spaniels and basenjis differed at multiple loci, each of which affected the character studied. Furthermore, the within-litter variances of the F_2 and backcrosses were not always greater than those of the F_1 and parents. Such a finding sometimes has caused consternation, but there appear to be three good explanations of why we should not expect the variances of the segregating generations to be larger than that of the parents and F_1. First, there is ample evidence that the basenjis and cocker spaniels were heterozygous at many loci which affect behavior so that the assumption of negligible genetic variation in these populations is unwarranted. Second, if many loci are involved, the proportion of extreme genotypes produced in the segregating generations will be low and the resulting increase in variance will be indistinguishable from sampling error in a small population. Third, the heritability of many characters is low enough so that a small increase in genetic variance is indistinguishable from sampling variance except in large populations involving many matings.

As suggested above, the findings indicate that a different experimental design involving a larger number of matings from a wider selection of breeds would have provided more precise knowledge concerning the genetic structure of dog populations. But such a plan would not have supplied as much information on the crossing of phenotypically distinct breeds which had been genetically separated for millenniums. The spaniels were, in general, superior performers on those tasks in which the terms "superior" and "inferior" are at all appropriate. In the training procedures, spaniels were more readily inhibited, which may be desirable or undesirable depending upon circumstances. On many performance tests, the hybrids were equal to or superior to the parent that was the more active or the better learner. We believe that a dual-process hypothesis will account for such findings. In the spatial-orientation test, the more reactive basenjis are strongly stimulated by small variations in the environment, and these variations interfere with such tasks as discrimination and the learning of a correct path. In these same tests some cocker spaniels show little distraction and also low spontaneous activity resulting in relatively poor performance. The hybrids of all degrees appear to be highly aroused like the basenji, and nondistractable like the cocker, so that performance is frequently superior on the average to either parent.

The characteristics of the two parental stocks are reasonably accounted for by past selection. Basenjis, living in villages in the African jungles, are given relatively little protection. Such circumstances place a premium on awareness of minor changes in the environment and excessive caution toward a strange object or individual. Cocker spaniels, on the other hand, have led a highly protected existence as English hunting dogs and as house pets. They were selected originally on the basis of great activity in hunting but ease of inhibition of attack on game. Though such selection has been relaxed in many strains, the popularity of the spaniel as a house pet is based in part upon its reputation for docility.

Determination of dominance relationships between the genetic systems responsible for these phenotypes (shyness vs. inhibition) could be made only if agreement could be reached on the scale for phenotypic measurement. Our data indicate that the shyness of the basenji may be dependent upon a relatively small number of genes with generally additive effects, and that a similar system involving different genes is associated with the inhibitability of the cocker spaniel. Such dual systems could explain the fact that on some measures the backcrosses were more basenji-like or more cocker-like than the purebred stocks. A basenji backcross homozygous or nearly so for shyness and also homozygous or nearly so for inhibition would perform less well perhaps than a purebred basenji which was less inhibitable.

We should stress that the basenji-cocker spaniel hybrids were physically and behaviorally excellent animals. The point is important because Stockard's (1941) statements regarding disharmonies in the F_2 of interbreed crosses have been cited as evidence for dangers inherent in racial crosses in man (George, 1962). Stockard, of course, started with bizarre types of dogs such as salukis, dachshunds, and bulldogs, which have achieved a rather perilous genetic adjustment as the result of long periods of selection. It is not surprising that jaws, limbs, and bodies of the F_2's did not fit together very well, although to a naïve Martian seeing dogs for the first time, the F_2's might appear no more strange than a purebred bulldog or a dachshund. As the photographs show, our backcross and F_2 groups included individuals which conformed to none of the types recognized as pure breeds, but which were nonetheless healthy, vigorous, and well adjusted.

GOALS FOR BIOMETRICAL GENETICS OF BEHAVIOR

This is perhaps a suitable place to record our change of attitude toward our experiments as we gathered data and subjected them to quantitative analysis. At the beginning of the hybridization experiment we were looking for genetic mechanisms to correspond with hypothetical traits. As data accumulated, it became clear that correlations between different tests given at different times and places were low—in other words, we found little evidence for pervasive traits affecting all aspects of behavior. To be sure, the factor-analyses of Royce (1955) and Brace (1961) uncovered many significant correlations, but these were highest between measures collected in the same test situation. Over the period of life with which we were concerned, trait structure was somewhat fluid, and genetic hypotheses regarding heritability were consequently specific to a particular age and type of measurement. Attempts to secure greater precision by substituting factor scores based on a combination of original scores were not successful. We somewhat regretfully came to the conclusion that classical biometrical models are not highly applicable to the inheritance of complex behavior in semi-natural populations (casually selected for behavior and not systematically inbred). If we can transfer to the dog the finding in mice that much of the genetic variation among hybrids is attributable to interlocus interactions (Fuller, 1964), we would expect relatively low parent-offspring correlations. Until better methods are devised, formal bio-metric analysis will be most fruitful in species in which numerous inbred lines exist, and in which large numbers can be bred at lower cost. In such situations the behavioral phenotype can be rigorously (and narrowly) specified, and the success of the biometrical approach has been amply demonstrated (Fuller and Thompson, 1960; Broadhurst, 1960; Bruell, 1962; Hirsch and Erlenmeyer-Kimling, 1962).

Broadhurst and Jinks (1961) have attempted the broader application of elaborate methods developed for the analysis of the effects of genetics on continuous quantitative variation in behavior, based on the techniques of Mather (1949). With these methods it is possible to estimate the amount of variation due to dominance, additive effects of genes, and interaction between them, as well as that due to environmental factors. Using a set of incomplete data from one of our problem-solving tests (Scott, 1954), they arrived at an estimate of heritability approximately twice as large as that obtained by us and based on breed differences alone. Such methods give us hope that the quantitative effects of heredity on behavior can be analyzed more exactly in the future. Because these methods depend upon empirical scaling of each individual test and hence become extremely laborious when applied to large bodies of data, they have not been applied here, although it would obviously be fruitful to use them on certain sets of these data and to compare the results with those obtained by other methods.

Can the results obtained in specialized laboratory populations be applied to larger, more heterogeneous groups such as dog breeds or mankind? In particular, we address ourselves to the interpretation of heritabilities. Heredity, in the broad sense, is usually assumed to explain differences between species. Occasional interspecies crosses are available for testing genetic hypotheses, but such results can have little application to natural species which are defined as individuals sharing a common gene pool. The variations in behavior between races or breeds of the same species are, parental influences aside, primarily genetic, but these are reflections of interallelic interactions (epistasis) as well as of additive effects of genes.

One should be cautious in applying quantitative figures for heritability outside of the specific context in which they were obtained. We question whether, once some genetic control of a character has been demonstrated, precise estimations of heritability have any great value except for selection programs with practical objectives. Here estimates of heritability for the criterion character obtained under conditions peculiar to the selection project have obvious significance.

To us, it appears that a species like the dog is most useful in behavior genetics for studies of genotype-life-history interactions. Such interactions need not be statistical abstractions from analysis of variance tables, but can be expressed in functional terms based upon observation and measurement of behavior at various stages in development. For this purpose, a group of purebred strains of divergent phenotypes (excluding monstrosities) can be extremely useful.

Results from such studies can be organized in the form of a matrix with varying numbers of rows (of breeds) and columns (of environments). Such a matrix is generated by any experiment comparing two or more breeds in two or more environments.

Our breed comparison study in one uniform environment formed a matrix with one column and eleven rows. For comparative purposes in future investigations, we recommend the addition of an additional row in the form of a 4-way cross, $F_1 \times F_1$, with each F, derived from a different pair of breeds. In such a population one obtains variable genotypes which facilitate correlational analysis. Furthermore, such a genetically diverse population can be used as a control for the generality of findings with the pure breeds.

For the future we also recommend additional columns of environmental treatments which will enable testing of the validity of various theories concerning the effect of early experience and training upon different genotypes. The standardized environment given our puppies was restricted in many ways, and different kinds of treatments might well result in the development of capacities hidden in this experiment.

CHAPTER 13

DEVELOPMENT OF PHYSICAL
DIFFERENCES AND THEIR RELATION
TO BEHAVIOR

INTRODUCTION

In previous chapters we have been concerned almost entirely with the development of behavior and the evidence that genetics produces behavioral differentiation and variation. This chapter is concerned with one aspect of the general problem of *how* genetics affects behavior: the effects of differences in physique. This is an important practical problem, for if there were a perfect correlation between physique and behavior, the task of the behavior geneticist would be enormously simplified. It is also an interesting theoretical issue, involving both the concept of pleiotropy, or manifold effects of the gene, and the application of the contrasting concepts of type and population to the analysis of variation.

Physique is the result of growth, a process which has only an indirect relationship to behavior. Some differences in behavior can be seen only when physical growth has reached a certain point, and differences in size may have important effects on certain motor capacities. However, the development of size and other physical characteristics can serve as a model for the action of heredity, partly because it is much closer to the physiological effects of genes, and partly because we know so much more about the inheritance of size differences.

PATTERNS OF GROWTH IN THE DOG

Growth in relation to periods of behavior development.—For the first 3 weeks of its life, a puppy normally receives all its nourishment by nursing from its mother. The milk supply does not increase as the puppy grows older, but reaches the limit of the mother's capacity a few days after birth. The general growth pattern reflects this limitation, and the neonatal puppy gains the maximum amount in the second week and slightly less in the third (Fig. 13.1).

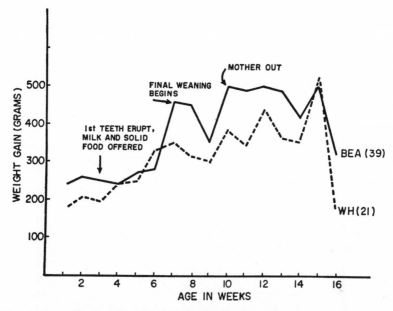

FIG. 13.1.—Absolute weight gain per week in male puppies of the largest and smallest breeds (beagles and terriers). Note that this measure is much more sensitive and variable than cumulative weight curves.

At 3 weeks of age, the puppies begin to take supplemental food. Under our special conditions of rearing, they now have access to an unlimited food supply, and this is reflected in their growth rates. From 3 to 7 weeks of age, there is an increase in the amount gained per week, the change taking place gradually or suddenly depending on how quickly the individual puppy or breed becomes adjusted to solid food. There is a temporary setback at the time of final weaning, but for several weeks thereafter the puppies gain a constant amount each week (about half a kilogram in an average male beagle). The dip at 16 weeks in Figure 13.1 is probably caused by reaction to distemper shots in some animals, as Corbin *et al.* (1962), report a constant increment up to 18 weeks of age.

These facts are reflected in conventional growth curves. There appear to be upper limits for the amounts of either liquid or solid food which a puppy can process, so that weight gain curves are flat up to 3 weeks of age and again from 7 weeks onward (Fig. 13.1). Since weight gain is not proportional to size, percentage curves decline throughout development. The cumulative weight curve up to 16 weeks, though superficially similar to a theoretical exponential growth curve, is actually composed of two straight lines connected by a curved portion reflecting increasing intake of solid food between 3 and 7 weeks (Fig. 13.2).

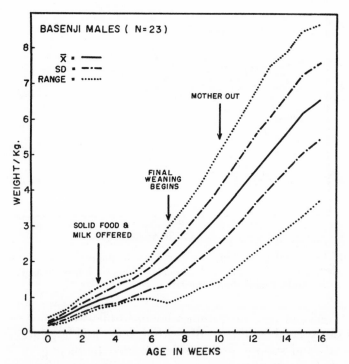

Fig. 13.2.—Combined growth curve for basenji males. Note how variation increases as the animals become larger.

For our puppies there was a second change in diet at 16 weeks, when they were put outdoors and no longer given supplements of milk and finely powdered dry foods. Most of the breeds showed a lessened rate of gain during the next two weeks (Fig. 13.4), but this may have reflected the fact that they were no longer drinking a quarter of a kilogram of milk before being weighed.

After 16 to 18 weeks, the increments of growth gradually decline.

By this time the puppies are more than half grown and beginning to approach adult capabilities, although they are just beginning to get their permanent teeth and would not under natural conditions be capable of effective hunting for another month or two.

This general pattern of growth is adapted to the life of a carnivore. A puppy grows as rapidly as possible until he reaches the size which enables him to be an effective hunting animal. This contrasts with the pattern of growth in human beings, in whom a period of rapid growth in early infancy is followed by a period of very slow and gradual growth up to shortly before puberty, when an adolescent spurt appears. There is nothing like the human adolescent spurt in the development of the dog, unless we consider the early period of rapid growth as such. It should be called a "puppyhood spurt," if anything.

Breed and sex differences.—There is always some overlap between individuals, but in every breed and at any age the males are on the average larger than the females (Fig. 13.3). By the time they reach one year, the males in beagles, cockers, and shelties are from 24 to 28 per cent heavier than the females. The differences in terriers and basenjis is much less, being 12 and 14 per cent, respectively.

These differences in growth curves suggest that the females may be maturing more rapidly than the males and hence slowing down in growth at an earlier age. However, if we compare the percentages of adult weight at 16 weeks (Table 13.1), it is obvious that the females are very little ahead of their brothers except in the beagles, where the figures are 61 per cent for females and 55 per cent for males. In basenjis the females are slightly behind the males.

TABLE 13.1

PERCENTAGE OF ADULT WEIGHT REACHED AT 16 WEEKS OF AGE

Breed	Males	Females	Adult Male/Female Weight Ratio $\times$ 100
Basenji..............	57	54	114
Beagle..............	55	61	128
Cocker.............	70	72	124
Sheltie..............	60	62	127
Fox terrier.........	62	63	112

There is one outstanding breed difference in rate of maturation: the cocker spaniels as a breed are considerably ahead of the other four, having already gained 70 to 72 per cent of their adult weight.

The shape of the curves in Figure 13.4 indicates that the dogs have not quite reached their maximum weight at 52 weeks of age. The

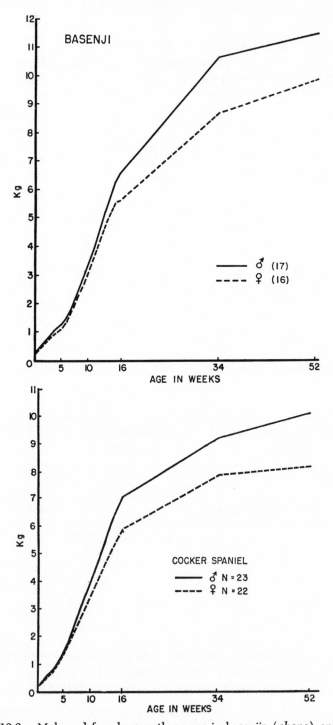

Fig. 13.3.—Male and female growth curves in basenjis (*above*) and cocker spaniels (*below*).

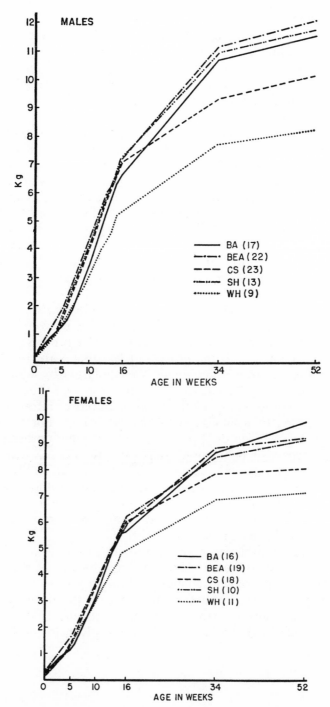

FIG. 13.4.—Growth curves of the five pure breeds. *Above,* males; *below,* females.

situation is somewhat complicated by estrus in the females, since a non-pregnant female may temporarily gain weight as a symptom of pseudopregnancy, a common occurrence in unmated females. This may account for the apparent discrepancy in the female basenji curve, since all females of this breed come into estrus around 10 months of age.

Differentiation of individual differences.—When we compare the weight curves of males and females, we see that there is very little difference in size at birth, and that they become more and more unlike as they grow older. This, however, is more a matter of change in absolute than in relative size, for if we analyze the variation of either sex within a breed, the standard deviation, a statistical measure of variability, fans out like a funnel as the animals grow older (Fig. 13.2). The upper and lower ranges of individual growth spread out even further. Initial differences in size become greater and greater as the animals grow, and individuals thus become increasingly differentiated from each other.

Can we take the differentiation of growth curves as a model for the differentiation of behavioral differences? The answer is that the model fits in some ways but not in others. The process of growth has certain similarities to the process of learning. Growth is an accretionary process, in which animals become larger by adding small amounts every day to that which they already have. In some respects learning is like growth, especially when small bits of information are gradually added to what has been learned before. For example, a child's learning new words is very much like a growth process. However, the process of behavioral adaptation is in some ways much more complex than growth, being complicated by the ability to make sudden jumps or reversals in understanding, and by the capacity to adapt to a uniform situation in a variety of ways. Furthermore, while animals may become increasingly different in certain behavioral traits, they may become more alike in others.

The partial resemblance between growth processes and learning has more importance than a simple analogy, for one of the ways in which heredity may affect behavior is through growth itself. As we saw in chapter 6, size has a definite effect on the development of dominance relationships. How important such effects may be is a question of fact, and the facts we have are given in the following pages.

THE RELATIONSHIP BETWEEN PHYSICAL
DIFFERENCES AND BEHAVIOR

There can be no doubt that physical differences can affect behavior. When differences are extreme, this statement hardly requires scientific proof. The short-legged dachshund runs more slowly than the long-legged greyhound, and the tiny Chihuahua is unable to climb upon a chair. The basic scientific problem here is whether or not physical traits *always* have an effect on behavior, and various theories have been proposed concerning this. One is the theory of pleiotropy, or manifold effects. In studying the genetics of the fruit fly, research workers discovered many cases in which there was one major effect of a gene and many minor ones. For example, Dobzhansky (1927) and Schwab (1940) studied the effects of various mutant genes on the shape of the spermatheca, one of the parts of the fruit fly skeleton. Both came to the conclusion that the majority, but not all, of the genes tested had an important effect on the spermatheca. From such facts several possible hypotheses can be drawn. At one extreme is the hypothesis that every gene will have some effect upon every process in the body. The logic behind this is that any gene is present in the body throughout life from the moment of conception and thus has the opportunity to affect all life processes.

At the opposite extreme we can state the hypothesis that each gene produces one specific primary effect and that pleiotropy, or multiple action, is directly related to primary gene action. In cases where primary gene action is well known, as in the synthesis of various chemicals necessary for life in the mold *Neurospora* and in bacteria, the usual situation is a long chain of chemical reactions, one gene being necessary for each reaction (Beadle, 1945). The gene can be thought of as one element in a genetic code whose function is to transfer information to a biochemical system. A change in a gene whose activity takes place early in the process may affect all the subsequent parts, and modification of this biochemical system may affect other systems dependent upon it, but all effects are traceable to a single primary effect.

To take another example, based on embryonic development rather than biochemistry: the dominant gene which in a homozygote produces the deformed polydactylous monster in the guinea pig affects almost every part of the body with such severity that the affected animals cannot survive beyond birth (Scott, 1937). However, all these effects can be traced back to an acceleration of growth for a few days in a particular stage of early development. All the organs growing rapidly at that time are affected. Thus the many actions of this gene can be traced back to a single primary effect.

These two theories are not necessarily mutually exclusive. There may be certain genes which have effects on all body processes and others whose effects are highly specific. The principal difference between the two hypotheses is a practical one. The first hypothesis assumes that the inheritance of behavior will be affected by every gene affecting a physical trait. This would make the study of behavioral heredity much simpler because the converse would also be true, and every behavioral difference would have an associated physical difference which could be used as a marker (Keeler, 1942). On the other hand, if the second hypothesis is correct, there will be many genes affecting physical traits which have little or no effect on behavior. Likewise, many behavioral effects could occur without visible changes in form or shape. Now let us examine some of the evidence as it emerged from our dog experiment.

The effect of hair length on behavior.—Hair length is a common variable trait in dogs and short hair usually has been reported to be inherited as a clear-cut dominant (Dawson, 1937). Cocker spaniels are uniformly long-haired dogs, and animals bred for the show ring are deliberately selected for long hair, particularly on the legs. On the other hand, basenjis are uniformly short-haired dogs. We measured the length of the hair on the back between the shoulder blades of all animals at one year of age. As can be seen from Figure 13.5, the results show considerable variability within the pure breeds, and particularly within the long-haired cocker spaniels. Part of this variability is produced by shedding the hair coat. One can get considerable differences in measuring the same animal, depending on whether the coat is measured at its maximum length or shortly after shedding has begun. Nevertheless, the shortest haired cocker spaniels had coats at least a centimeter longer than the longest haired basenjis.

The picture is also complicated by the fact that males tend to have longer hair than females, particularly in the long-haired cocker spaniels. This is true also in the wild Canidae, where the males develop a longer ruff than the females.

The F_1's of the cocker spaniel-basenji cross turned out very much like basenjis except that a few individuals had slightly longer hair. On the other hand, there was only one case of overlap between F_1's and cocker spaniels, that of one F_1 male with the same hair length as a cocker spaniel female. There was no overlap between like-sexed

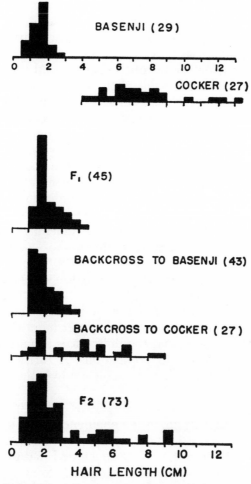

Fig. 13.5.—Inheritance of hair length. A single dominant gene will account for most of the observed variation, although there are probably modifying genes present in the cocker breed. Hair length was measured on the back between the shoulder blades.

individuals. Using these relationships as a guide, we classified the animals in the other hybrid populations. In the backcross to the cocker spaniels there were 14 long-haired and 14 short-haired dogs, corresponding exactly to the expected 50–50 ratio. In the F2 generation there were 16 long-haired dogs and 57 short-haired ones, corresponding quite closely to the expected ratio of 18.25 to 55.75. We concluded that the data strongly supported the findings of other authors that the inheritance of hair length is determined by a single gene in which short hair is dominant over long hair. As indicated above, the recessive gene may have some effect in the heterozygote, a few F_1's being longer haired than their basenji parents, and there is also an indication that other genes having a minor effect on hair length are present, since none of the long-haired cocker backcrosses developed coats as long as some of the purebred cockers.

This gave us the opportunity to see whether short hair had any effect on behavior. We had strong a priori reasons for supposing that it would do so, since a short-haired dog is obviously more affected by cold and less protected against physical injury than a long-haired one. We therefore examined the records of long- and short-haired dogs in the backcross and F_2 generations in relation to the 10 physical measurements and 40 behavioral measurements which Brace (1961) had used in his factor analysis.

In assessing such results, one has to remember that differences can occur simply by chance when finite numbers are used. We would expect that at least one difference out of 20 would be statistically significant at the .05 level simply as a matter of chance. In 50 variables chosen at random, we ought to get on the average, 2.5 such results purely by chance. On the other hand, if there is a real and important effect of the gene, we should get more than this number of significant differences, and the differences should be in the same direction in the two hybrid populations. A further test would be whether a reasonable mechanism could be hypothesized for the observed effect.

Among the behavior variables there was one significant difference in the backcross population, in the obedience test. However, the difference in the same test in the F_2 generation was in the opposite direction and not significant, and we concluded that this apparent difference was due to an accident of sampling. Among the F_2's, which we had in larger numbers, there were two significant differences (Fig. 13.6). One of these again showed an opposite difference in the backcross generation and was probably accidental. However, there was one highly significant difference, in the confinement test. Although the difference was not significant in the backcross, it was fairly large and in the same direction. In the confinement test, the dogs in a litter were confined outdoors in small wire-bottomed pens and graded on all forms of activity while a litter mate was removed, handled, and fed. The results indicated that short-haired dogs were more active under these circumstances than long-haired dogs, which is a reasonable conclusion, because the short-haired dogs would have been considerably less comfortable in cold weather. This appears to be a real effect of a physical trait on behavior.

We also compared the long- and short-haired dogs on their physical measurements. The long-haired dogs averaged larger in every measurement in both backcross and F_2 generations. All the differences were in the same direction, but only one of these was statistically significant. Such a result could be easily explained by the fact that all the measurements were made without shaving the dogs, and thus the measurements of long-haired animals would be slightly larger, even in the case of body weight. Another possibility might have been that there were accidentally more males in one group than another, males being both larger and having longer hair. This

BACKCROSS TO COCKER

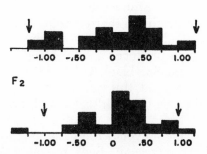

FIG. 13.6.—Distribution (in stanines) of differences between long- and short-coated animals in relation to 34 behavioral variables (omitting those confined to 2 categories). Arrows indicate approximate standard error of the difference for most variables. Long or short hair has little effect on behavior.

was not the case in the F_2 generation, in which males and females were almost evenly distributed. We can conclude that long hair may possibly have an effect on size measurements, but if so it is probably an artificial one resulting from slightly less accurate measurement of the long-haired dogs. In any case, the effect is not an important one.

Despite these negative results, there is no doubt that long hair may have an effect on behavior in special situations; e.g., a short-haired dog will hesitate to sit down on snow or ice. The experiment demonstrates that there is no important *general* effect of the gene for short hair upon behavior. If it is true that every gene affects all activities of the body, the effects of this gene must be so small as to have little, if any, importance and in fact be unmeasurable. The result supports the second hypothesis, that of specific primary gene action.

The effect of hair color on behavior.—While we were conducting our experiments, Dr. C. C. Little (1957) was making extensive breeding tests to determine the inheritance of color in a large number of breeds, including those used in the behavioral experiments. While various colors are permitted by the basenji breed standards, the animals of our strain were all pure red as adults (except for white spotting), but the puppies were always born with black hairs mixed with the red, the black being lost at about 5 or 6 weeks of age.

Crosses with other breeds established the fact that this color is produced by the same gene as that which produces the color commonly called "sable" in collies, whose long hair is reddish or yellowish at the base and black at the tips. Modifying genes in basenjis cause the black tips to be lost in adults.

The gene a^y which produces the red color in basenjis is an intermediate member of an allelomorphic series, the most dominant gene being A^s, a gene which produces a solid black color, and the least dominant being a^t, producing black with reddish or tan points. While some of the cockers were red, this was the effect of another recessive gene, e, which produces a clear red color with no black at any time in life.

Consequently, the basenjis had the genetic constitution $a^y a^y$, the cockers $A^s A^s$, and the F$_1$'s, $A^s a^y$. Backcrossing the F$_1$'s to basenjis gave two colors in approximately equal numbers: black ($A^s a^y$) and red ($a^y a^y$). This gave an opportunity to compare the behavior and physique of two groups of animals, one having the gene A^s and one without it.

This analysis gave no statistically significant differences in any of the physical variables, although the group of reds averaged slightly larger than the blacks on most measurements. This result was probably accidental. There was one significant difference in the behavioral variables ($P = .05$), but the same test in the F$_2$ population showed a difference in the opposite direction, and not significant. We may conclude that there is no general pleiotropic effect produced by the gene A^s.

This does not mean that it is impossible for color genes to have effects on behavior. There is a color mutation in the house mouse which its discoverer named "varitint waddler" (Cloudman and Bunker, 1945). This animal has a peculiar multicolored spotting combined with a limping gait resulting from some defect in the nervous system. Likewise, the "merle" gene, found in several dog breeds, has the effect of changing black color to a light gray with black spots. This color is much admired among the fancy, particularly in shelties, and is produced by a dominant gene, M. Homozygotes are produced which are almost pure white and often have defective eyes and ears. These defective sense organs, of course, cause differences of behavior. We reared one such animal in our testing program. It had small eyes, one of them being almost completely sightless, so that it did not respond to vision on that side. It was also deaf, but this seemed to have the effect of making it less easily distracted, so that in many tests it actually did better than normal animals.

Since the pigment cells originate in embryonic development from the crest of the neural tube, it is likely that the primary effect of the gene is to alter the development of the neural tube and related sense organs, this in turn affecting the distribution of the neural crest cells which eventually produce pigment.

We can conclude from these data that at least some color genes have no demonstrable effect on behavior. Although certain other color genes may have effects on behavior, these effects are associated with defects in the development of the nervous system.

The effects of body size and proportions on behavior.—Physical build and behavior have long been supposed to be associated. According to tradition, fat people are jolly, thin people are sour and dyspeptic, and heavily muscled men are quick to anger. The most ambitious attempt to investigate this tradition was undertaken by W. H. Sheldon, who developed a technique for estimating body proportions which has been widely used in anthropology and by departments of physical education. Sheldon measured the body in terms of three components, each measured on a 7-point scale. The first, endomorphy, is based on the amount of body fat. The second, or mesomorphy, is a measure of heavy bones and muscles; and the third, ectomorphy, is the tendency to have a high proportion of skin (actually, to be thin) with light bones and muscles. The technique of measuring body types, or somatotypes, in this way is to take photographs of the nude body in standard positions, blacking out the face for reasons of anonymity. The body type is then established on a 7-point rating scale for each characteristic.

Sheldon's technique can be criticized on the grounds of both subjectivity and logic (Hammond, 1957). The accuracy of the scales depends a great deal on the experience and skill of the person using them. There are logically only two independent components, since ectomorphy is essentially an absence of the other two. Howells

(1952) has shown by a factorial analysis of numerical measurements made of men designated as extreme somatotypes, that there is actually only one scale, endomorphy–ectomorphy, and one component, fat, that can be related to Sheldon's system. However, the chief interest to us lies in the fact that Sheldon and Stevens (1942), on the basis of many case histories, stated that they found a correlation between somatotype and temperament. As we have said before, physique obviously affects behavior. A boy with strong, heavy muscles, for example, is more likely to enter a sport like football or boxing than a boy with slight and wiry build. Sheldon and Stevens went beyond this, implying that the same factors which produced physical build affected behavior independently, producing traits which had nothing to do directly with physical build. This is essentially the same theory of general pleiotropy which we have discussed before.

Our long-term experiment with dogs seemed like an ideal opportunity to test Sheldon's theory of the effects of somatotypes on behavior. Charles Shade, of the Harvard University Department of Physical Anthropology, spent a summer devising a system for making body measurements on dogs. He found first of all that Sheldon's system of photographs was impractical with dogs, since the fur makes so much difference in physical appearance. Every animal would have to be shaved for this system to work. Furthermore, it seemed desirable to try to improve on Sheldon's system and devise more objective measurements. Shade finally developed a series of seven general measurements which could be done with reasonable accuracy, and we later added three special measures of bone and muscle size to make a total of ten.

The first two of these were over-all measurements of body size: weight, and body length from the anterior end of the breastbone to the posterior point of the hips. The accuracy of this last measure depended on being able to get the dog to stand with its body in a straight line and in a natural position. Then there were three measurements of the chest region of the body: brisket depth (measuring the chest from top to bottom), chest breadth, and chest circumference. The abdominal region was measured by waist circumference. Heaviness of bone was measured by taking the diameters of the femur and humerus, the large bones of the hind- and forelimbs. Heaviness of muscles was measured by taking the circumference around the thigh, thus measuring one of the largest muscle groups in the body. Finally, leg length was taken by measuring the standing height at the shoulder. Thus, if there were dogs with unusual

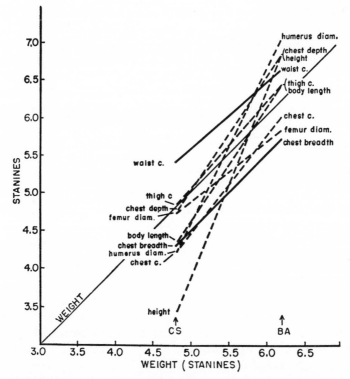

FIG. 13.7.—Relative proportions of cocker spaniels and basenjis. If the two breeds had the same proportions, all the lines would be parallel to that of weight. Actually, only waist circumference and chest breadth are approximately similar.

If the basenjis were simply thinner dogs, all lines would be parallel but run upward with respect to weight. Actually the lines run in different directions and the measurements occur in different orders in the two breeds (rank order correlation = −.17). Basenjis are relatively taller, and have a thicker humerus, deeper chests, and longer bodies than spaniels.

amounts of fat, muscle, or bone, these measurements should reflect such traits.

We began taking the full set of measurements at 17 weeks of age, and repeated these at 34 and 51 weeks. We took the last measurements to represent the final physical build of the adult dog, and analyzed them in various ways.

When we take the average of each one of these measurements for each pure breed and compare it to the average weight of the breed, we find that there are indeed differences in proportions (Table

13.2). The shelties show the greatest number of extreme deviations and could be described as having large and deep chests, thin waists, small thighs, small bones, long bodies, and long legs. In many respects they correspond to the ectomorphs of Sheldon's somatotypes.

TABLE 13.2

RANK FROM LARGEST (1) TO SMALLEST (5) IN BODY MEASUREMENTS
RELATIVE TO WEIGHT, IN THE FIVE PURE BREEDS

	CIRCUMFERENCES			LINEAR MEASURES						TOTAL EXTREME RANKS
BREED	Chest	Waist	Thigh	Body Length	Height	Chest Breadth	Brisket Depth	Femur Diam.	Humerus Diam.	
Basenjis.....	3	2.5	2	3	2	4	3	3	1	1
Beagles......	4	4	4	2	5	1	4	4	4	2
Cockers.....	5	2.5	3	5	4	5	5	5	3	4
Shelties.....	1	5	5	1	1	3	1	1	5	8
Fox terriers..	2	1	1	4	3	2	2	2	2	3
Range (stanines)	1.0	0.9	0.9	1.1	1.9	1.3	1.3	1.3	1.1	...

None of the other breeds shows such extreme deviations. Wire-haired terriers are thick waisted and have a heavy thigh and femur, indicating that they have large muscles and thick bones which would be useful to a breed used for attack. They might be labeled mesomorphs, though not of an extreme type. Cockers are relatively small chested and short-bodied in proportion to weight. Beagles are relatively broad chested and short-legged, but otherwise undistinguished. However, this may have been caused by one of the foundation males having these proportions, rather than being characteristic of the breed as a whole. Basenjis, although they have a distinctive appearance, are outstanding only in having a heavy humerus.

To compare the two breeds used in the cross, basenjis are larger in almost every dimension than cockers, both in absolute size and size relative to weight. This last means chiefly that basenjis are relatively thin and cockers relatively fat. The two breeds are most different in respect to height; the next most important differences are in the depth and size of the chest, and body length. Thus basenjis are tall, thin, deep chested and long-bodied, while cockers are relatively short, fat, shallow chested and short-bodied. The differences in proportion are not extreme, and in Sheldon's system the basenji would probably be a "mesectomorph," and the cocker a "mesendomorph."

After measurements were collected for all hybrids as well as pure breeds, another physical anthropologist, Loring Brace (1961), began his detailed study by the method of factorial analysis. He first converted all measurements into stanines. As described in chapter 7, this is a method often used in psychology to convert measurements into a form in which they can be easily analyzed by the methods of normal curve statistics, including standard deviations and correlation coefficients.

Brace chose 40 behavioral measures to compare with the 10 measurements of physical size, selecting those which seemed to show the most clear-cut breed differences and indications of Mendelian inheritance. Some of these measurements came from the same tests, but these were ones which appeared to be at least partially independent of each other. There were 13 separate behavioral test situations represented. Brace thus had 50 measurements on each dog in five different pure breeds, plus six hybrid populations between the basenjis and cockers, as basic data.

Factor analysis is a mathematical method based on correlation. The logic behind its use was that if there were several different behavior traits or physical measurements affected by the same genes or group of similar genes, then each trait should be correlated with all the others, appearing together in the same animal and being measurable by correlation coefficients. The method itself is designed to sort out groups of measurements which have a common correlation, and the presumed cause behind such a group is called a factor. If there is any truth in Sheldon's contention that somatotype is related to behavior, we should get factors including both physical measurements and behavioral measurements.

The first mathematical step was to correlate each of the 50 variables with each other one. This meant calculating a total of 1,225 correlation coefficients, along with the means and standard deviations for each measurement. Brace decided to calculate such tables for each breed and hybrid population, and by using modern calculating machines, he was able to calculate 50×50 correlation tables on 16 different populations of animals in a single night's work. This meant a total of 19,600 correlation coefficients. This, however, was only the beginning. The communality of correlation still had to be determined by the process of rotation, which is equally laborious. This, however, could be done by programming computing machines, and after several months' work, Brace was able to get the machines set up so that they would do the job accurately. In so doing, he was able to use a mathematical method (that of principal axes) which had hitherto been impractical because of its laborious nature.

We may now consider factor analysis in relation to genetics. The method is based on correlation, which in turn is based on the assumption of additive effects of causal factors. Studies of growth and physical size have generally indicated that genetic factors affecting size are additive; that is, each gene makes the animal a little bigger. On the other hand, as we have shown in other chapters of this book, complicated interaction between genes is a very common phenomenon affecting the expression of other kinds of physiological and behavioral traits. The mathematical logic behind the use of the correlation coefficient assumes that the same gene always produces the same effect, and this may not be true in the case of genes affecting behavior.

Assuming, however, that we can get good additive effects, there are two ways in which heredity can produce correlation in addition to the case of a single gene's effect on different traits. One of these is the presence of two genes on the same chromosome, producing the genetic phenomenon of linkage. The other is the simultaneous selection of several different traits in a single breed. We would expect that in any pure breed there would be a number of different genes brought together by selection. Hence, in a population of several pure breeds we would expect to get groups of artificially correlated genes which would appear in a factor analysis as breed factors.

In segregating populations such as the backcross or F_2 generations, we should get correlations between all of the effects produced by each segregating gene, or by each group of closely linked genes. The F_2 population would provide the best opportunity for such correlations to show up, because the effects of all segregating genes would appear, as they might not in a backcross to a dominant parent population.

We would also expect to get different factors in the different segregating populations because of dominance, which would completely suppress the action of a recessive gene in a backcross to a dominant. There should, therefore, be fewer common factors between backcrosses than between backcrosses and F_2's.

On the other hand, if we are dealing with non-genetic correlations, the same factors should appear in all populations. We would expect that at least some of these non-genetic factors would turn up because of the fact that more than one measurement was taken from certain tests. Some measurements could be correlated simply because they came from the same tests and were done on the animal at the same time.

The correlation between physical size and behavior.—The F_2 population provides the best chance for detecting correlations due to common genetic factors. Without going through the entire factor analysis, we can get some idea of the effect of physical measurements on behavior by summarizing the correlation matrix (Table 13.3). In interpreting these results, we must remember that a certain number of correlations may occur simply by an accident of random sampling. In our population of 72 F_2 hybrids there is 1 chance in 20 that correlations larger than .23 will appear simply by accident.

TABLE 13.3

SIGNIFICANT CORRELATIONS BETWEEN PHYSICAL MEASUREMENTS AND
BEHAVIORAL VARIABLES IN PURE BREEDS AND HYBRIDS

BREED OR HYBRID	NUMBER	.05 LEVEL OF SIGNIFICANCE	CORRELATION BETWEEN PHYSICAL VARIABLES		CORRELATION BETWEEN PHYSICAL AND BEHAVIORAL VARIABLES	
			Observed	Expected	Observed	Expected
Basenji............	22	.420	15	2.25	26	20
Beagle............	24	.405	41	2.25	12	20
Cocker............	23	.410	35	2.25	40	20
Sheltie............	16	.496	45	2.25	47	20
Terrier............	15	.510	19	2.25	43	20
Total........	100	...	155	12.25	168	100
All Pure Breeds....	100	.194	45	2.25	137	20
BCS F_1...........	24	.404	37	2.25	22	20
Backcross to cocker	28	.374	42	2.25	42	20
BCS F_2...........	32	.339	45	2.25	28*	20
CSB F_1...........	28	.374	42	2.25	28	20
Backcross to basenji	41	.304	41	2.25	37	20
CSB F_2..........	41	.304	45	2.25	29*	20
All F_1........	52	.270	43	2.25	44	20
All F_2........	73	.229	45	2.25	24	20
P and F_1......	97	.200	45	2.25	138	20

* Correlated tests are not identical in these two groups.

The physical measurements in the F_2's are all highly correlated with each other. The 45 possible correlations range from .90 to .40, with most of them toward the higher end. There is no doubt that size measurements are correlated with each other.

On the other hand, the correlations between physical and behavioral measurements are all smaller, the greatest one being —.36. Each of the physical measurements could be correlated with the 40 behavioral measurements merely by chance 1 time out of 20, or a total of 2 times. The 10 measurements should therefore show 20 significant correlations merely by chance, and the observed number is 24. Although the possibility exists that some of these are real correlations, there is no evidence that there is any strong or important effect of physical size on the 40 behavior measurements chosen for analysis.

Since the physical measurements are correlated with each other, we should expect them to be correlated with the same behavioral measurements whether or not this was a matter of chance. This was indeed the case, and 18 of the 25 significant correlations relate to 4 behavioral tests: habit-formation, vocalization in the reactivity test, number of errors in the barrier test, and playful aggressiveness in the handling test. Some of these, like playful aggressiveness, might reasonably be effects of size differences. However, the low correlations and the small excess of significant correlations beyond chance expectations lead to the conclusion that the effects of physical size on the behavior tests are minor ones, if they exist at all. In short, variation in physical size has little if any effect on the variation of behavior in the F_2 population.

When we look at the results of correlating physical and behavioral variables within the pure breeds, we get a somewhat different impression. As Table 13.3 shows, the basenjis show slightly more than

TABLE 13.4

FACTOR LOADINGS OF PHYSICAL VARIABLES IN COMBINED F_2 POPULATION, PRINCIPAL AXIS ROTATION*

Measure	Factor Loading	Quality Measured
Body weight..............	.94	General size
Chest circumference......	.94	General size of thorax (largest part of body, much of viscera)
Thigh circumference......	.90	Size of hind limb (muscle, bone)
Waist circumference......	.86	Size of abdominal region (fat, intestines, lumbar muscles and vertebrae)
Brisket depth............	.83	Linear measure of thorax (vertical)
Chest breadth............	.82	Linear measure of thorax (horizontal)
Body length.............	.82	Linear measure of entire trunk
Humerus diameter........	.82	Thickness of bone, anterior limb
Femur diameter..........	.81	Thickness of bone, posterior limb
Standing height..........	.64	Linear measure of forelimb (vertical)

* The loadings indicate the relative efficiency of each variable as a measure of general size and fall into 4 groups.

the expected number of significant correlations and the beagles definitely less. The other three breeds each show twice the expected number of significant correlations. Within certain of the pure breeds, it looks as if physical size might have some effect on behavior.

Such correlations might in part be caused by the fact that the physical measurements themselves are correlated. We can test this possibility by examining the sample of physical measurements independently. Actually, any one of the physical measurements shows more than the number of correlations expected by chance. There is thus a strong possibility that physical size has an important effect on behavior against the genetic backgrounds peculiar to certain breeds. A difference in size might be expected to produce a relatively consistent effect on animals showing similar behavior and thus show up as a correlation. However, an alternate explanation of the results might be that there are individuals or strains within a pure breed in which physical measurements and behavioral measurements are accidentally correlated, and this possibility cannot be entirely ruled out by the data.

Relationships between the physical variables.—The factor loadings in the F_2 populations (Table 13.4) show how each measurement is related to general size. Weight has the highest loading, followed by the three measures of circumference, with all linear measures falling below these. This is a reasonable result, since weight combines all dimensions of the body, a circumference combines two, and a linear measurement only one.

The loadings thus fall into four groups relative to each other. Body weight and chest circumference have the highest loadings on the general size factor, forming the first group. Body weight measures the entire size, whereas chest circumference measures the size of the thorax, the largest part of the dog's body, which includes many of his internal organs such as the heart, lungs, liver, etc. In the next group are thigh circumference, which measures the size of the hind limb, and waist circumference, which measures the size of the abdominal region. In the third group are five measurements which have almost identical factor loadings. The two chest measurements are highest and the two bone measurements the lowest. Finally, standing height, which measures the length of the forelimb, falls into a group by itself, with a very low loading compared to the rest.

The factor analysis indicates that any measure of the body is strongly related to general size, and thus that all the dogs in the F_2 population have much the same proportions. The only exception is the measure of height, which may be disproportionate to the rest. There is no indication of disproportion in any special tissues, such as fat, bone, or muscle, in this population.

General results of the factor analysis.—Brace used two methods of factor analysis on each population. The principal axis method is the ideal mathematical method (Thurstone, 1947) and produces a group of measurements related by common correlation and weighted according to the importance of each. This weighting, or factor loading, can be either positive or negative, and can vary from zero to one. Because so many measurements appear in each factor, the results are sometimes hard to interpret. The varimax method (Kaiser, 1958) is used to emphasize those measurements having the largest effect and to eliminate the unimportant ones, and so aid in the interpretation of the factors. Only the results of the principal axis method are considered here.

As we stated above, the F_2's provide the critical test of the hypothesis that physical measurements are associated with behavior, since in this population genetic variation should occur freely in all respects, with a minimum amount of correlation due to linkage and a minimum effect of dominance. As we might expect from the basic correlation table, only one behavioral measure has a loading of any appreciable importance on the general-size factor. Good performance on running to a goal in the habit-formation test is weakly associated with large size (.29). This is a reasonable effect, but so small that it may well be accidental. Furthermore, none of the physical measurements have loadings on any of the other four behavioral factors. We can only conclude that the physical measurements have no important connection with behavior in the cross between these two breeds.

TABLE 13.5

Factor Loadings of Factor 1 (General Size)
in Combined Pure Breeds (N = 100)

Variables	Factor Loading
Physical:	
Chest circumference......................	.88
Brisket depth...........................	.87
Weight................................	.77
Body length...........................	.76
Standing height........................	.74
Humerus diameter......................	.73
Waist circumference....................	.73
Thigh circumference....................	.71
Chest breadth.........................	.65
Femur diameter.......................	.64
Behavioral:	
Few errors, **U**-shaped barrier.............	.48
Fast speed, long barrier.................	.44
Few errors, long barrier.................	.43
Few noises, **U**-shaped barrier.............	.37
Success, manipulation, string pulling.......	.36
Success, manipulation, moving box.........	.36
Fast speed, maze.......................	.35
Active while weighed....................	.30

When we look at the other populations, we find that in every one the physical variables are strongly correlated and that the factor analysis consequently assigns them high loadings on a factor of general size. The segregating populations are of particular interest in determining the relationship of this factor to behavior. In the back-cross to the cocker, the heart rate—both of puppies and adults—is associated with size, large dogs having a slower rate. The same thing is true in the backcross to the basenji, but this effect does not appear in the F_2's. The most logical explanation is that it is an effect of linkage upon two traits, neither of which is affected by domi-nance. In the backcross to the basenji, there are three other be-havioral variables associated with size, but the loadings are low and lack a logical explanation. These last results are therefore probably accidental.

In contrast to the segregating hybrids, the pure breeds, both to-gether and separately, show a large number of behavioral variables associated with size (Tables 13.5 and 13.6). These seem to be differ-ent in every breed, and the varimax rotation has little effect upon them. It looks as if size may have an effect on the behavior of animals which are similar to each other otherwise, but that the effects are not consistent. In other words, size does not produce the same effect on the behavior of a sheltie as it does on that of a fox terrier.

TABLE 13.6

FACTOR LOADINGS OF FACTOR 1 (GENERAL SIZE)
IN SHETLAND SHEEP DOGS AND FOX TERRIERS

Variables	Shelties	Fox Terriers
Physical:		
Chest circumference........................	.98	.83
Femur diameter........................	.98	...
Brisket depth........................	.97	.93
Weight........................	.95	.66
Chest breadth........................	.94	.57
Humerus diameter........................	.89	.68
Waist circumference........................	.89	...
Body length........................	.87	.80
Thigh circumference........................	.81	.62
Standing height........................	.77	...
Behavioral:		
Manipulation, moving box, success.........	.64	...
Manipulation, string pulling, success........	.63	...
Little vocalization, leash control...........	.63	...
Little vocalization, reactivity..............	.60	...
Much vocalization, U-shaped barrier........	.56	...
Fighting leash...........................	...	.68
Active in confinement....................	...	.60
Obedience, little docility..................	...	.72
Few errors, long barrier test..............	...	.65

Of course, it may be that these correlations are produced by accidental association of size and temperament among the offspring of different matings within the same breed. This possibility can be tested by comparing the results of reciprocal crosses between the two breeds, since the parents in a cross between male basenjis and female cockers are necessarily different individuals from those in the reverse cross. In the two F_1 populations there are 13 behavior tests which have moderately high loadings on factor 1. Of these, only two are found in both populations. This would mean that most of the apparent effects of physique on behavior are probably caused by accidental association but that some effects may be real.

These results correspond to those of Humphrey and Warner (1934), who correlated 42 measures of physique with 9 measures of behavior within the German shepherd breed. They obtained only 15 correlations out of a possible 378. The most that we can say is that there is a possibility that size has varying effects on behavior depending on the genetic constitution of the individual involved, but that this is at present only an interesting hypothesis.

With regard to the other results of the factor analysis, one basic hypothesis to be tested was that there were a few basic genetic traits of behavior which might affect the results of many different tests and so run through the whole life of an animal. If such traits represented the action of different genes, then one would expect to get different factors in different populations, depending both on the presence or absence of these genes and also upon their masking by dominance. Instead of this, Brace got the opposite result: much the same factors appeared in every population. There was one factor emphasizing physical measurements, another emphasizing physiological measurements, and two others which included behavior measurements. Of the latter, one included most of the performance tests, and the other was strongly related to tests of emotional reactivity.

All this suggests that similar kinds of behavior and physiological activity are closely related only to themselves. Here again the data disagree with the type concept; the behavior of an animal in one situation is not "typical" for his behavior in another situation. A more detailed study of the behavioral aspects of the factor analysis will be given in another chapter.

Thus the results of this experiment are almost completely negative with regard to the hypothesis that a "somatotype" is strongly correlated with temperament and behavior. It can, of course, be argued that such effects might appear in a cross between two breeds contrasting more sharply in physical build than do cockers and basenjis.

However, there are good theoretical reasons for believing that genetic factors affecting body build and behavior are largely independent of each other.

Evaluation of the somatotype concept.—We can make a basic theoretical criticism of the somatotype approach to body build and behavior on the ground that it represents a primitive and now almost discarded method of biological study. The typing method was used by the early naturalists in their description of animal species. They first selected an individual for very careful description and called this the "type" of the species. Often the individual was selected on the basis of intuition or simply because it was available. Then they attempted to describe the species as a whole as compared to the individual "type." This method has obvious drawbacks, and it resulted in the artificial creation of many species which did not exist in nature and were later shown to be part of the same population.

The typing method can give a very misleading picture of a population. For example, the type specimen selected for Neanderthal man turned out to be an old man probably suffering from arthritis, with knobby bones and stooping posture. Later collections of other specimens show that many Neanderthal skeletons are indistinguishable from those of modern-day Europeans (Brace, 1962).

The modern approach of naturalists is to measure and describe the population first and then to describe the individual in terms of the population. This means the collection of measurements on a reasonably large sample of the population, from which estimates of the average, standard deviation and variance can be made. In this experiment we have used the population method, but in such a way that types can be recognized if they do in fact exist. If a population is divided into four or five types instead of showing continuous variation, we should get clusters of correlated traits by the factor-analysis method. If physical type is associated with behavioral type, then some physical measurements should be associated in one group by themselves, with different sets of physical measurements in other groups. This did not happen in the factor analysis, and hence all the evidence which we have collected is against the existence of such types.

Dog breeders, in changing and improving their breeds, have used the type method, selecting an individual as a grand champion and then attempting to breed for this particular type. If the characteristics of an individual are in fact produced by a large number of variable genes, producing a uniform type should be an extremely difficult task, as the dog breeders have indeed found. Much variability remains in every breed. However, the breeds have to some extent been standardized by selection, and when the combined pure breeds are analyzed by the factor-analysis method, there should consequently appear certain factors which correspond to the group of traits peculiar to the "type" of each breed. As indicated above, this was not the case, and we must conclude that the association of traits produced by selection for a type is quite weak.

The early geneticists also used the type concept in describing mutations and, in fact, did not even bother to study hereditary variation unless they found a gene which produced a clear-cut "phenotype." As with taxonomists, the modern tendency is to use the population approach, but the type concept may still be useful in describing certain extreme traits or combinations of traits.

This brings us down to the underlying theory of genetic effects on adult characteristics. All our evidence is consistent with the hypothesis of specific primary effects of genes. While pleiotropy may exist, it certainly does not have universal importance. Typing may be used to describe the effects of a single gene which has either important pleiotropic effects or one major effect which carries the phenotype outside the normal range of variation. However, the majority of evidence presented in this and previous chapters is on the side of the conclusion that the differences between the dog breeds are produced by large numbers of independent genes, each having a highly specific effect on behavior, physiology, or structure.

CONCLUSIONS AND HUMAN APPLICATIONS

Both breed and sex differences affect the patterns of growth of young puppies. There is little difference in size at birth, but as the puppies grow older they begin to diverge and differentiate from each other, the amount of difference being proportional to age. Can this differentiation be taken as a model for the differentiation of behavior? Growth is an accretionary process, with a little weight being added each day to what was there before. To a certain extent, this is also true of the learning process, as each day the puppy has new experiences and presumably learns a little more. In this way, the two types of differentiation are similar.

However, in solving a particular new problem the puppy does not invariably make use of all that he has learned previously. In fact, if the problem is new enough, he may have to start from the beginning and will be able to use very little of his old knowledge. Furthermore, the process of learning often proceeds by sudden jumps or saltation rather than gradual accretion (Hebb, 1949). While one can weigh a dog and measure all that he has grown, it is difficult to devise any test situation which will measure all that he has learned, or even to give him many different tests and add the results together in any meaningful way. Our conclusion is that where learning is truly accretionary, the model of growth will apply, but that there are many cases of learning which do not have this property and for which growth is hence a poor model.

Physical traits have obvious effects on behavior. The short-legged dachshund cannot run as fast as a normal-legged beagle, and the heavy muscles of a terrier have obvious advantages in fighting. This kind of effect hardly needs scientific proof. A further problem is whether or not physical traits have more subtle effects on behavior, and this is related to the genetic theory of pleiotropy. Proponents of its most extreme form have argued that every gene must have some effect on every body process. Opposed to this we have the theory of specific primary effects of genes—that a gene has one primary effect which may or may not affect other processes. Under this latter theory some genes might have pleiotropic effects and others not. We have tested these hypotheses by measuring the correlation between known inherited physical traits and behavior. If the general theory of pleiotropy were true, such genes ought to affect almost every behavioral test. As a matter of fact, the genes controlling hair length and the red coloring of basenjis had no appreciable effect on a wide variety of behavioral tests, or even upon measurements of body proportions. The evidence is definitely in favor of the theory of one specific primary effect of a gene. This does not exclude pleiotropy in special cases, and we know of at least one dog gene having pleiotropic effects, that for merle spotting. Even in this case, however, there is evidence for a specific primary effect.

Related to the theory of pleiotropy is the theory of body and behavioral types; i.e., that all the manifold effects produced by a pleiotropic gene can be summed up as a type. Brace conducted a very extensive test of the type theory through the method of factor analysis. While the dog breeds have been selected as types, the crucial genetic test in the F_2 population shows that the breed types do not reappear as such in the F_2's, either in body conformation or in other physical characteristics such as hair length and color. The evidence is in favor of a large number of individually segregating genes rather than a few major genes having pleiotropic effects and producing separate types. The type concept has only limited use in modern biology and is being replaced by the study of variation according to populations.

These conclusions and ideas have an important relationship to human affairs. Fresh from using the type method to classify animals, many biologists in past generations thought of the human races in terms of types. As we have pointed out above, the type concept is a useful one wherever a single major cause consistently produces several effects. The earlier classification of human races was based on the idea of one major cause producing skin color and all other associated characteristics, not only physical but cultural traits as well. The modern study of races as populations shows that each group is enormously variable in itself, and that populations unlike in many respects may have the same skin color. In those genetically determined characteristics which can be measured, such as blood groups, human populations differ in the average distribution of certain genes, but the differences between averages are usually small compared to the differences between individuals within the same population.

What does this mean in terms of interpreting human behavior? It does not mean that physical differences have no effect on behavior, but simply that these differences are of a simple and understandable nature. American Negroes have on the average longer legs in proportion to the trunk than do American whites (McKusick, 1960). One would expect that on the average they might be better performers in jumping and running, as indeed their performance in recent Olympic events would indicate. In the case of skin color, there is no evidence that it has any effect other than what might obviously be expected, namely, that Negroes endure severe physical exertion better in a hot moist climate than do white people, presumably because they are able to radiate body heat more rapidly (Baker, 1958). All the evidence is against any association between skin or hair color and behavioral traits other than the above. Wherever comparable intelligence tests can be given under conditions of equal opportunity, there have been no differences between different races (Anastasi, 1958). The kinds of behavioral differences which we now find between populations reflect the combined effects of cultural and environmental advantages. We should remember that civilizations of a type at least comparable to those of Babylon and Egypt were in existence in the New World when Columbus discovered it. A two-thousand-year start gave a tremendous advantage to the older culture when the two came into competition, but all indications are that another advanced civilization would have been developed on this continent if it had been left alone under favorable environmental conditions. There is likewise every evidence that human beings who live under extremely unfavorable environmental conditions such as the far north or the tropics are basically just as capable as those coming from more temperate regions, but that the severe environment provides a strong handicap to the development of these capacities.

In short, all the weight of modern scientific evidence, both from direct studies on human beings and from animal studies such as this one, is on the side of the conclusion that the human race is divided into populations rather than types. These populations are becoming larger and larger as they absorb former small local populations, and as even large continental populations begin to merge with each other.

In dealing with individuals, whether they are dogs or men, we need to remember that each is a unique collection (or population) of genetic traits, not a type representing a race. Likewise, in dealing with a group, we need to remember that we are dealing with a population: a group of variable individuals rather than a single type.

CHAPTER 14

THE EFFECTS OF HEREDITY
UPON THE BEHAVIOR OF DOGS

So far in this book we have been considering the results of individual tests, many of them especially selected for analysis because genetics appeared to be important and its mode of action readily explainable. In this chapter we shall consider all the tests together and so obtain an over-all and reasonably unbiased viewpoint. This will give us information on three principal points: general evidence for the existence of Mendelian segregation, evidence for the extent to which heredity affects behavior, and, finally, evidence for the existence of specific and general behavioral traits.

GENERAL EVIDENCE OF SEGREGATION
AND DOMINANCE

The detailed analyses of individual tests have revealed numerous ways in which complex interactions on the behavioral level can modify the expression of hereditary variation. Nevertheless, on the basis of Mendelian theory, one would expect greater variation in the hybrid populations in which segregation would occur, than in the F_1 and parental populations. In order to get an over-all picture, we combined the results from the cocker spaniel and basenji breeds and the two F_1 hybrids in one population, and combined those from the two backcrosses and the two F_2 generations in another. Each of the two populations was thus composed of four subpopulations having approximately equal numbers. One of them included all the segregating hybrids, and the other all the non-segregating populations (Table 14.1).

TABLE 14.1

Composition of Combined Populations

Non-segregating		Segregating	
Cocker..........	31	BCS × CS.......	28
Basenji..........	33	CSB × BA.......	42
BCS F_1.........	24	BCS F_2.........	32
CSB F_1.........	28	CSB F_2.........	41
Total........	116	Total........	143

Three statistical operations were performed. In the first, the sub-populations were simply combined and the total variance calculated. In the second, only the variance within populations was combined; and in the third, only the variance within litters was combined. The last should, of course, give us the best picture of the effects of segregation, since within a litter the effects of random variation in environmental factors are greatly reduced and the effects of accidental selection of different sorts of parents are eliminated. Within-population variance should include more environmental effects, as well as effects of differences between strains within a population, while the total combined variance of all the subpopulations includes variation from all sources, including that from differences between the subpopulations themselves.

If ideal Mendelian populations having equal numbers are combined in the above ways, there should theoretically be no difference in the means between the segregating and non-segregating populations, whether or not dominance is present. Actually, there were a large number (26 out of 44) of significant differences between the two. This result may in part be due to the fact that the component populations were not actually equal in numbers, and it also may be related to accidental selection of different individuals for the parent animals.

With regard to variation, the theoretical genetic variance of the *combined* parent and F_1 populations should be greater than that of the combined segregating populations, 47 per cent larger in the case of no dominance and 18 per cent in the case of dominance. The actual figures (Table 14.2) indicate that the variance of the non-segregating populations was greater in slightly over half of the variables. This small excess would argue that dominance is a common phenomenon in these variables.

On the other hand, the theoretical within-population genetic variance of the non-segregating populations should be zero, while

TABLE 14.2

NUMBER OF VARIABLES IN WHICH THE COMBINED VARIANCE OF SEGREGATING
POPULATIONS EXCEEDS THAT OF NON-SEGREGATING POPULATIONS

	Combined Variance Greater	Combined Variance Equal or Less
Total variance...............................	20	24
Within-breed or within-hybrid variance.........	28	16
Within-litter variance........................	31	13

that for the segregating populations should, on a scale of 1.0, average
.083 in the case of no dominance and .146 in the case of dominance.
Either the variance within populations or the variance within litters
should show an important increase in the segregating populations. A
glance at Table 14.2 indicates that this is indeed the case. As we
progress from the combined total variance to the within-litter variance, there is an increase in the number of variables in which the
segregating populations are more variable than the non-segregating
ones. The ratio of 31 to 13 is significantly different from that expected
by chance ($P = .005$). This ratio is approximately the same in the
variables which show significant differences between the backcrosses
and hence show important differences between the two parent
breeds.

Thus the over-all evidence is consistent with the assumption that
Mendelian segregation increases the amount of variation in hybrids,
although this may not be true in certain individual measurements
subject to special behavioral interactions (see chap. 12). Likewise
there is some indirect evidence that dominance is probably a common phenomenon in this data.

THE IMPORTANCE OF GENETIC VARIATION: RESULTS OF ANALYSIS OF VARIANCE

From the very start we realized that even within pure breeds
there was a great deal of variation between individuals, and our
original mating plan, described in chapter 1, was intended to minimize individual genetic differences by first developing strains based
on the offspring of a single mating and, second, by crossbreeding between individuals coming from a single mating in each of two pure
breeding stocks thus developed. This plan should have had the effect
of reducing variation due to individual differences relative to variation between breeds. This fact should be kept in mind in interpreting
the results.

With the help of Dr. Gunther Schlager and the Jackson Laboratory Biometric Service, we made an extensive analysis of variance of the 50 variables used in Brace's (1961) factor analysis. For each of these variables, the variance was computed: (1) between breeds or hybrid populations, (2) between litters within populations, and (3) within litters. The first of these components can be assumed to be largely genetic, as every effort was made to rear all breeds under uniform conditions. Likewise, the third component of within-litter variance should reflect a large amount of genetic variation, that between individuals. Since all members of a litter underwent testing at the same time and had very similar environments, it can be assumed that this part of the variance has a large genetic component. On the other hand, the between-litter variance should include a large part of the environmental effects, especially where repeated litters from the same mating were tested. Where different matings are involved, it also reflects genetic differences between the parents.

These computations were made for all pure breeds, the two parent breeds (cockers and basenjis), the two F_1's, the two backcrosses, and the two F_2's (Table 14.3). Six of the variables were distributed in a dichotomous fashion so that variance estimates could not be made. In these cases, the chi-square analysis was used to measure the importance of differences between populations.

A further and more detailed analysis was done for selected backcross and F_2 populations. In these selected groups there were two litters from each of two matings in each population. The parents in each case were brothers and sisters from the reciprocal crosses, so that the whole group was descended from four sibling basenjis and four sibling cocker spaniels. In these populations, the between-litter variance could be separated into a between-mating component (largely genetic) and a between-litter–within-mating component (largely environmental).

Since litter sizes were variable, variance components were computed according to the method of Gower (1962). In this method, the within-litter variance value is assumed to be correct, but the remaining two components are calculated with some degree of approximation. The method tends therefore to slightly reduce the estimates of between-breed and between-litter variance; that is, these are conservative estimates.

The general result is to give us estimates of the relative importance of two largely genetic components, between-breed and between-mating, where the latter is somewhat minimized. It also gives us a

TABLE 14.3

PROPORTION OF VARIANCE ATTRIBUTABLE TO POPULATION DIFFERENCES

Variable	All Pure Breeds	BA & CS	F₁'s	Back-crosses	F₂'s	Selected Back-crosses	Selected F₂'s
Physical Measurement:							
Weight	27**	42**	18*	5*	0		*
Body length	59**	76**	11*	33**	0		
Shoulder height	52**	82**	62*	53**	0	**	
Chest circumference	21**	37**	2*	21**	0		
Waist circumference	24**	21**	48**	11**	0		
Brisket depth	27**	65**	11	33**	0		
Chest breadth	27**	27**	0	0	0		
Femur diameter	0*	4*	3	0	0		
Humerus diameter	21**	58**	15*	0	0		
Thigh circumference	16**	19**	0	21**	0		
Mean	27	43	17	18	0		
Physiological-Emotional (Heart Rates):							
Average rate, 11–16 weeks	39**	27**	0	24**	2	**	
Adult rate	46**	37**	7*	17**	7*	**	
Change, quieting	27**	42**	0	16**	0	**	**
Change, bell ringing	10*	11*	9*	18**	0	**	
Arrhythmia index	66**	0	9	0	0		
Mean	38	23	5	15	2		
Emotional:							
Distress vocalization while weighed, 1–4 weeks	26**	24**	0	0	0		
Activity in confinement	13**	8*	16*	0	7	**	**
Responses to stimulation during reactivity change, alone and with experimenter	1	8*	29**	0	0		
Activity during bell ringing	34**	11*	5	0	0		
Posture	14**	43**	0	0	13*		*
Muscular tremor	44**	0	22**	40**	0	**	
Vocalization^d	**				*	...	...
Investigative behavior	24**	46**	11*	13*	12		
Escape attempts	28**	56**	7	29**	0		
Mean	23	25	11	10	4		
Social relationships:							
With humans:							
Avoidance and vocalization, 5 weeks	29**	59**	3	24**	0		
Playful fighting, 13–15 weeks	15**	42**	0	10	0		
Approach and following, 13–15 weeks	18**	27**	9*	30**	30**	**	**
With dogs:							
Dominance score^d	**	*	*	**	*	...	...
Mean	21	43	4	21	10		
Trainability:							
Forced training:							
Inactivity while weighed, 5–16 weeks^d	**	**	**	**		...	...

TABLE 14.3—Continued

Variable	All Pure Breeds	BA & CS	F₁'s	Back-crosses	F₂'s	Selected Back-crosses	Selected F₂'s
Leash fighting...............	55**	77**	12*	58**	1	**	
Leash, interference...........	33**	0	0	9*	21**	*	**
Leash, vocalization...........	26**	0	0	0	2		*
Docility, sit training..........	18**	48**	13*	24**	0	*	
Reward training:							
Goal orientation.............	7**	15**	9	0	0		
Motor skill.................	27**	0	0	0	0		*
Mean......................	28	23	6	15	4		
Problem solving:							
Detour test:							
Errors, first trial.............	8**	13*	0	17**	4		
Errors, long barrier...........	33**	39**	0	1	0		
Errors, U-shaped barrier^d....	**	**		**			
Final running time, long barrier	55**	78**	0	46**	0	**	...
Vocalization, U-shaped barrier	12**	47**	0	37**	1	**	
Manipulation time, string pulling^d.....................	**	**	*	**		...	...
Manipulation, uncovering food^d......................	**	**		**		...	...
Maze, corrected time score first day..................	34**	53**	0	31**	0	*	
Maze, average error score.......	22**	0	23**	14**	0		
Trailing......................	20**	33**	10*	0	0	**	
Spatial orientation, errors.......	40**	31**	4	0	0		
Spatial orientation, time........	35**	35**	5	6*	0	*	
Spatial orientation, persistence	33**	31**	0	0	0	*	
T-maze, motivation.............	27**	16**	0	49**	0	**	
T-maze, errors.................	8**	0	0	37**	3	**	
Mean......................	27	31	4	20	1		
Average, all behavioral measures...	27	28	6	16	3		
Average, all measures.............	27	31	9	16	2		
** Significant differences .01 or less.	47	35	5	28	2	13	4
* Significant differences .05–.02.	2	7	15	4	4	5	3

^d — dichotomous distribution, variance components not calculated.

picture of the change in distribution of the between-breed component as a result of hybridization.

Results with five pure breeds.—As Table 14.3 shows, there is only one variable out of the 50 which does not show significant differences between the breeds, at least on the 5 per cent level. The great majority of the differences are significant to far beyond a probability of .01. Since these variables were selected to represent all important behavioral tests, it is obvious that breed differences in behavior are both real and important in magnitude.

With respect to the relative amounts of breed differences, these vary from zero to 66 per cent of the total variance. The largest between-breed component occurs in the arrhythmia index, produced almost entirely by the fox terrier breed, which is highly different from all the others. A near zero percentage is found in the variable of change in overt activity during the reactivity test, but only in the part in which the dog is left alone and later comforted. The computations also reduce a small but significant difference in one of the physical variables (femur diameter) to zero.

As shown at the end of the same table, the portion of variance attributable to breed differences (comparable to the intraclass correlation figures in previous chapters) averages approximately 27 per cent for both physical and behavioral variables. Even if the physiological measures are omitted, the remaining behavioral variables still average 25 per cent. We can conclude that behavioral differences between the five pure breeds are as important as differences in physique, even though physical measurements are usually considered to be more accurate than behavioral ones.

Results with basenjis and cocker spaniels.—As might be expected, there are fewer significant differences between these two parent breeds (Column 2, Table 14.3) than when all five breeds are considered. One notable example is that of the arrhythmia index, in which there is no difference between cockers and basenjis or any of their hybrids. Nevertheless, there are still 35 out of 50 variables which are significantly different at the .01 level or less. These two breeds are obviously different from each other in many respects.

Indeed, certain of the between-breed variances even exceed those calculated for all five breeds. Those for leash fighting and running time on the long barrier of the detour test are 77 and 78 per cent respectively. Almost all the physical variables are higher, that for height being 82 per cent. Nevertheless, there is a great deal of individual overlap between the two parent stocks in most of the variables, even those which show the greatest differences. This overlap might be caused either by environmental variation or by genetic variation within the breed, a question answered by further computations described later in this chapter.

Results with the F_1's and F_2's.—Differences between the reciprocal crosses in these two populations could be caused by accidental selection of different parents, by maternal inheritance, or by sex-linked inheritance. As shown in an earlier chapter, there is no evidence that sex-linked inheritance affects any of the variables, and, indeed, this is not surprising. There are 39 pairs of chromosomes in the dog, and therefore the chance of finding a gene carried by the X chromosome is quite small. Maternal inheritance is a likely possibility only in the case of approaching and following behavior exhibited toward a human handler, where both the F_1's and F_2's show large differences and where there is a logical means of passing on a trait from generation to generation by following the maternal example.

There are many more differences between the reciprocal F_1 crosses than would be expected purely by chance. Out of 50 variables, 2.5 would be expected by chance at the .05 level. Actually, 20 variables showed differences significant at the .05 level or less. Since two of the three alternative explanations listed above have been eliminated, there remains the possibility that the differences between the F_1's were caused by accidental selection of the parents. This is especially likely where the differences between the F_1's are opposite from those in the parents, as in all six of the physical measures and three of the behavioral measures. The results therefore indicate that individual genetic variation within a breed is important.

However, the number of differences between the reciprocal crosses is greatly reduced in the F_2 generation, there being only six differences significant at .05 or less. Some of this reduction is undoubtedly produced by selection of a small number of F_1 parents from the offspring of the parents selected in the previous generation, which would have the effect of decreasing variation. It is also possible that some breed characteristics are fully developed only in the maternal environment of the pure breeds; that is, the development of breed characteristics may in part depend upon an environment produced by association with members of the same breed. For example, fox terrier mothers appear to be considerably more aggressive during weaning than do other mothers, and this may aid in the development of excessive aggressiveness among their pups. This general possibility is reinforced by the fact that 11 out of the 14 significant differences in behavior are in the same direction as differences between the purebred parents.

Results from the backcrosses.—In these populations, genetic segregation takes place. The result is that genetic variation is redistributed, being decreased between populations and increased within populations. On any theory of inheritance involving one set of alleles, there should be differences between reciprocal backcrosses. Selecting those variables which show highly significant differences should therefore give a group of variables in which the between-breed component is greatest. The number of variables significant at .01 is reduced from 35 in the parent strains to 27 in the backcrosses, and for the most part the variables showing most significant differences correspond to those which showed the maximum differences between the parents.

There are three exceptions. One of these is the average error score in the maze test. In this variable there is a large difference between the two F_1's, but no difference between the two pure breeds. The significant difference between the backcrosses probably results from accidental selection of parents which are different from the breed averages. The second exception, tremor during the reactivity test, has the same explanation. Finally, there is the error score in the T-maze, for which there is no obvious explanation except that the original similarity between the parent strains may have been produced by different genes which segregated separately and thus produced differences between the backcross generations.

Results from selected backcrosses and F_2's.—In the above calculations it is impossible to differentiate between the between-mating variance and the between-litter variance within matings, as various accidents prevented our obtaining repeated litters from every mating. The following calculations are based upon a selected group of matings in which it was possible to obtain two litters from each. The original parents were all obtained from basenji mating 4 and cocker spaniel mating 7, so that differences between the matings represent genetic differences between siblings.

Descendants from these four matings produced two litters each in the backcross and F_2 generations, so that for each population there were four litters and 19 to 21 animals. It is therefore possible to measure the variance produced by differences between matings in a direct manner. It should be recalled that in the backcross and F_2 generations, the variance due to segregation is maximized, producing a large amount of within-litter variance. The differences between matings, however, should not be affected by segregation but only by accidental selection and small numbers.

As seen in Table 14.3, the number of significant differences between the selected populations is still further reduced because of the reduction in numbers and the restriction of differences between matings. The variables which still show significant differences are the same as the variables which showed the largest differences between the backcrosses, with three exceptions: the confinement test, the trailing test, and the spatial-orientation-persistence test. These three show significant differences between the pure breeds but not between the total backcrosses. The fact that the selected backcrosses are significantly different probably results from an accidental selection of parents which showed greater differences than the average.

There should be a slightly higher proportion of within-litter variance in the F_2's than in the backcrosses, because of reduction in the latter of a genetic variation due to dominance, and this in fact is the result. As seen in Table 14.4, the within-litter variance in each F_2 population is greater than in the corresponding backcross population.

TABLE 14.4

VARIANCE COMPONENTS OF EACH SELECTED BACKCROSS AND F_2 POPULATION

POPULATION	BETWEEN MATINGS	BETWEEN LITTERS	WITHIN LITTERS
Mean Variance Components of 10 Physical Measures, in Percentages			
BCS × CS.........	18	16	67
CSB × BA.........	4	22	74
BCS F_2.............	20	3	77
CSB F_2.............	5	0	95
Total..........	12	10	78
Mean Variance Components of 34 Behavioral Measures, in Percentages			
BCS × CS.........	13	15	72
CSB × BA.........	10	14	77
BCS F_2.............	13	11	76
CSB F_2.............	13	9	78
Total..........	12	12	76

The between-mating variance averages approximately 12 per cent on both physical and behavioral variables. The between-litter–within-mating variance is very close to the same amount. It should be recalled that the former represents genetic differences between brothers and sisters rather than a random sample of the breed and is therefore a minimum estimate. It must be concluded that genetic differences within breeds are important. If we compare these results with those from the pure breeds, about 75 per cent of the variance occurs within breeds, including about 21 per cent due to differences between litters. If we assume that one-half of the last figure, or 10.5 per cent, is caused by differences between matings, the figures are quite comparable.

In Table 14.5 the variance components of certain selected variables showing the greatest differences between the backcrosses are shown. On the average, there is no difference in the relative size of the variance components between these and the total of all variables. There are, however, great individual variations between the tests. For example, height shows almost no between-mating variance,

TABLE 14.5

MEAN VARIANCE COMPONENTS IN VARIABLES SHOWING DIFFERENCES
BETWEEN THE BACKCROSSES WITH A $P < .01$

Variable	Between Matings	Between Litters	Within Litters
Height...............................	2	12	86
Ave. heart rate, 11–16 weeks..............	26	35	39
Adult heart rate.......................	11	9	80
Heart rate change, quieting..............	0	4	96
Heart rate change, bell ringing............	7	15	78
Activity in confinement..................	4	31	65
Muscular tremor.......................	21	20	59
Approach and following, 13–15 weeks.......	7	0	93
Leash fighting.........................	22	7	71
Final running time, long barrier...........	5	1	94
Vocalization, U-shaped barrier............	7	18	75
Trailing..............................	9	3	88
T-maze, motivation.....................	11	4	85
T-maze, errors.........................	20	1	79
Mean total........................	11	12	77

whereas infantile heart rate shows a great deal. This would mean that in the selected sample, parents were accidentally chosen that were uniform in height but variable in heart rate, although it will be recalled that in the total sample there was a great deal of parental variation in height.

In the case of approach and following, there is no between-litter variance within matings. This, it will be recalled, is a variable in which maternal inheritance is a likely possibility. In the T-maze error score, which is one in which the parent pure breeds showed no differences, there is a large between-mating variance. In this case there was probably accidental selection of different sorts of parent animals from the two breeds, although the breeds themselves are similar on the average. Finally, in the leash-fighting test a large between-mating component confirms conclusions from the detailed analysis of this test in chapter 12.

Conclusions.—Reviewing this complex material, we can come to certain firm conclusions. One is that the differences between breeds are important, both in physical measurements and behavioral measurements. The variance averages between 25 and 30 per cent of the total and ranges as high as 66 per cent. In only one behavioral test and one physical test are breed differences estimated as being close to zero.

At the same time there is a great deal of overlap between the breeds. Although this overlap is in part caused by random environmental factors, it has a substantial genetic component. The general picture which emerges from the study is that in most traits there are average breed differences. Taking any two breeds at random, one would expect a large number of significant differences between mean scores. However, there are genetically overlapping individuals in each of these breeds, so that in many traits one breed could be made like the other within a very short time by selection. This means that the dog breeds have retained a great deal of genetic flexibility.

For example, fighting the leash is characteristic of basenjis and is shown by no cocker spaniels. Nevertheless, there are a few basenjis which lack this trait, and it would be possible by selection to produce a strain very much like cockers in this respect. This type of genetic overlap is not necessarily reciprocal, as it would probably be much more difficult to select cockers in the opposite direction.

In a few cases selection has apparently produced traits for which certain breeds are very nearly genetically homozygous, such as the low capacity for developing fighting behavior in the hound breeds and the high capacity for the same trait in the terrier breeds. Such instances are, however, relatively rare.

Within a breed, individual capacities are likely to be highly variable; but most individuals may be able to perform the tasks required of them by bringing different capacities into play. Outstanding performers within a breed probably have special combinations of certain capacities which are largely the result of accidental selection.

EVIDENCE FOR THE EXISTENCE OF GENERAL TRAITS: RESULTS OF FACTOR ANALYSIS

Alternate theories of inheritance.—We have now confirmed the reality of a large number of differences between the behavior of basenjis and cocker spaniels. The basenjis had a tendency to fight the leash, to show a great deal of playful aggressiveness, and to be highly fearful as young pups. In contrast, the cockers almost never fought the leash, had less tendency to playfully fight and chew on their handlers, and showed very little fear as young puppies. We might suppose either that all of these differences are caused by one basic behavior trait, or that each of these is a special trait and unrelated to the rest. If the first theory is true, then the test results should be correlated even in the segregating hybrid generations. If the second theory is correct, the test results should show little if any correlation.

The traits are, of course, correlated in the original pure breeds, no matter what the genetic situation might be, and a few independent traits might be correlated even in the segregating generations if they were caused by two genes which are close together on the same chromosome and hence closely linked. However, since the dog has 39 different pairs of chromosomes, the chances of close linkage are not very great.

The theoretical result is also complicated by the fact that a particular behavior trait might be caused either by a single major gene or by several genes with minor effects. The first condition would produce very strong correlations between tests. The second could produce such correlations, but only if the effects of minor factors were additive. If a particular trait is produced only by a special combination of genes, it will disappear during segregation and be found only in a very small number of individuals, not enough to produce any strong correlation. For example, if a trait was produced by a combination of three recessive genes, this combination would appear in the F_2 generation in only one out of 64 individuals. In any case, one way to test these various theories is through analysis of correlations.

The theory of factor analysis.—Dr. Joseph Royce, who was at the time we began these experiments a graduate student working with the late Professor Thurstone, strongly advocated the use of factor analysis on our data and has presented the theory involved in several papers (1950, 1957, 1963). If the hypothesis of basic general traits is correct, factor analysis should give a clear-cut indication of the nature of these traits and of the situations in which they appear. The method does have one limitation. Since it is based on correlations, it will work only where the basic mathematical assumptions of correlation are met. These assumptions are that causal factors produce certain specific effects, and that these effects are additive and distributed in normal curves.

The factorial method has proved very useful in analyzing the results of human intelligence tests. Each test score is the result of adding together several test items. It therefore fulfills the criterion of additive effects. Since many intelligence tests were designed by devising a number of questions and problems, and afterward selecting those to which students gave different answers, the original intelligence tests were a mixture of many different sorts of questions. The factorial method was successfully used to sort out the common factors in various questions and to produce tests which could be interpreted in terms of special capacities. Thurstone and Strandskov (Strandskov, 1953) tested the hypothesis that these purified tests represented basic genetic differences in capacities by giving the tests to identical twins and determining the correspondence between

them. In most cases the correlations were low, indicating that heredity was not highly important in determining these capacities, although some of the more strictly physiological tests showed higher
resemblances. Judging from the evidence on transfer of training,
which indicates that people are able to transfer material learned
on one situation only to closely related situations, it is probable that
the factors obtained by Thurstone represent problems with a common recognizable similarity rather than problems solvable by a
single common capacity. As we have seen from the results of performance tests with dogs, many capacities may be involved in the
simplest performance test, and animals can organize these capacities
in various ways.

A *factor analysis of emotional and physiological tests.*—Royce
(1955) did a factor analysis of tests made on 53 purebred dogs distributed in six different breeds. Thirty-two variables were included
in the analysis. These variables were obtained from only five different test situations, and some of them would be expected to be correlated simply because they occurred in the same test. The basic
correlation table, however, shows many more correlations than might
be expected from this source alone. Of the 396 correlations of unrelated tests, 85 were significant at the .05 level, as opposed to the
19.8 which might be expected by chance. Real correlations between
the different tests must exist, although these may be caused by the
fact that the animals tested belonged to pure breeds and hence had
common traits produced merely by breed selection. Royce was able
to identify several factors, two of which were highly physiological.
One was characterized by high blood pressure and arrhythmic heart
beat and may have been associated with emotional reactions of
timidity. The second showed a high degree of change in heart rate
to social stimulation, associated with a high degree of external activity. Four factors were largely behavioral. In one of these, there were
high loadings for both playful aggressiveness toward human handlers and confidence while being weighed. A second was characterized by a high degree of timidity in reaction to people, along with
some emotional reactions. A third test included a high degree of reaction to the noise of a bell ringing, and a fourth was called general-
activity level.

One of the most interesting results was the indication that timidity and aggressiveness were not opposite ends of the same continuum
but were found in different factors. The results also indicate that
there are several different kinds of timidity and thus argue for the
theory of highly specific effects of genes, but since any of the fac-

tors might have been caused by accidental breed association, this last conclusion is not a final one.

Royce's factor analysis enabled us to select certain variables which seemed to have important general effects and to eliminate others which seemed to be either inconsistent or specific. Thus the testing program, and particularly the scoring of certain tests, was considerably improved. Another result was the demonstration that the same test scores given at different ages came out in different factors. The timidity-confidence rating on the fifth week, or first testing, appeared to be largely independent of the average ratings obtained from 5 to 10 and 11 through 16 weeks. The last two, however, were consistently associated.

Factor analysis of performance tests.—Anastasi and Schmidt (1955), did a similar preliminary analysis of performance tests. The dogs tested included all five pure breeds, totaling some 50 animals, plus 23 of the F_1 hybrids between the cockers and basenjis. The group tested was therefore somewhat similar to that used by Royce. The results of eight different test situations were included, but several of these included more than one score, making a total of 17 variables to be analyzed. The method was the same as that used by Royce, the centroid method.

The analysis produced five factors. One of these was clearly a "test" factor, involving two scores from one test, and meant little else than that these two scores were similar to each other. Another factor chiefly included three scores indicating poor performance on the T-maze test, plus scores associated with good performance on the second barrier (maze) test. Since the principal cause of error in the T-maze is a tendency to always run to the same side and thus show position preference, this factor was called "persistence of positional habits." It is interesting that a later detailed analysis of the second barrier test revealed breed differences in the tendency to form such habits, the beagles being quite different from the rest of the breeds in being slow to form fixed habits.

The other factors were interpreted as largely emotional rather than intellectual in nature. Factor 1 was called "impulsiveness"; Factor 2, "docility"; and Factor 4, "visual observation."

Some of the different scores on the same test situation appeared in different factors, particularly those on the barrier test, but in most cases all scores from one test appeared in the same factor. On this basis we could select the most important score from each test for detailed analysis and future work.

Whether these factors were the result of the correlations arising from selection of breed types, or whether they represented general behavior traits in dogs, was not apparent from the results.

Considering the two preliminary analyses together, it looked as though it might be possible to account for most of the differences in dog behavior on the basis of a few emotional and temperamental traits having an effect on performance as well as on emotional expression. The test of this theory was to combine both emotional and performance tests in a single factor analysis and to further determine whether or not these characteristics were really basic traits by applying factor analysis to the segregating hybrids as well as to the pure breeds and F_1 hybrids (Brace, 1961).

A general factor analysis.—Brace's extensive factor analyses have already been described in chapter 13 in connection with the effect of physique on behavior. Brace used the 40 behavioral and 10 physical variables listed in Table 14.3 as a basis for separate factorial analyses on each of the pure breeds and each hybrid population, as well as on several combined groups. Of these, results from the F_2's were most important, since they were based on the largest numbers and on the population where maximum effects of segregation might be expected.

It will also be recalled that all the physical measurements came out as a single factor of general size in the F_2 generation, indicating that there was no genetic basis for physical types. In both cocker spaniels and basenjis, however, the physical measurements did appear in two factors, indicating that there were long- and short-bodied types of animals within the pure breeds, probably relating to strains within a breed. In this case there may have been variation in one respect only, namely, body length, producing two types of dogs. Thus the method is capable of identifying types if they do exist.

We can again ask the most important question: is there any indication of the existence of true behavioral types? An affirmative answer would be related to the general theory of pleiotropy discussed in the last chapter: namely, the existence of major genes, each of which produces effects in many different situations. The F_2 population gives the most clear-cut test of any hybrid group, since the effect of all genetic factors should appear, and since there should be a minimum effect of genetic linkage upon correlation.

In the first analysis, we can examine a completely null hypothesis: that none of the behavioral variables are related to each other aside from those which are obviously correlated because they were derived from the same test situation. This hypothesis can be discarded, because the correlations between variables in different behavioral

TABLE 14.6

FACTOR 2, ACTIVITY-SUCCESS IN THE COMBINED
F₂ GENERATION (BRACE)

Variable	Loading
Active during bell ringing	.57
Spatial orientation, fast	.56
Trailing, success	.54
T-maze, few errors	.54
T-maze, fast	.52
Maze, fast	.50
Motor skill, success	.50
Active escape, many attempts	.46
Active investigation, many attempts	.42
Much vocalization, reactivity test	.42
Few vocalizations, U-shaped barrier	.42
Active in sit training	.41
Erect posture, reactivity test	.40
Spatial orientation, persistence	.39
Active while weighed	.39
String pulling, success	.38
Much playful fighting	.34

tests far exceed the number expected by chance. The different tests are definitely related to each other in some way. However, the correlations between them in the F₂ generation are uniformly low. If there were major genes for basic traits segregating in the F₂'s, we would expect groups of tests with high correlations between them.

Looking at the factors themselves, we see that Factor 2 in the F₂ generation includes a large number of performance tests and certain measures of high activity (Table 14.6). Brace therefore labeled it "activity-success." Factor 3 includes all of the measures of heart rate and a few which may indicate timidity (Table 14.7). Beyond this, Factors 4 and 5 include only miscellaneous variables with low

TABLE 14.7

FACTOR 3, HEART-RATE, IN THE COMBINED
F₂ GENERATION (BRACE)

Variable	Loading
High adult heart rate	.68
Low arrhythmia rate	.56
Many errors, U-shaped barrier	.43
Little playful fighting	.43
Much distress vocalization while weighed	.41
Active during bell ringing	.37
Great docility, sit training	.34
Slow on T-maze	.33
Much tremor, reactivity	.32
Few errors, long barrier	.32
Erect posture, reactivity	.31
High infantile heart rate	.31
Much vocalization, U-shaped barrier	.30

loadings. There are therefore only two explainable behavioral factors in the F_2 hybrids.

The other groups of segregating hybrids show similar results. In the backcross to the basenji there are two factors very similar to those in the F_2, and possibly a third which consists of a group of tests indicating a high rate of vocalization. In the backcross to the cockers, Factor 2 is again marked by success on a variety of performance tests, although these are a different group from those emphasized in the other two hybrids. Likewise, the next factor is one which can be labeled "heart rate," although not as definitely as in the other two.

Another analysis in which genetic differences would be expected to be highly important is that of the combined pure breeds, in which breed differences are maximized. Here Factor 2 is again one of good performance on the same series of tests as in the F_2 and the backcross to the basenji. Factor 3 is once more a recognizable heart rate factor. Thus there appear to be only two recognizable behavioral factors derived from the populations in which genetic differences should be important.

Comparing this result with that in the F_1 generation, in which genetic differences are theoretically less important, we find an entirely different situation. Here there is only one recognizable behavioral factor, which includes all of the 11 measures from the reactivity tests, each loading indicating an active animal in overt behavior or in heart rate (Table 14.8). Thus, in a population in which genetic

TABLE 14.8

FACTOR 2, REACTIVITY, IN THE COMBINED F_1 HYBRIDS (BRACE)

Reactivity Tests	Other Tests	Loading
Active investigation...............	....	.71
High adult heart rate..............	....	.63
Erect posture.....................	....	.61
Little change in heart rate.........	....	.59
Active during bell ringing..........	....	.57
Much vocalization.................	....	.51
Active escape.....................	....	.50
	Little vocalization, leash training	.49
	Little fear of human handlers	.46
	Much interference, leash training	.45
Little change in activity...........	....	.42
	Much distress vocalization while weighed	.40
Little cardiac arrhythmia...........	....	.40
	Active while weighed	.39
	Much playful fighting	.38
Little change in heart rate, bell ringing	....	.38
	Fast on T-maze	.33
	Few errors, long barrier	.31
Little tremor.....................	....	.30

differences are less important, the greatest amount of correlation appears to come from a common test situation.

Brace came to the conclusion that he was dealing with four major factors in all the dogs: body size, activity-success, heart rate, and general activity. Instead of a variety of general traits, the evidence shows that there are only a limited number.

We may now take a second look at the most important behavioral factor, activity-success (Table 14.6). It includes a very large number of significant factor loadings, none very high, but most of them indicating good performance on a variety of performance tests. In fact, almost all of the intelligence tests are included. Along with this there are high loadings on certain reactivity tests, indicating high activity and general confidence. We can therefore label this factor "General good performance and active, confident behavior."

A similar factor involving good performance appears in the backcross to the basenjis. Factors involving several performance tests are also found in other hybrid populations and in the pure breeds, but there is less consistency about their composition. Looking at the correlation tables of the F₂'s and basenji backcross, we immediately notice a large number of significant correlations with the trailing test (Table 14.9).

TABLE 14.9

CORRELATIONS OF POOR PERFORMANCE ON THE TRAILING TEST
WITH PERFORMANCE IN OTHER TESTS IN TWO HYBRID POPULATIONS

Test Performance	Backcross to Basenji	F_2
Many errors, T-maze	.61	.42
Slow on motor-skill test	.54	.32
Slow, spatial-orientation test	.48	...
Little approach and following behavior	.47	...
Slow on T-maze	.44	.36
Little playful fighting with handler	.41	.32
Poor performance on string pulling, manipulation test	.38	...
Little persistence, spatial orientation	.38	...
Many vocalizations, U-shaped barrier	.33	.33
Few escape attempts, reactivity test	.33	...
Fast, goal-orientation test	.31	...
Little activity during bell ringing	...	.32
Little investigation, reactivity test	...	.32
Many errors, U-shaped barrier test	...	.29
Slow on maze test	...	.26

As we saw in chapter 10, the trailing test shows a high degree of variation in every population, but the differences in variance between these populations are almost entirely accounted for by fear responses to the apparatus. A large amount of variance due to this sort of timidity occurs in the backcross to the basenjis, suggesting that we may be dealing with a backcross to a recessive.

Assuming that we have such a relatively simple Mendelian mechanism, we would expect certain results in a factor analysis. The factor-analytic method is based on correlation, and correlation is possible only where there is variation. According to Mendelian theory, there should be genetic variation only in the backcross and F_2 generations. Therefore, we should get clear-cut genetic factors from only these populations. This expectation is borne out by the results, at least in part. In the backcross to the basenjis, the trailing test has the highest loading on the general factor mentioned above. We could therefore call this factor "timidity, or fear," particularly of strange apparatus but also involving some fear of human beings because of the negative correlation with playful aggressiveness and attraction on the handling test. Fear or timidity is, of course, the opposite end of the scale from confidence, which was the original label of the factor and an equally appropriate one. We can now specify the kind of confidence—that evident in dealing with strange objects.

We may conclude that this is a case of a similar effect on performance in many different situations. This fits the concept of a general behavior trait produced by a pleiotropic gene. However, with the possible exception of the heart rate factor, which might itself be related to another sort of timidity, apparatus-fear is the only case of this sort which turned up in the factor analysis. There are no other well-identified factors involving large numbers of tests, and we must conclude that the vast majority of genetic effects on behavior are highly specific and restricted to one or two situations.

Re-examining the correlation tables, we would expect that if general traits were common, there should be many high correlations in the backcross and F_2 generation as a result of genetic segregation. With the exception of the trailing test scores and certain others derived from common tests, and hence dependent on each other for other reasons, there are no such high correlations. There are many more correlations between the independent tests than might be expected by chance, but all of them are low. This indicates that there may be traits which have major effects on one test and only minor effects on others. Evaluating the whole study, we can say that the majority of behavior traits appear to be of this latter kind, specific in their effects and inherited independently of each other, and that the only verified general trait is an emotional one.

Value of the factorial method for genetic analysis.—The value of the factorial method is that it selects, on an entirely objective and impersonal basis independent of the preconceived ideas of the experimenter, those measurements which have common correlations. The same objectivity comprises its chief limitation for genetic analysis, because the method makes no distinction between those measurements which are correlated because of heredity and those which are correlated because of various environmental agents. The distinction between factors which are primarily genetic and others must therefore be made on the judgment of the experimenter.

The essential characteristic of a genetically determined factor is that it should appear only in populations in which there is important genetic variation. In a Mendelian cross, it should appear in the segregating populations, i.e., the backcrosses and F_2's. If the genetic system involves dominance, the factor will appear only in the backcross to the recessive. As we have seen above, only one factor had this characteristic in any clear-cut way.

Factor analysis will also select measurements made in situations which for some reason appear to be similar to the animals, whether or not the similarity has a genetic basis. For example, both the motor-skill test and spatial-orientation test required climbing on a ramp. Correlation may occur because tests are actually similar, as in this case, or because measurements are made at the same time and hence are not completely independent of each other. Still another possibility is the effect of previous success, which may carry over from one test to another, whether or not the tests are similar. These kinds of correlations may in practice be difficult to tell from those produced by genetic correlation. Genetically based correlations should appear in only the segregating generations, while non-genetic correlations might appear in non-segregating populations as well, inevitably producing some confusion between the two sources of correlation.

Analyzing genetic variance by correlational methods.—Analyzing genetic results by the method of factor analysis is a difficult process without additional clues based on a direct knowledge of the tests themselves. This suggests another possible method of analyzing inheritance—one based on correlation. If we take two tests which show high correlation in the segregating generations and low correlations in other populations, we should be able to make an estimate of the amount of variance due to genetic factors, using the principle that the percentage of variance due to correlation is given by the square of the correlation coefficient.

As shown in Table 14.10, this method can be used to transform the variance of the trailing test, the key variable in the "active confident success" factor, by computing the common variance with errors in the T-maze, the test with which it is most highly correlated. The result is a distribution of variance which is consistent with the

TABLE 14.10

COMMON VARIANCE IN TRAILING TEST AS INDICATED
BY CORRELATION WITH PERFORMANCE IN T-MAZE

Population	Variance, Trailing Test (σ^2)	Proportion of Common Variance with Errors in T-Maze (ρ^2)	Common Variance $\rho^2\sigma^2$
BA..............	4.42	.03	.13
CSB F_1...........	2.94	.00	.00
CSB × BA.........	4.07	.37	1.51
CSB F_2...........	3.91	.09	.35
CS..............	3.19	.10	.32
BCS F_1...........	3.69	.11	.41
BCS × CS.........	3.52	.03	.11
BCS F_2...........	3.56	.31	1.10
All F_1.........	3.64	.04	.15
All F_2.........	3.78	.18	.68

hypothesis of a recessive gene or genes for timidity in the basenji breed. Furthermore, the results are very different from the original raw data, which looked as if there were many complicated interactions involved with the maternal environment, etc. In other words, the correlation coefficient can be used to "strain out" that portion of the variance which conforms to the mathematical assumptions of the correlation coefficient; namely, those of simple linear effects and a normal distribution of variation.

The method should work well only if the two tests or measurements used are situationally independent of each other, so that there are no common environmental factors and correlation is due chiefly to the hereditary organization of each individual. It should also work best where numbers large enough to insure the accuracy of estimates of the correlation coefficient are used.

This again brings up the need for new methods of analysis. As indicated in chapter 12, presently available quantitative methods are inadequate for analyzing the effects of heredity upon intricate problem-solving behavior, except to indicate that complex interaction is important. Progress in the future will in part depend on the development of statistical methods whose assumptions more accurately meet the facts of behavior inheritance. However, there is already abundant evidence from our own and other work that heredity has a quantitatively important effect on behavior, and future research should also be directed toward the important question of *how* heredity affects behavior—a problem which is still almost untouched with respect to the kinds of genetic differences in emotional reactions and social behavior which are so important in the dog.

CONCLUSION

In this chapter we have attempted an unbiased general survey of the effect of heredity upon the behavior of dogs. While many of the tests were set up without any previous knowledge of behavioral differences, many others were devised to measure the breed differences in behavior which we had seen in casual observations. In short, we tried to design our experiments so that there would be a favorable chance of revealing genetic differences. The data are biased in this direction, but the statistical methods are not.

In general, the results show that heredity is an important quantitative determiner of behavior in dogs and that genetic differences in behavior can be as reliably measured and analyzed as can hereditary differences in physical size.

Considered as a group, the results of the various tests are consistent with the principle of Mendelian segregation, with some indication that dominance is an important factor in most cases. At the same time, the results of individual tests are frequently difficult to interpret upon a simple Mendelian basis because of the complex interactions with the environment which are so often involved in the process of behavioral adaptation.

There are relatively few general behavioral traits. Rather, each breed shows a combination of many special characteristics which are often related to the special behavioral tasks for which the breed has been selected.

Furthermore, there are relatively few behavioral traits for which any breed is actually homozygous. Even within the restricted samples chosen for this experiment there was a great deal of individual genetic variability.

This large amount of genetic variation, both within and between breeds, leads to the conclusion that it is impossible to generalize about any one breed from experience with one dog or even one strain of dogs, and that it is likewise impossible to generalize about all dogs from experience with one breed. Furthermore, their great range of genetic variability makes dogs highly reactive to selection.

Selection of certain kinds of behavior can greatly modify a breed within a few generations, and increasing the range of variability still further by breed crossing makes it possible to readily create new and unique combinations of behavioral traits to meet the specialized needs of changing human societies.

PART IV

GENERAL
IMPLICATIONS

CHAPTER 15

IMPLICATIONS FOR THE ART
OF DOG BREEDING

INTRODUCTION

Our primary purpose in this study was to discover as much as possible about the effect of heredity upon behavior, and we chose the dog as one of the most favorable species for demonstrating the magnitude of such effects. To achieve our objective it was necessary to study the genetic background of dogs, their behavioral development, and their social organization. All this information, together with the findings from our genetic experiments, has implications for the practical dog breeder as well as the scientist and the student of human affairs.

We have used the word "Art" advisedly in the title of this chapter because any practical application of scientific knowledge involves a certain amount of individual judgment and adaptation to special circumstances. The scientific findings must be interpreted, and when this is done we are entering the realm of the arts. Applications must be designed to individual circumstances. Our own work was necessarily limited to only a few out of the more than one hundred breeds now recognized by the American Kennel Club. In the following pages, we shall outline the major findings of our study and indicate their general implications.

RESULTS AND THEIR IMPLICATIONS

Basic dog behavior.—All present evidence indicates that dogs were first domesticated from wolves and that both species possess the same basic patterns of behavior. Much dog behavior can be understood in terms of the social life of wolves, and dog breeds can be thought of as more or less specialized populations of wolves. The raw materials of the dog breeder are the behavior patterns of dogs and wolves which are listed in chapter 3. He can modify the frequency of expression of these patterns in all sorts of combinations except those which are incompatible with life and reproduction. This general repertory of behavior likewise imposes certain limitations upon breeding, in that it is difficult, if not impossible, to go beyond the basic capacities of dogs.

Combined action of environmental and hereditary factors.—We found breed differences in behavior beginning at birth and extending throughout the first year. During this time, particular differences might wax and wane and environmental factors could affect behavior at all stages of development. The important fact is that behavior is never wholly inherited or wholly acquired but always *developed* under the combined influences of hereditary and environmental factors. The conclusion for the practical breeder is that it is almost always possible to modify behavior by modifying environment as well as heredity. Since the former can be done so much more rapidly than the latter, it is always a good idea to supplement a breeding program with one of improved training and upbringing.

This brings up the practical question of whether it is better to rear dogs in a poor environment, with the idea that only the best animals will overcome it and will be chosen for breeding, or whether it is better to set up a favorable environment and select animals which do best under these conditions. This, of course, is a matter of practical judgment. The former plan obviously leads to a great deal of waste, and dogs may be selected which do well in poor environments but in no others. In general, the best plan would be to raise the animals under the best possible conditions and base a selection program on increasingly higher standards.

Periods of development.—Designing a good environment for the development of a puppy depends upon knowledge of the periods of development. It is, of course, important at all ages to provide adequate nutrition and to prevent disease. Beyond these requirements ideal care varies from period to period. In the neonatal period a normal mother will provide optimal care for her puppies, and attention should be concentrated upon making sure that the bitch is well nourished and allowed to care for her puppies undisturbed. The most that the owner needs to do is to inspect the puppies once a day for possible illness or accidents. This inspection may lead to secondary benefits. Although we have no direct evidence, experiments with other species (Levine, 1962; Denenberg, 1962) strongly indicate that young animals benefit from the stimulation of handling.

A similar regime can be continued through the transition period, and it is not until the beginning of the period of socialization that additional care becomes important. The first thing is to provide additional food, beginning at 3 weeks of age or possibly earlier if the mother is short of milk or has a large litter.

In the period of socialization there are two basic rules for producing a well-balanced and well-adjusted dog. The first of these is that the ideal time to produce a close social relationship between a puppy and his master occurs between 6 and 8 weeks of age. This is the optimal time to remove a puppy from the litter and make it into a house pet. If this is done earlier, especially at 4 weeks or before, the puppy has little opportunity to form normal social relationships with other dogs. It will form close relationships with people but may have difficulty adjusting to its own kind even in mating or caring for puppies. On the other hand, if primary socialization with people is put off to a much later period (the outside limit being about 12 weeks), the social relationships of the puppy with other dogs may be very good, but he will tend to be timid and to lack confidence with people. Although all dog breeds have the capacity to develop a close social relationship with people, the importance of this relationship varies with the dog's future use. A strong relationship is highly important with pet dogs, working dogs, and those hunting dogs which work under close direction. It is probably not so important in most hounds, with which the dog-human relationship is not so essential for successful hunting.

The second general rule is that the young dog should be introduced, at least in a preliminary way, to the circumstances in which it will live as an adult, and this should be done before 3 or 4 months of age. The young puppy from 8 to 12 weeks is a highly malleable and adaptable animal, and this is the time to lay the foundation for its future life work. Dogs left in a kennel until 4 months of age or older are frequently poorly adapted to any other life. This conclusion is strongly supported by the work of Krushinskii (1962) as well as our own. Although an exceptional animal or breed may do fairly well when these two rules are not followed, their observance will bring out the full genetic capacities of a maximum number of animals.

The importance of breed differences.—There are important breed differences in almost every aspect of behavior and physique, and even in the development of social relationships. At the same time, the capacities of each breed are a great deal broader than most people realize. A breed may have one highly specialized ability, but in most cases it also retains broad and general capacities as well. For the practical breeder and even for the pet owner this finding has relevance to the choice of a breed for any particular purpose. Some breeds are rather narrowly specialized in their behavior, particularly the working dogs, which often make poor household pets unless they are given a rather high degree of training and regular tasks to perform. On the other hand, even these breeds retain wider abilities. German shepherds can be taught tracking, and collies retain an ability to hunt deer, as many collie owners have found to their cost.

In general, hunting dogs do better in a kennel than do the working breeds. In our own experiments, basenjis, beagles, cockers, and terriers all adjusted well to the laboratory conditions. The same hunting breeds adapt themselves well as house pets. Beagles are a favorite for this purpose because of their extreme lack of aggressiveness. With a minimum of training, a beagle will become a bearable house pet, and the chief difficulty with the breed under modern conditions is inattentiveness to cars, with a resulting high death rate. On the other hand, they do poorly as trained performers. Other kinds of hunting breeds, such as the bird dogs and retrievers, also adapt well to household living and usually are much better at accepting inhibitory training than the hounds. The general principle is that many breeds have a wider range of adaptability than is ordinarily recognized, and that the scope of adaptability may be much narrower in some breeds than others.

Heterogeneity within pure breeds.—For our purposes it was desirable to minimize variation within breeds by starting with one or two foundation pairs in each group. The procedure was successful in demonstrating significant variation between breeds in almost every measure, but the preceding pages also document the wide range of behavioral and physical variation within each breed. We may recall such examples as the weight of Shetland sheep dogs at one year (mean = 11.7 kg.; range, 4.2–31.6 kg.) or leash-training demerits of basenjis (mean = 57; range, 20–97). Obviously a great deal of phenotypic variation is still present in the pure breeds of dogs, and the genetic source of some of this variation is shown by the general finding that matings within a breed tended to be consistently different from each other. This means that the practical breeder can usually effect major changes in a breed by means of a selection program. Whether or not he will be successful in achieving the changes he desires will depend upon the nature of his objectives and the details of the breeding program.

Importance of emotional differences.—Differences in emotional reactions are very common, both between and within breeds. This has several consequences of interest to the practical breeder. Most of these emotional reactions are specific in nature. For example, a basenji is afraid of strange apparatus but shows little timidity with respect to other dogs. There appear to be several different kinds of fearfulness in dogs, and selection for confidence in one particular situation will not necessarily affect confidence in another. The problem of unwanted timidity presents itself in almost every breed. Its very universality suggests that it may be a trait necessary for existence. An animal which is afraid of nothing is not likely to live long or be an amenable companion for man, and selection should probably be directed toward a desirable balance between timidity and overconfidence rather than complete elimination of either trait.

Emotional reactions have important effects upon performance, so that selection for ability to learn particular tasks is likely to affect emotions first and true cognitive abilities later. It is sometimes very difficult to separate the two in a practical situation.

Motivational differences.—Closely related to emotional differences are variations in degree of motivation. In terms of psychological learning theory, repeated reinforcement of an act by reward or punishment leads to increased motivation. Our five breeds differed in their preferred modes of reinforcement. Beagles, for example, were reliably motivated by food; Shetland sheep dogs and basenjis would sometimes merely sniff at the food rewards and were typically hesitant about eating. An opportunity to explore, particularly in an odor-rich environment, appeared more rewarding to beagles than to Shetland sheep dogs. Differences in reaction to unpleasant or aversive stimulation were also manifested. Physical restraint evoked more struggling and distress vocalization in basenjis than in wirehaired terriers, although in their pens both breeds appeared to be active and aggressive. Shelties are highly sensitive to punishment. Hayes (1962) has gone so far as to suggest that the effects of genes upon intelligence are exerted primarily through modifications of the motivating properties of particular classes of stimuli.

For practical dog breeders, the most important motivational characteristic of a dog is its response to praise and blame from a trainer. Breed variations in attraction to humans developed early in our puppies, as shown by the handling-test results and were obvious when the dogs lived in outside runs. Although all the dogs had the same amount of previous handling, spaniels and beagles would typically swarm over an experimenter during the weekly examination, while basenji and Shetland sheep dogs stood aside. Undoubtedly selection for social responses to man has played an important part in dog breeding. When an animal can be reliably motivated, in either a positive or negative way, its behavior can be molded by controlling reinforcement, so that this aspect of hereditary variation in behavior has the greatest significance in choosing a breed for a particular task.

Heredity and intelligence.—The inheritance of the tendency to perform well in a particular situation, insofar as it can be separated from emotional and motivational reactions, appears to be highly complex. The general principle here is that problem solving is a process of adaptation, and the animal uses whatever capacities it has in order to solve the problem. This was apparent in the different solutions of the problem of box climbing in the motor-skill test, in which the cocker and basenji breeds used very different capacities to achieve the same result. In a practical program of selection this might mean that two dogs having very different basic capacities but equal performance might be selected and mated together, giving rise to offspring with still other combinations of abilities which might be inferior to either parent. Any real progress in such a situation would depend on analysis of the basic capacities involved and indedendent selection for each.

In dogs we found nothing like the general-intelligence factor sometimes postulated for humans. This failure might have been due to inadequate scope of testing, since the dimensions of intelligence were not the major goals of the study. Our results are, however, compatible with those of Searle (1949), who found that rats selected for good performance on one type of maze performed no better than the average on other learning tasks. Provisionally, we have adopted the view that genetic effects in specialized tasks are mediated through numerous independent pathways, and that selection for good performance on one task will improve only very closely related tasks. The unity or multiplicity of intelligence (from the genetic point of view) was not critically tested in these experiments, and more research is needed to settle the issue. It seems possible that since certain psychological processes have been shown to be critically dependent upon the integrity of localized portions of the brain, these or other psychological processes might be controlled by a particular group of genes. It should eventually be possible to describe any specific problem-solving ability in the dog in terms of a few basic capacities, but at present we can specify only a few of these relating to emotional and motivational responses.

Trait independence: lack of correlation between physical and behavioral measures.—One of our major findings was a negative one, the failure to find correlations between physical and behavioral measures. The F_2 basenji-spaniel hybrids demonstrated this best. These animals showed a wide range of variability in appearance, none showing exactly the same combinations of characteristics as the parent breeds. The physical phenotype yielded no clues to the results of the behavioral tests. The dog with basenji coat and tail carriage often was docile like the average spaniel. Although some heritable variation in behavior obviously rests upon heritable structural characteristics, such as length of legs and lop ears, most if not all behavior is unrelated to the color and form characteristics which define the common image of a basenji or cocker spaniel. Therefore, attempts to characterize a breed biologically and psychologically with respect to a type embodied in a particular individual are unsound. The definition of a type has some value in the show ring, because it gives judges an objective basis for scoring a competitive event, but it should be recognized as an abstraction. A breed is not defined by conformity to type, but by common ancestry and absence of outbreeding. If conformity to type is established as the goal of a selective breeding program, certain difficulties are bound to appear because of the complexity of genetic determination of the many traits which enter into a type definition.

Inheritance of coat color and body structure.—Closely associated with the behavior research project C. C. Little (1957), building on the previous work of others, worked out the major outlines of color inheritance in the dog. In most cases color, hair length, and hair texture are determined by major genes with clear-cut dominance and epistatic relationships between them, with some minor modifying genes. This makes the selection of a desired color a fairly easy process. Size and physical conformation are apparently determined by multiple factors, but since these characteristics can be easily quantified, this does not interfere with selection.

The inheritance of structural defects is less well known. The pure breeds usually contain genes which produce physical defects. Cleft palate, hip dysplasia, blindness produced by progressive retinal atrophy, and hemophilia are all known to be at least partially caused by heredity. More extensive lists have been made by Burns (1952) and Fuller (1954, 1956a, 1960). In our own particular strains we found cases of undershot jaws and a tendency to inguinal hernia in the basenji, otocephaly ("pig jaw") in the beagles, low-degree hydrocephaly in cockers, obesity and monorchidism in shelties, and club feet in wire-haired terriers. Infertile individuals produced by various physiological defects showed up in all the strains.

Some of these conditions are rare and are seldom seen by the ordinary breeder; others are common enough to cause concern among breed associations. Their genetic management depends upon the mode of inheritance, the frequency of the defect, its economic importance, and its effect upon viability and fertility. Most breeders will find it desirable to consult a geneticist for advice on specific problems, but a few examples will illustrate the difficulties involved.

Our Shetland sheep dogs illustrate a simple type of problem and its genetic management. Some of them carried the dominant M gene, which changes black hair to gray with black spots. In the proper combination with other genes, it produces the highly admired blue merle color. A merle is always heterozygous (Mm), since MM animals are not blue but almost completely white and partially blind and deaf. Prevention of the defect is easily achieved by not allowing any two animals carrying M, whether merle or white, to mate with each other. As long as one parent is mm, no defective whites will be produced.

Other deleterious genes are not as readily managed. Scattered among the dog breeds are many recessive genes which produce no obvious phenotypic effects except in double dose. Many of these genes are quite rare and the defects correspondingly infrequent, but under special circumstances a particular gene may become common. Progressive retinal atrophy leading to blindness is such a condition. In certain breeds a few popular stud males apparently carried the gene and disseminated it so widely among numerous descendants that blindness has become a serious problem for the breed associations. The only way to eliminate the gene is through breeding tests. Without them, it is impossible to distinguish between carriers and non-carriers. Test matings to a blind animal must be made and the production of six or seven normal-sighted puppies taken as presumptive evidence of the gene's absence. Such a procedure is inevitably time consuming and expensive.

Some physical defects consistently appear in certain strains but follow no regular pattern of inheritance. Thus occasional cases of cleft palate appeared in our beagles and club feet in our wire-haired fox terriers. Such cases can be accounted for by a genetic constitution sensitive to harmful environmental influences; the more sensitive the genotype, the less environmental disturbance is needed to produce the anomaly.

The published data on hip dysplasia indicate that it is an environmentally influenced defect with important hereditary determinants (Fuller, 1960). Vulnerability runs in families, but the degree of affliction is so variable that precise genetic predictions cannot be made. A survey of three breeds being reared as potential guides for the blind showed that offspring of parents with good hips practically never had dysplasia as judged from X-ray photographs; on the other hand, parents with poor hips might have satisfactory offspring but produced more than their share of defective pups. In addition, it appears likely that more than one gene is concerned.

The practical genetic management of such a multiple-factor defect of moderate heritability is, however, reasonably simple. It consists of breeding from the best stock available, selecting those lines in which the defect appears infrequently, and disregarding rare cases of defects which may result from environmental causes. The method is not infallible, but it should produce desirable results unless all the available breeding stock are carriers of the deleterious genes.

For the practical breeder the widespread occurrence of inherited physical defects means that any program of selection must be based on multiple criteria. It is of no use to produce a highly intelligent dog if he is also infertile.

BREED AND STRAIN IMPROVEMENT

Selection.—The major resource of the animal breeder is selection, which implies both a desired objective and the provision of a standardized rearing procedure to minimize environmental variation. Selection has been treated by a number of authors from both practical and theoretical points of view (Lush, 1943; Lerner, 1958; Falconer, 1960). Basically, selection is an attempt to bring about the differential reproduction of genotypes, and so change the relative proportions of alternate alleles from generation to generation.

The results of selection depend chiefly upon the degree of heritability and the mode of inheritance of the desired trait. As shown in chapters 12 and 14, heritability estimates based on the five pure breeds and crosses between two of them range between zero and 66 per cent, with an average of 27 per cent. Heritability estimates on traits valuable in livestock production, such as wool length or egg production, rarely run over 60 per cent and often are much lower (Falconer, 1960). On the average, therefore, behavior in dogs should be as responsive to selection as are physical traits.

Of course, the average breeder is more likely to be working with a pure breed than with hybrids. Even here the indications are favorable, since the between-mating variance averaged 12 per cent in a sample in which this kind of variance had been deliberately reduced. The amount of variance available for selection in an unrestricted sample of a pure breed must be much greater.

The other factor on which the success of selection depends is the mode of inheritance. If inheritance is simple, results will be achieved very rapidly. For example, selection for an effect produced by a single recessive gene, such as the non-merle condition, can produce maximum effects in one generation. On the other hand, if an effect depends upon a combination of several genes, which may interact with each other, results will come much more slowly. Our evidence indicates that the mode of inheritance of differences in emotional reactions and related simple behavior patterns is relatively simple, whereas the behavior involved in problem solving and other complex tasks, while equally heritable, is governed by complex modes of inheritance. To the numerous possibilities of interaction in gene physiology are added the enormous possibilities of interaction on the behavioral level; the animal organizes its capacities in different ways depending on the task at hand. Selection in such a case will be based on the best *combination* of genes present in the original population, and Lush (1943) makes the suggestion, based on the earlier work of Wright (1935) that it may be very difficult to get away from this combination and make further progress, without adding new combinations of genes through outcrossing.

All in all, selection for behavior traits in the dog should give most rapid results when applied to emotional and motivational traits and simple behavior patterns. The final problem of selection is, of course, that of selecting for many traits at the same time, and this is an important practical problem for any breeder.

Hybridization.—The well-known effect of hybrid vigor can be utilized in dogs as well as corn. Hybridization is essentially a technique of reliably producing a highly desirable combination of genes. It depends upon maintaining pure strains and, at the outset, upon having a large number of parent strains from which to choose, so that the best combination of parents can be selected. The technique is comparable to that of "nicking," a phenomenon in which a cross between two specific individuals produces unusually desirable offspring. The hybrid effect cannot be maintained beyond one generation, as is also true with nicking between individuals.

The pleasing quality of F₁ hybrids from cockers and basenjis suggests that breed intercrosses might be used to produce superior working animals, just as hybrid corn is used in agriculture to produce superior grain. If there are objections to crossing breeds, separate lines within a large breed might be developed to serve the same purpose. The method would probably be more effective, however, with the wider difference between parents, and we see no reason for the non-acceptance of planned hybrid matings once their purpose is understood.

Methods for the individual breeder.—The use of the above tools will of course depend a great deal upon the objectives and resources of each breeder. An individual owner usually has a limited number of animals at his disposal and cannot afford the 10 or 15 years necessary to accomplish a selection program. As stated above, most of the behavior concerned with complex performance depends upon a combination of a large number of genetic capacities. The individual breeder will usually find that the best chance of getting such combinations will be to try out various matings within his own stock and other stocks accessible to him. Once a mating is found which produces a high proportion of desirable puppies, the breeder can then repeat this mating throughout the lives of the parent animals and obtain a large number of offspring. This general plan, based on "nicking," is suitable for the breeder with limited resources and time.

Methods for breed associations.—The breed associations were founded in order to overcome the limitations imposed upon the individual breeders, and it is possible for them to accomplish much more through long-continued selection programs. In the future, the breed associations can accomplish more than they have in the past by modifying their objectives and making use of newer genetic theories and techniques. First of all it should be realized that a breed is a *population* of individuals showing a limited but still important degree of genetic variability. If selection is confined to one narrowly defined type, the result will almost inevitably be the accidental selection of various undesirable characteristics. Breed standards should include regulations relating to health, behavior, vigor, and fertility as well as body form. These can perhaps best be accomplished by introducing tests of performance and emotional reactions as well as appearance. Obedience trials and field trials are a valuable step in this direction.

The desirability of multiple standards makes the practice of breeding a champion to a large number of females within a breed a questionable one. Almost every animal carries some sort of injurious recessive genes, and this practice insures that they will be spread throughout the whole breed, with resulting disappointment as the descendants of these champions are eventually bred together and the recessive traits begin to show up in large numbers. The breed objectives should not be the development of a single, fixed type—something which is only possible by strict inbreeding—but rather for the development of a population varying within desirable limits and within which new and more valuable combinations of genes will always be possible.

The breeding of working dogs.—Dogs still perform major services as stock herders, as guide dogs, and as police and guard dogs. Breeding improved animals for such duties involves selection based on multiple criteria, including behavior, fertility, vigor, and special physical attributes of size and strength.

Another important consideration is the maintenance of uniform rearing and training conditions, which should be the best available. Otherwise, variation between individuals may depend as much or more upon the early environment of the animal as upon its heredity. As a first consequence of a selection program for good performance, we would predict rapid changes in emotional reactions. Second, because selection must be made on the basis of performance of a complex task, and because such a performance is likely to rest on a particular combination of genes, one would expect that continued selection on this basis would soon lead to a standstill in progress, and that there would be great difficulty in fixing the desired genetic combination. Progress should be made most rapidly if the ability to perform can be broken down into specific capacities having a reasonably simple genetic basis. Otherwise, the breeder should combine selection with occasional outcrosses in order to bring in new combinations, repeating those matings which appear most successful. A large number of animals must be available to compensate for waste from unsuccessful matings.

Dogs for research.—The many advantages of pure bred dogs whose entire life history is known are obvious, and the use of such animals is increasing rapidly (Scott, Ginsburg, *et al.*, 1962). We have already pointed out the necessity for the socialization of kennel-raised animals. Laboratory-raised puppies must also have the opportunity for socialization at the proper time if they are not to become fearful and untrainable. In addition, they have the same requirements for foundation training and early introduction to the environment in which they are to live as do other dogs. As a minimum, such puppies should be given elementary obedience training, including leash and sit training, and an introduction to the laboratory rooms, before the age of 3 or 4 months.

The choice of breeds depends upon the research objectives. Beagles have been widely used, and some workers have advocated that this breed be adopted as *the* laboratory dog, in the same way that psychologists used to use Wistar albino rats as the standard laboratory strain. There are good reasons for the choice. As hunting dogs, beagles do well in kennel conditions. They are of medium size and inexpensive to feed. Having been developed primarily as a hunting rather than a show breed, they retain a great deal of vigor. They have short hair, and therefore require little grooming. Finally, they are one of the least aggressive of dog breeds, so that they can be handled safely with a minimum of training. At the same time it should be remembered that beagles are not a universal dog any more than any other breed, being specialized in many ways. To adopt one breed as a standard laboratory animal would be to throw away one of the main scientific advantages that dogs possess—their enormous genetic diversity. For purposes of experimental surgery, for example, some of the larger breeds are more useful, and for purposes of studies of behavior the highly trainable working breeds are more interesting. In long-term experiments which require a high degree of co-operation and control, trained working dogs would have obvious advantages, and it should be remembered when selecting a breed that dogs in general are highly adjustable and adaptable animals. Almost any breed can do well in a laboratory environment provided certain adjustments in care and training are made for that particular breed.

The vigor of F_1 hybrids between the pure breeds recommends their use in many laboratory situations, and their nature needs to be explored. Each new hybrid is essentially a new breed, as it will possess new combinations of characteristics depending on whether the parental traits are inherited as dominants or recessives. Our F_1 cocker-basenji hybrids were like basenjis in their ability to climb, like cockers in their response to food, and intermediate with respect to barking. New characteristics are also possible through interaction between new combinations of genes.

Finally there is a possibility of creating new breeds especially selected for research purposes. In this case it would be well to start with hybrids between two or more breeds in order to obtain the maximum advantages from variation.

SUMMARY

The tools of the animal breeder are improvement of the environment, genetic variation, selection, and hybridization. The results of the experiments described in these pages have been to sharpen these tools rather than create new ones. Our experiments have led to important ways of improving the early social environment, particularly by taking advantage of critical periods in development. We have found that genetic variation is highly important within breeds as well as between them, so that it should be possible to transform a strain within a breed into a very different sort of animal within a few generations. The most important genetic differences in behavior appear to be emotional and motivational rather than those of cognition or basic intelligence. Such traits have important effects upon any type of performance and should respond rapidly to selection, provided the animals are raised under uniform conditions minimizing environmental variations. Superficial appearance is no guide to behavior, and the use of "marker genes" consequently has little validity. Finally, any selection program leading to the improvement of behavior must of necessity be one which involves multiple criteria including general physical and reproductive fitness.

Our two major experiments, breed comparison and breed intercrossing, do not in themselves yield a blueprint for dog breeding, but they do furnish guides for persons interested in selecting for behavioral characteristics or in producing working dogs. The authors hope that this by-product of their research will be of value to the large group who find pleasure in trying to understand man's four-footed companion. As will be seen in the next chapter, an understanding of dogs also helps those whose chief interest is understanding man.

CHAPTER 16

THE EVOLUTION OF DOGS AND MEN

INTRODUCTION

When dogs were first domesticated, some eight to ten thousand years ago, they became a part of human society, and they have since undergone most of the cultural and environmental changes that have affected their masters. Dogs, however, have shorter lives than men, and they go through new generations at the approximate rate of one in every two years instead of man's one in twenty. We can roughly estimate that dogs have gone through some four thousand generations since their domestication while man has gone through only four hundred. These facts suggest a hypothesis: the genetic consequences of civilized living should be intensified in the dog, and therefore the dog should give us some idea of the genetic future of mankind, always assuming that there are no radical new changes in the conditions of human living. In short, the dog may be a genetic pilot experiment for the human race.

HISTORICAL PERSPECTIVE

Wolves and men had much in common even before domestication took place. Just as the lion in the southern hemisphere occupied the ecological position of the dominant carnivore preying on the herd mammals, so did the wolf, *Canis lupus,* in the northern one. Once early men had invented the spear and the bow and arrow, they too could successfully attack and prey upon the large ungulates such as deer, mountain sheep, and bison. Stone Age hunting tribes soon found themselves in direct competition with wolves for similar prey.

Furthermore, as we have pointed out, both wolves and men are social species, showing a great deal of co-operation and mutually

397

helpful behavior within their own groups. Finally, as an ecologically dominant animal, with no other competitors except man, the wolf shows some polymorphism, as humans do, in addition to the usual division of social animals into males, females, and young. Wild wolves show enough variation in form, color, and behavior that Murie (1944) could easily recognize each member of a wild wolf pack.

Adaptive radiation.—When an animal species moves unopposed into a new environment, there is an opportunity for very rapid genetic change, because each small group populates a large area with its own descendants, which in turn reflect the individual genetic peculiarities of their parents. At the same time, there is a rapid selection for those animals best adapted to the new areas. The whole process is called adaptive radiation and is the sort of thing which once happened to the whole class of mammals after the reptiles had been eliminated as possible rivals.

Adaptive radiation on a smaller scale seems to have taken place soon after the dog became domesticated. Within the various human societies, dogs found a whole new habitat. The dog, as one of the first domestic animals, was a remarkable social invention, both for protection and as an aid to hunting, and every tribe must have wanted to get hold of one. In this way dogs spread rapidly over the world, differentiating as they moved, and so produced the southern short-haired varieties like the dingo and, at the opposite extreme of their range, the northern Eskimo dogs which are almost like wolves (see Fig. 2.2). A further multiplication of habitats was provided when the herd animals were domesticated. Now dogs were needed to protect these herds against their own close relatives, the wolves, which found the domestic beasts easy prey.

The domestication of animals and plants produced another major environmental change. Hunting was no longer a primary occupation, but a sport and frequently a luxury of the rich. Under these circumstances dogs became specialized for the hunting of many different animals. Still another expansion of habitats occurred after the industrial revolution, with the use of dogs as pets and social companions for city dwellers. However, none of these later developments had the importance of the first major expansion into the new environment provided by human societies.

Period of isolation.—As Wright (1950) and other geneticists have pointed out, an ideal situation for rapid genetic change is one in which a population is divided into small subpopulations with contacts permitting occasional gene exchanges between them. When dogs were first domesticated, this was still the situation in the majority of human societies, which were divided into small semi-isolated tribes, still reflecting the conditions which had led to the very rapid biological evolution of modern man in the previous fifty thousand years.

Divided among tribes, dogs were maintained in even smaller social groups than their masters, with only occasional opportunities for mixing with animals from adjacent tribes. Under these conditions we would expect rapid differentiation between subpopulations, and this was probably the foundation of the major breed differences. Historical records of American Indian tribes show that each one had its own particular breed of dog, perhaps not very different from that of the neighboring tribe but greatly different from those at long geographical distances (Allen, 1920). Under similar conditions in ancient Eurasia there originated the large war dogs of one tribe, the long-legged greyhounds used for hunting gazelles on the deserts, and the shepherd dogs used to guard domestic flocks.

Period of panmixis.—The period of local isolation was succeeded by one in which new methods of transportation brought explorers and immigrants together from all over the world. Wherever the European explorers went, whether for trade, discovery, or war, they brought their own breeds of dogs, which readily interbred with the native varieties. Sometimes the native breeds disappeared, as they did in America and South Africa. Sometimes the Europeans brought new breeds back to their shores, as the crusaders brought the greyhounds back from the Middle East. In either case, this was a period of mixing and crossing.

Period of genetic control.—The idea of intensifying and preserving desirable characteristics in dogs through scientific breeding is hardly more than a century old, the first dog show having been held in Newcastle, England, in 1859 (Ash, 1924). Before this time dogs were mainly produced and bred on the basis of results, no matter what their ancestors might be. In the latter part of the 19th century people began to appreciate the possibilities of artificial selection and began trying to keep dog populations completely separate in the modern pure breeds. These attempts at genetic change and improvement were not always scientific or completely effective, but they have resulted in a new diversity of dog populations. Many of the modern breeds were originated within the past century, having been produced by crossing older breeds and then selecting their descendants for special characteristics.

Genetic changes in human populations.—The first three of the

above periods can also be found in the history of human populations, although extended over greater lengths of time. The modern
human races, with lighter skinned people concentrated toward the
northern polar region and the darker ones toward the equator, probably represent the result of an ancient period of adaptive radiation.
The period of isolation into small tribes was one which began to
disappear with the dawn of civilization some eight or ten thousand
years ago. As people moved into cities, they began to live in larger
populations, and these populations began to be continuous with and
mix with adjacent ones. In modern times this process of panmixis is
still accelerating, with people moving rapidly over continents and
the entire world, leaving their descendants everywhere.

A period of scientific breeding of human populations is yet to
come, but the beginnings are already being made (Hammond, 1959;
Reed, 1963). As we accumulate authentic information about the inheritance of genetic defects and other qualities, prospective parents
are being given advice affecting their decisions to have children.
Moreover, whether it is based on scientific information or not, the
cultural habit of free choice in marriage leads to assortive mating.
There is consequently a strong correlation between body size, appearance, and even intelligence tests between husbands and wives.
What does the future have in store? One possible answer is given by
the canine pilot experiment.

THE RESULTS OF GENETIC CHANGE

Diversity of form and behavior.—Our first problem is to assess
what has happened to the dog as a result of long membership in
human society. The most obvious thing is an extraordinary diversity
of form and behavior. We now have dwarf breeds which weigh less
than four pounds at maturity, and giant breeds like the Great Danes
and Saint Bernards which weigh at least as much as the largest
northern wolves. There are all sorts of variations in hair structure
and color. Density ranges from near absence in the Mexican hairless
breed to heavy curls in the ever growing coats of poodles. Hair may
be wiry or silky, curly or straight, and appear in almost any color of
the visible spectrum with the exception of green. Ears can be erect
or drooping and all gradations between. Tails can be short or long
and vary from a sickle shape to a tight curl. Legs also can be short
or long, and the skull can be deformed and shortened as in bulldogs,
or long and narrow as in the greyhounds. The only thing which remains relatively invariable is general body form. Aside from occa-

sional obese individuals, all breeds retain the large chest and thin waist of their ancestors.

Such wide variability is also characteristic of human beings, among whom there is wide variation of skin and hair color, shape of facial features such as ears and nose, and, unlike dogs, in general body proportions. If human beings are more variable than dogs in some ways, they are less so in others. For example, a Great Dane may weigh forty times as much as a Chihuahua. Nothing like this extreme diversity is found in the human species, where the largest adults may be only four or five times heavier than the smallest, except for cases of extreme obesity and dwarfism.

This genetic diversity of men, dogs, and other domestic animals contrasts greatly with what we see in most wild animals. It is almost impossible to tell one squirrel from another of the same species by outward appearance. In fact, this difficulty of telling individuals apart is one of the perpetual problems of studying animal sociology under natural conditions. We assume that uniformity is a result of natural selection; that mutations which produce white spotting, for example, make a squirrel so conspicuous that he is much more likely to be taken by a predator than are his standard-colored relatives.

In a species not subjected to intensive selection by predators, a greater variety of individuals can survive without difficulty. Furthermore, in a highly social animal many sorts of new variations will be protected by group action. In short, the condition of being a highly successful social species automatically results in a relaxation of natural selection, permitting a greater variety of individuals to exist.

We may wonder why the tendency toward diversity has gone to greater lengths in the dog than in man. There are two possible reasons. One is that the dog has had more genetic time, in terms of generations, to show the effects of relaxed natural selection. The other is, of course, that people have deliberately preserved and multiplied harmful canine mutations as objects of curiosity, in a way which has never been done with human mutants. We might then conclude that the dog is genetically weak compared to the ancestral wolf, and that we ought to attempt to restore the more uniform appearance and greater vigor characteristic of these ancestral animals. Is this conclusion really justified?

Modification of fertility.—Compared with the wolf, dogs are generally more fertile than their wild ancestors. Wolves mature at two years of age and sometimes later, and the average litter size is four or five. This probably represents the number of animals which can be successfully reared to maturity under natural conditions, where food must be provided by hunting. Dogs of most breeds, on the other hand, become sexually mature before one year of age and have developed a twice-yearly breeding cycle instead of an annual one. This means that a pair of dogs has the potentiality of producing twice as many offspring as a pair of wolves. Medium-sized breeds usually produce litters about the same size as those of wolves, but the larger breeds exceed the production of wolves on the average. Against this, the dwarf breeds produce smaller average litters than do wolves. Also, while we have no statistics on this, there are probably more cases of complete infertility among dogs than among wolves.

However, we can say in general that domestication and becoming a part of human societies has not only extended the range of fertility in dogs but actually made dogs more fertile than their ancestors. Such high fertility would be a disadvantage under natural conditions because of the difficulty of raising the puppies, but in a human society, where the puppies are partially raised by human beings, it is actually more efficient to rear a large litter than a small one, since the amount of work required is roughly the same.

Reduction of wildness.—If a wolf puppy is taken from its parents before its eyes are open and is adopted into a human society, it will develop very much like a domestic dog. However, adoption becomes rapidly more difficult as the wolf cubs grow older, because of the quick development of fear reactions and fighting behavior. An older wolf cub bites in frantic fear when caught. Although Ginsburg's (1963) recent work has shown that even adult wild wolves can be socialized eventually, this is a long and delicate process. There has evidently been selection in dogs toward the extension of the critical period during which socialization is easy. There may also have been some reduction of alertness and wariness of strange sounds and movements, but this is doubtful, because one of the principal values of dogs is just this capacity—to alert the householder to possible dangers. In fact, a perennial problem in most breeds is the development of more timidity than the breeders consider desirable.

The specialization of behavior.—In almost any behavioral characteristic there are breeds of dogs which surpass the capacities of wolves in either direction. The terrier breeds, as evidenced by their incapacity for living in large groups, are more aggressive than wolves, whereas the hound breeds like the beagles are far less aggressive. Greyhounds are faster runners than wolves, but many dogs are much slower, especially those from short-legged breeds such as the dachshund. Some breeds like the trail hounds appear to be better trackers than wolves, whereas others like the terriers appear to be poorer. The sheep dogs are better herders than wolves and, of course, many breeds are much poorer. The setters, pointers, spaniels, and other bird dogs appear to have much more interest in birds than do wolves, a special capacity useful only with human co-operation. Wolves themselves are poor bird hunters compared to foxes.

Although many breeds of dogs consistently exceed wolves in their capacities, none of them can be considered super-wolves. Their special capacities have been achieved at the price of sacrificing others. The greyhound has achieved speed by sacrificing the heavy muscles and jaws which enable a wolf to live on bones if need be. It is inconceivable that any particular domestic breed could compete with wolves under natural conditions, with the possible exception of the Eskimo dogs. Even these would be at a disadvantage because of their smaller teeth and jaws. In fact, dogs have successfully gone wild again only where there is an absence of such competition, as on the continent of Australia or around towns and cities.

None of the dog breeds are super-wolves. A wolf is a rugged and powerful animal adapted to life under a variety of adverse conditions. Consequently, no one of his behavioral capacities can be developed to a high degree. Compared with wolves, dogs are a group of specialists. But as they are co-ordinated and sheltered by human society they can perform their functions more efficiently than any group of wolves.

We can conclude that the development of all-round capacities to meet a variety of environmental conditions leads not to a superb development of these capacities but to a balance between them. An all-round development means great adaptability, but not superb performance.

This suggests that the idea that natural selection will produce a super-man or super-animal of any sort is an unobtainable myth. As far back as the Greeks, and possibly long before, people have dreamed of godlike or heroic figures who combined the physical strength of a gorilla with the intelligence of an Einstein, the musical gifts of a Beethoven with the beauty of a Greek God. From the historical and genetic evidence afforded by dogs, this appears to be an impossibility, and we are now beginning to realize why. On the other hand, a complex society can use all these superb qualities in the form of specialized individuals, so that the society as a whole has capacities far beyond those enjoyed by a uniformly developed group. The super-man is not to be found as an individual but as a well-developed human society.

THE ACCUMULATION OF MUTATIONS

Almost as soon as Darwin put forward his theory of natural selection, scientists began to reason that if natural selection would cause the improvement of a species, then the relaxation of selection would automatically cause degeneration. The argument seemed plausible enough because, if unfit individuals were not eliminated, there ought to be more of them in the next generation, and if this process continued generation after generation, the result would eventually be genetic disaster. However, this idea first became current before the discovery of Mendelian genetics and long before we had any idea of the process of mutation. As we shall see below, the relaxation of natural selection does not necessarily lead to degeneration.

The Hardy-Weinberg law.—Once the mechanism of Mendelian heredity had been discovered, it was possible to mathematically calculate its effect upon future generations. All chromosomes exist in pairs, and there is an equal probability that one or the other of each pair will be passed along to an individual in the next generation. In an infinitely large population with no selection and no assortive mating, genes carried on these chromosomes will therefore be passed along in exactly the same proportion as that existing in the parent population. The Mendelian mechanism has the function of recombining these genes in all possible combinations without changing their frequency. This principle, known as the Hardy-Weinberg Law, may be briefly stated as that of "the constancy of gene frequency."

Therefore, there is nothing in the ordinary mechanism of inheritance to cause degeneration. If natural selection ceases to act upon a large population, the heredity present in the group at this time will be passed along without change from one generation to the next, from that time forward. The result is neither degeneration nor improvement, but stability.

The situation is slightly different in a small population. Here the accidental selection of one kind of parents may produce a shift in heredity from one generation to the next, but the change is as likely to be in the direction of improvement as toward degeneration. In a large population composed of a number of small subgroups, we would expect that some would be improved and some worsened, with the whole population remaining exactly as before.

Thus neither the relaxation of natural selection nor the operation of the Mendelian mechanism of heredity will lead to progressive genetic deterioration. The one remaining possibility of degeneration lies in the process of mutation.

The effect of mutation on the dog.—The dog provides an extreme example of the unfavorable effects of mutation. Not only has selection been relaxed because dog owners protect and feed their pets from the kennel to the grave, but fanciers have practiced breeding systems which actually increase and multiply injurious mutations. In the first place, they have selected certain freakish traits, such as the bulldog head, which are definitely inferior to the natural conformation. Second, they have selected for neutral traits such as coat color without regard to viability and fertility. The best looking show animal in a litter may accidentally be a poor breeder or subject to various sorts of sickness. Third, in recent times most dog breeders have deliberately followed a policy of outbreeding, trying various sorts of matings between non-related dogs in the hope of obtaining an unusual or desirable type. For practical purposes this is a good way to change or improve a breed. However, continuous outbreeding also takes away the natural method of reducing the number of injurious recessive genes, i.e., through death or infertility of homozygous individuals. Under these breeding methods such recessive genes will not even be recognized unless they become so frequent in the population that carriers begin to mate with each other. According to the Hardy-Weinberg Law, there should be no spread of such mutations through the population, but the dog breeders have provided a method for doing just this. If a male becomes a great champion, everyone wants one of his puppies, and he may be bred to several hundred females. If he carries even one recessive gene, it will be spread throughout the whole breed in such numbers that it will be almost impossible to eliminate when his descendants eventually begin to be mated with each other and it is finally recognized as a serious problem. Thus current dog breeding practices can be described as an ideal system for the spread and preservation of injurious recessive genes.

The seriousness of this effect is well illustrated in Table 16.1. In the pure breeds, the neonatal death rate averages about 15 per cent. This is, if anything, an underestimate of mortality rates in the total breed populations, since females with unusually good breeding records were selected for our experiment whenever possible. The F_1 hybrids showed only two neonatal deaths, both stillbirths, and the mortality rate in the segregating hybrids is intermediate between the F_1's and the pure breeds. Assuming that the F_1 deaths were caused by

TABLE 16.1

MORTALITY RATE PER 100 BIRTHS IN PURE BREEDS AND HYBRIDS

Population	No. at Birth	Neonatal Mortality (before 3 weeks)	Total Mortality (1 yr)
Basenji...................	57	3.5	8.7
Beagle.....................	83	13.2	16.8
Cocker....................	90	18.9	21.1
Sheltie....................	43	20.9	23.2
Fox terrier................	54	14.8	22.2
Total..................	327	14.4	18.3
BA + CS...................	147	12.9	16.3
All F_1......................	59	3.4	3.4
All backcrosses..............	90	7.8	16.6 (11.1*)
All F_2......................	87	5.8	5.8
All segregating hybrids........	177	6.8	11.3 (8.5*)

* Excluding five animals which died in distemper epidemic.

environmental accidents, we can estimate that the genetic death rate is 10 to 12 per cent. Estimates from other colonies give neonatal death rates of anywhere from 10 to 30 per cent (Corbin *et al.*, 1962), depending upon the strain of animals involved, so that 15 per cent should be a good rough estimate of neonatal genetic mortality for pure breeds in general.

In addition to these lethal effects, there is widespread occurrence of crippling defects and deformities in purebred dogs. In our experiments we began with what were considered good breeding stocks, with a fair number of champions in their ancestry. When we bred these animals to their close relatives for even one or two generations, we uncovered serious defects in every breed. As mentioned in the last chapter, our African basenjis showed a high frequency of undershot jaws in both sexes and of inguinal hernia in the males. Beagles showed an opposite defect—a reduction of the lower jaw. Until corrected by selection, wire-haired fox terriers produced many puppies with club feet. The Shetland sheep-dogs produced individuals with hereditary obesity. All breeds showed a tendency to decline in fertility.

In addition, the cocker spaniels showed an outstanding defect which had apparently been established throughout the strain before we received it. This breed has been selected for a broad forehead with prominent eyes and a pronounced "stop," or angle between the nose and forehead. When we examined the brains of some of these animals during autopsy, we found that they showed a mild degree of hydrocephaly; that is, in selecting for skull shape, the breeders had accidentally selected for a brain defect in some individuals. Besides all this, in most of our strains only about 50 per cent of the females were capable of rearing normal, healthy litters, even under nearly ideal conditions of care.

Among other dog breeds, such defects are quite common. One which has recently received a great deal of attention is hip dysplasia, particularly in German shepherds, but also fairly commonly in other breeds. In this defect the hip joint is not properly developed, with the result that the animals become lame and crippled. In the German shepherd breed it is associated with the fact that the show breeders have preferred a sort of "downhill" carriage (associated with "angulation" of the hindlegs) in which the shoulders are higher than the hips. In selecting for this trait they may have unwittingly selected for lame dogs.

Another example is hereditary hemophilia, which has been discovered in English setters. Indeed, almost any important constitutional disease of human beings has its counterpart in dogs, which are consequently highly useful for medical study. All the evidence is in favor of the conclusion that injurious recessive mutations have accumulated in the dog and become extremely common. At the same time there is no indication that dogs are becoming extinct as a result of genetic deterioration. If anything, there are more dogs today than ever before.

The theory of genetic equilibrium.—The hereditary composition of any wild species tends to reach a condition of balance, insofar as the various factors producing genetic change work against each other. For example, the increase of injurious recessive mutations reaches a state of balance with the death rate of homozygous individuals, and the frequency of appearance of homozygotes depends upon the degree of inbreeding in the population as well as on the mutation rate.

In the case of the domestic dog a similar balance seems to have been reached. One reason why dogs are not disappearing is that, although about 50 per cent of the females are unable to produce living young, reproduction by a fertile animal is approximately double that of the ancestral wolves. The canine reproductive rate far exceeds that for human beings, as a good bitch may easily produce 50 living offspring by the time she is 6 years old. This increased reproductive capacity more than compensates for the genetic loss of fertility.

In spite of the accumulation of recessive mutations, there is still selection against abnormal individuals. Extremely defective individuals, such as dogs with hereditary obesity, are often completely infertile. Those with minor defects such as undershot jaw are frequently not chosen for breeding stock, and dog breeders are becoming more and more aware of the necessity of selecting for good health and vitality as well as particular show points. In short, even in a highly protected species, natural selection is not completely inoperable, nor does artificial selection always work against it.

We could make another point in passing. A dog which is infertile or unsuitable for breeding may still make a highly satisfactory house pet or even occasionally be an outstandingly successful performer as a guide dog, hunter, or herder. In a social animal, or an animal such as the dog which is actually part of human society, reproductivity is not the sole criterion of usefulness.

We can conclude that the relaxation of natural selection in the dog has not produced a continuing process of degeneration but rather has set up a new condition of genetic balance which permits a wider degree of variation.

The theoretical danger from mutations.—From the study of fruit flies and mice, geneticists have been able to calculate mutation rates. Some genes mutate fairly frequently and others almost never. The average rate seems to be about one mutation in a particular gene for every million individuals. Mutations can also be in the reverse direction; a new mutation will occasionally revert to the original. The average rate is presumably the same in either direction, and we can hypothesize that if a mutation had no effect on survival and was therefore not removed by natural selection, the process of mutation would go on until the reverse mutation rate balanced the original rate, which should be at the point where there would be equal numbers of the new mutation and the original gene (Wright, 1955).

This kind of mutation may be very common, but usually escapes notice because a mutation which has little effect on survival is, of course, one which has little effect of any kind. The mutations which we notice are those which produce large effects, and these are almost universally injurious. If such an injurious gene should occur in as high a ratio as in the hypothetical example above, in which 50 per cent of the genes in the population were of the mutant sort, there would be 25 per cent of defective individuals in every generation. If these defective individuals were as capable of reproduction as normal animals, the same high proportion of defectives would be maintained in future generations. The consequence would be a continuing heavy burden on the population.

However, all our evidence indicates that defective individuals, while they may occasionally be able to reproduce, are always selected against and frequently are completely infertile. If the whole 25 per cent of defective individuals in a population were incapable of reproduction in the next generation, one-half of the normal genes would have to mutate in each generation in order to maintain an equal ratio of mutant and non-mutant genes. This is a tremendously high mutation rate compared with one in a million or, indeed, with anything that has been produced artificially by radiation.

Many scientists are concerned with the genetic effects upon human populations of increased exposure to radiation, either in medical practice or through the use of atomic weapons, or even by exposure to radiation in outer space. According to Crow (1961), any increase in the mutation rate over the natural one is a bad thing, as it will in the long run increase the number of seriously defective individuals in our population. To give some idea of what this increase might mean, Morton (1960) has calculated the effect of doubling the mutation rate for certain defects, until a condition of equilibrium is reached. The mutation rate can be doubled with the surprisingly low radiation dose of 25 roentgens per generation. When an irradiated population of one million people had reached stability, there would be an increase of at least 60 cases of limb-girdle muscular dystrophy (a disease in which the muscles waste away), 450 cases of deaf mutism, and 1,370 cases of low-grade mental deficiency—an increase of some 1,880 defective individuals in just these three categories.

An increased mutation rate would, of course, greatly increase the prenatal and postnatal death rates of human infants. We are still a great deal better off than purebred dogs. In 1960 the combined deaths at birth and up to one year of age were less than one in every twenty live births, while the corresponding figures in purebred dogs were about one in five (Table 16.1). With increased mutation rates, we could expect the neonatal and infantile death rates to rise to a much higher level than at present.

We have been thinking of the dog as being a pilot experiment for the genetic future of the human race. However, human heredity need not follow the same course, because we can at least avoid the mistake of special selection of defective individuals for parents. The example of the dog is somewhat hopeful, in spite of all the unfavorable changes that have taken place, because the species continues to flourish. By analogy, we may predict that human heredity will eventually come into a new condition of balance, depending on the nature of the forces for change.

With regard to fertility, one of our most important world problems is that of increased human fertility produced by superior nutrition and medical care. There is no indication that there has been any genetic increase in fertility, as in the dog. On the other hand, we know that at the present time about 10 per cent of marriages in the United States are confronted with the problem of sterility. We may expect that if mutation rates increase, the percentage of sterility will gradually rise and will in time reach a balance which would cancel out the increased survival rate. However, it might take centuries before such balance is actually achieved.

The changes produced by increased mutatation rates must be slow, for even if natural selection is relaxed in any population, it is certainly not eliminated and may even be increased in some respects through the special stresses of civilized living (Dobzhansky, 1962). There is always selection against sterility. More than this, the improvement of the environment achieved by social and civilized living has made very little change in the prenatal environment. The usual estimates indicate that there is a prenatal mortality rate of approximately 20 per cent, that one out of five pregnancies never comes to term. The real rate may be even higher, since many embryos die so soon that their existence is ordinarily never detected. Figures for some of the domestic animals indicate a prenatal mortality rate of 30 to 40 per cent. Prenatal life is therefore the major screening ground for injurious mutations and one which is not likely to be seriously interfered with in the future. The prenatal death rate also indicates that there are actually a large number of injurious mutations already present in human populations (Morton estimates 3 to 5 per individual), and that there is strong selection against them.

To get the whole matter into perspective, the relaxation of selection chiefly affects postnatal life, after the most important part of selection has already taken place. Even in postnatal life, natural selection has not been eliminated. Parents who have a defective child do not recklessly have many more, being strongly deterred by the inevitable tragedy and expense. It used to be thought that we were selecting in favor of feebleminded individuals, and that morons would breed like rabbits. As Reed (1963) has shown, the actual figures indicate that while there may be a few exceptional feebleminded persons with many children, the average feebleminded individual has far fewer children than the rest of the population. Furthermore, the near relatives of such persons also have fewer children.

In short, we may conclude that the relaxation of natural selection in human societies has not set into motion a continuing process of degeneration but has rather set up a new condition of equilibrium which permits a wider degree of variation. In addition, we should remember that some of these variants may well have special abilities superior to the ideal of an all-round man, instead of being invariably below the average.

We have already shown that selective breeding has not produced a super-wolf among dogs, and we might predict that if anyone should attempt to select a breed of men having all-round ability, health, and vigor, he would probably produce not a group of supermen, but a very average group of individuals with no exceptional qualities. In essence, such people might indeed be very similar to Stone Age men, but we have no indication that primitive men were anything other than extremely tough individuals who could survive under many conditions. By contrast, our present situation with its relaxation of selection permits the trying out of almost infinite numbers of gene combinations, and some modern individuals may well be superior in all-around abilities to the best that Stone Age man could offer. However, under our present circumstances of large populations and panmixis, these generally superior individuals are unlikely to produce a new race of men. If they appear, they will make a social rather than a genetic contribution to the future. Even in this respect, their social value is likely to be relatively small compared to that of individuals with markedly superior special talents.

Present conditions permit the survival of socially valuable but genetically specialized individuals, lacking the all-round survival capacities of Stone Age man, but possessing extremely superior special qualities which can be used to good advantage in a highly developed civilized society.

Human society is thus basically polymorphic (Dobzhansky, 1962). As with other societies of mammals, we are biologically differentiated into three types: males, females, and young. A high degree of social organization encourages the development of a further diversity of individuals within these main classes. Even in a society of wolves, it is an advantage to have both extremely timid and extremely courageous animals. The timid animal may scent danger from afar and withdraw to warn the others, while the bold animal may press forward to investigate and obtain food for the pack when the danger proves to be small. In a human society a diversity of individuals is even more useful.

A human society thus rests on a genetic basis different both from that in the insects, where polymorphism is expressed in a small number of genetically determined castes, and from that in dogs, where genetic diversity is organized into breeds. Human societies are founded on the genetic basis of many individual differences within the three biological types of males, females, and young, and the ideal human society is consequently one which recognizes, respects, and uses human individuality and variation.

CHAPTER 17

TOWARD A SCIENCE OF SOCIAL GENETICS

The science of heredity is primarily concerned with the action of genes or basic hereditary units. At first, the principal problem was their transmission from generation to generation by means of the chromosomes. Later, scientists became interested in how genes work, and this problem developed into a subscience of physiological genetics. Another problem was how genes are distributed and passed along in natural populations, as opposed to the controlled experiments of the laboratory. In answer to this question the subscience of population genetics was born.

The science of heredity has thus been subdivided according to its effects at different levels of biological organization. In order to complete the table, we need to add two more divisions at the individual and social levels. One concerns the effect of genes on the activity of an entire individual. This division, which we call "behavior genetics," is treated in chapters 7–14 of this book. Another involves the effect of genetics on social groups, or "social genetics." Finally, the science of developmental genetics cuts across all levels of organization (Table 17.1).

TABLE 17.1

THE SUBDIVISIONS OF GENETICS

Level of Organization	Unit of Organization	Subdivision	
Ecological	Population	Population genetics	
Societal	Society	Social genetics	Developmental genetics
Behavioral or Psychological	Organism	Behavior genetics	
Physiological	Organs, tissues, etc.	Physiological genetics	
Cellular	Cells	Mendelian genetics	
Molecular	Gene	Chemical genetics	

413

For the most part population genetics has dealt with two units: the individual and the whole population. Research workers sometimes assume for theoretical purposes that a population is unorganized, with its members moving about and mating at random, just as a group of gas molecules are unorganized. This assumption may be justified in many cases, particularly in some of the lower animals, but the higher animals are always organized into some sort of social groups in which they neither move nor mate completely at random. Furthermore, in such social animals there is a new basic unit, the social relationship, which means that two or more individuals form a unit. A number of relationships in turn make up an organized group, which may range in size from a single pair to several hundred individuals. In many animals each group constitutes an entire society in itself, but in many others the groups have at least some connection with each other and are organized into a larger whole. Among human beings, such groups of groups are more complexly organized and make up institutions, and it is a group of institutions that becomes a society. Finally, a population may consist of a single society or a group of societies, depending on the size of the geographical unit on which it is based.

These differences between various kinds of societies mean that the first problem of social genetics is one of basic information. We need to know the anatomy of the animal society with which we deal: the composition of groups, the nature of the relationships within a group, and the important types of basic relationships. Within this framework, there are two major theoretical problems. One is how genetics affects a relationship and the other is the reciprocal question of how the social environment affects genetics and the process of evolution.

THE ANATOMY OF CANINE SOCIETIES

All members of the family Canidae show the same basic patterns of social behavior, just as they show the same basic type of structure (Scott, 1950). Even foxes (Tembrock, 1957) go through much the same patterns of sexual and agonistic behavior as dogs and wolves. Such differences as exist are chiefly matters of degree rather than kind. A fox's bark can easily be told from that of a dog, but it is still recognizably a bark.

The principal differences between these canine societies, therefore, lie not in basic social behavior patterns, but in the complexity of social groups. For foxes, the mated pair is the largest social group aside from the temporary association of a mother and her young litter, and fox breeders have great difficulty in getting males to mate with more than one female. Likewise, the mated pair is the typical group in coyotes and jackals, although they are occasionally seen in small packs. At the opposite extreme are the dogs and wolves, whose typical organization is the pack of several adult males and females. In wolves, a group composed of a single mated pair is the exception rather than the rule.

The comparative behavior of canine individuals suggests that behavior is a conservative trait in evolution just as their comparative anatomy suggests that general body form is resistant to change. In spite of all the human selection to which dogs have been subjected, we still do not have any two-legged breeds of dogs or dogs that employ human speech. Conservatism in both form and behavior probably results from the fact that it is difficult to change any highly organized system without disrupting it.

In contrast to the anatomy and behavior of individuals, the structure and function of canine societies is relatively labile and changeable, there being wide differences in the composition of groups. The basic reason for this lies in the transition from one level of organization to another. There are greater changes on the social level of organization than on the behavioral level because a new type of organization on the higher level can be constructed without greatly modifying that on the lower. The behavior patterns of wolves are different from other Canidae in only minor ways, but these make possible the development of larger social groups and all that a large group can do in contrast to a small one.

The typical dog or wolf society is not a highly organized one. The number of individuals in a pack is usually small, and the relationships between the members are simple. Food is divided through a dominance organization but can be obtained either by combined or individual effort. As might be expected in a group of animals which are likely to be all the same age, there is no strong system of leadership.

When socialized to people, both dogs and wolves transfer to human beings the social relationships which they would normally develop with their own kind insofar as this is permitted by their owners. Dogs do not acquire human behavior through this association, but continue to show the dog-like patterns of behavior common to all the Canidae.

Novelty does arise in the development of the relationship between dog and human. People transfer their social relationships to dogs and attempt to teach them human social customs and new skills. The result is a relationship which is neither typically human nor typically canine. The dog is no longer an equal member of a pack, although he will develop along these lines if allowed. Rather he becomes a perpetual follower and dependent. Nor is he like a human child, who eventually becomes more independent and capable of caring for others of his kind. The dog brings bones back to his yard as he would to a pack, but these are not shared by his master.

In our experiments we have principally studied the effect of genetics upon these two simple social relationships: that between the puppy and his litter mates and that between dog and master. The simplicity of these relationships has made the task somewhat easier than it might otherwise have been and also has had the advantage of making certain general principles more obvious.

The development and differentiation of social relationships.—One of our most important discoveries was that the development of puppies is divided into definite periods, each characterized by an important process. The neonatal period is dominated by the process of infantile nutrition. The transition period is one of rapid acquisition and development of adult behavior patterns. The period of socialization is one of development of primary social relationships. The juvenile period is one of growth and the acquisition of skills which eventually make the puppy potentially independent of its parents or caretakers. Finally, sexual maturity marks the development of sexual relationships, and the consequent birth of young makes possible the development of the parental relationship with a new generation.

Social development correlated with social organization.—Canine development is so organized that the period of socialization is one in which the mother normally leaves puppies for long periods while the litter stays together and waits to be fed. Consequently, the strongest social relationships are formed not with older animals but with litter mates, and this appears to be the foundation for pack formation. We still do not have complete information on the natural formation of new packs in wolves, but Murie (1944) observed a young litter moving about as a unit as it grew up, already acting as a pack. He also saw a single adult attempting to join an established pack and being rejected by it. It seems likely that most wolf packs are formed by litters which stay together as adults.

Thus the development of social behavior in dogs is directly related to the type of social organization typical of adult wolves and dogs allowed to run wild in packs. When we compare canine development with that in other species, we can see that we have good evidence of a general law. In sheep (Scott, 1960 and 1962*a*), the young are born capable of moving around actively; the process of socialization begins immediately after birth; and lambs begin to follow their mothers within a few days. Thus the stage is set for the formation of permanent flocks led by older animals in contrast to the canine pack of a similar age group.

Human beings, though different from either sheep or dogs, follow the same general law. They are more like dogs in that babies are born in an extremely immature state, but unlike them in that physical development is much slower. Along with this, the process of primary socialization begins before the transition to the adult form of locomotion, at a time when the infants still require constant care. (Scott, 1963a). This means that babies form their primary social relationships with older persons. Human societies, and indeed most primate societies, are characterized by long associations between younger and older members of the group and in this way are more like a sheep society than a canine one where the members of a pack may all be the same age. However, human beings are sharply different from sheep because of their slow development, which means that the human infant is helpless and must be carried while the process of socialization is going on, whereas a lamb is immediately capable of independent locomotion and at once starts to form a habit of following its mother. Theoretically this would make the human infant develop somewhat more independently, since training in following begins relatively much later in life.

If this generalization is correct—that the development of social behavior is directly related to the type of social organization typical of adults—it should follow that the disturbance of social development would cause the disturbance (or at least the modification) of social organization. The point at which the maximum change in social development can be most easily produced is during the process of socialization. This is therefore a critical period for the determination of adult behavior and social organization.

Our evidence on the puppy reinforces the general conclusion that in any highly social species of animal there is a relatively brief period, usually early in development, when primary social relations are established, normally to members of the same species. During this period, it is easy to transfer social relationships to another species by experimental means. This law has great significance in explaining the domestication and taming of wild animals. It also has many implications for the modification of social organization in human beings and the disruption of organization by disturbances in development.

Genetic variation in development.—Now, how does genetics modify these general developmental processes? In the first place, all dog breeds go through the same sequence of developmental stages and processes at approximately the same time. The variations due to selective breeding have produced no fundamental disturbance of development. However, there are variations in the speed of development of various behavioral capacities, and these can vary independently of each other. Fox terriers, for example, are quick to develop the capacity for hearing but at 3 weeks of age are slower than the other breeds in most respects. This perhaps foreshadows the future behavior of these animals, as fox terriers are highly responsive to sound. There is no indication that such incompatibility of development produces abnormality. In studies of human development it has been suggested that an individual who has slow physical development but rapid social development, or vice versa, may be maladjusted. Our work with dogs suggests that such variations may be related to individual differences but that the result is eventually a well-integrated individual.

In human development, Bayley (Bayer and Bayley, 1959) has shown that variations in speed of development may have important effects on children in our society. Early sexual maturity usually has favorable effects on boys and unfavorable effects on girls in regard to social adjustment to their classmates. We have not observed any important effect of this sort in dogs raised under our conditions except possibly the late maturation of the sheep dogs, which performed relatively poorly on many tests. The general tradition among dog owners is that these animals develop confidence more slowly than some other breeds. However, there are other possible explanations of their poor performance, and we can only say that genetic differences in the speed of development do not seem to produce important effects in the early life of dogs, although they may do so later on. One reason for this is the fact that the possible range of such differences in the period of early development is quite small compared with that in later life.

Heredity and social relationships.—We may now consider the effect of hereditary differences upon the development of a social relationship. This kind of development goes along with the behavioral development described above, but involves in addition the behavior of a second individual. Also, while behavioral development is to a large extent dependent on growth, the development of a social relationship may take place very rapidly and at any time in development.

Our best example of a social relationship is the dominance-subordination relationship between litter mates. We studied the development of this relationship in detail under controlled conditions and found that within it the breeds developed important differences in agonistic behavior. In fact, one of the biggest genetic modifications of basic dog behavior patterns has been the selection of certain breeds for aggressiveness and others for peacefulness. As described in an earlier chapter, one of the most obvious effects was the limitation of group size in fox terriers. Fox terrier puppies were unable to work out any kind of livable relationship in litters larger than three individuals.

The situation poses an interesting theoretical question. Presumably the members of each breed are nearly equal in aggressive abilities, but the nature of fighting makes it impossible for both animals to win and continue to be equally aggressive. One result of this situation might be a fight to the death in the highly aggressive breeds, and this is exactly what happens if fox terriers are left to themselves. Another possibility is that the more aggressive breeds would work out systems of complete dominance while the more peaceful breeds would show little or no dominance among themselves.

The breeds did indeed show this kind of difference. Fox terriers and basenjis show many more instances of complete dominance than do beagles and cocker spaniels, while shelties are intermediate. Examining the situation in detail, we find in every breed more completely dominant males than females, but the differences between the sexes in this respect are far greater in the two most aggressive breeds. That is, in aggressive breeds the great majority of relationships with complete dominance are cases where males are dominant over females.

Here we see the result of a hereditary difference within breeds —that between males and females. When we compare dominance between males in one breed with that in another, the differences in the amount expressed are not great. The big difference occurs in the male-female relationships; these are more highly differentiated in the more aggressive breed than in the more peaceful one. In a dominance-subordination relationship the expression of a genetic difference in aggressive behavior depends upon the two individuals also being different from each other with respect to sex. This would not necessarily be true in every social relationship, but it does illustrate the complexity of the problem of how a difference between genes is finally expressed in a social relationship between two unlike animals.

"Personality" measured by social relationships.—Human beings are often concerned with the effect or impression that one person

produces on another and speak of his "personality." One of the descriptive terms applied to human personality is that of dominance. What shall we say about the "personality" of the dogs in this experiment? Obviously we cannot say that all fox terriers are dominant and all beagles subordinate. This might be true if they developed social relationships between the two breeds, but within its own group the same fox terrier may be dominant in one relationship and subordinate in another, and the same thing may be true of a beagle. Neither can we say that one breed is aggressive and the other submissive. Again this depends upon the social relationships concerned. All that we can say is that a fox terrier has a greater capacity than some other breeds to develop aggressiveness in a relationship where the capacities of the two individuals are unequal. That is, the development of the capacity depends not only upon the possession of latent ability but on the behavioral development of the other individual concerned.

This is a new concept of personality or, more properly speaking, a more accurate way to measure it. What we are basically interested in is the characteristic social behavior of an individual. Since it varies from relationship to relationship, we can only describe it in terms of the most important relationships in an individual's life. For a dog in the Jackson Laboratory School for Dogs, the most important relationships were those with his litter mates, with the human caretakers and experimenters, and with occasional human strangers. The same fox terrier might be dominant with all his litter mates, subordinate to the caretakers, and fearful with strangers.

Among human beings we can easily observe the same kind of contradiction. It is a rare individual who appears the same to his wife as to his business acquaintances. The small boy who attempts to treat his playground acquaintances in the same way that he manages life with his parents is quickly disillusioned. We have all seen cases of individuals who produce a wonderful impression upon strangers but who cannot get along with their everyday acquaintances. All this suggests that the only realistic way to get a practical and measurable picture of personality is in terms of the important social relationships of each individual. Most current "personality tests" are based on information regarding a mixture of many relationships and often depend upon the individual's view of himself rather than his effect on others. Even the clinician sees an individual only in one relationship, that of doctor and patient. Some relationships are highly specific and others broad and general. Sometimes they may resemble each other but, as our experimental evidence shows, some can be highly independent of others. At the very least we must keep in mind the

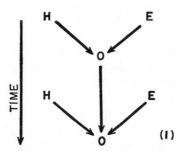

Fig. 17.1.—Functional differentiation of form during growth. Hereditary (H) and environmental (E) factors both act on physiological processes within the organism (O), and these processes by their own function modify further processes.

distinction between the *capacity to develop* a relationship, which may depend largely on heredity, and the actual relationship, which depends on the heredity of the other individual as well.

The prediction of social development.—Figures 17.1 to 17.3 graphically represent the effect of heredity upon the development of a social relationship. The diagrams are somewhat artificial in that they are inevitably more simple than reality, and because it is difficult to express the fact that several processes are going on simultaneously. The arrows are intended to indicate action of one process

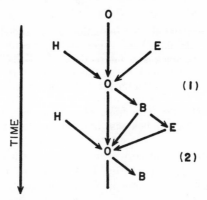

Fig. 17.2.—The functional differentiation of behavior in relation to the physical environment. The organism previously differentiated by growth processes at (1) is now capable of behavior (B), which may change the nature of the environment, as well as the nature of the organism (O) through the process of learning. Heredity, on the other hand, remains unchanged but acts on a continually changing organism. Just as in the previous diagram, the activity of the organism modifies itself, but this time the arrows lead through behavior and the physical environment before feedback takes place.

on another, and the arrows all point in one general direction, indicating the passage of time.

The diagrams reflect two fundamental effects. One is that an organism undergoes continual change and development, so that neither behavior nor physiology and structure are the same at one point in time as at an earlier one. The second effect is that an organism changes itself by its own activity. This is the fundamental concept of functional differentiation. Early in life these changes occur mostly through the process of growth. By differential growth and other

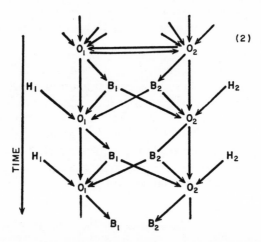

Fig. 17.3.—The functional differentiation of a social relationship. Both organisms (O_1, O_2) stimulate each other at (2), resulting in an extremely complex diagram of feedback and interaction. The behavior of an individual in a social relationship (B_1 and B_2 at the bottom of the diagram) depends upon the heredity and past environment of the other individual as well as his own. Further, his own heredity and environment affect him through the other organism as well as directly. The effect of a developing social relationship is thus a mutual modification of behavior and the creation of something new.

physiological processes, an embryo changes and modifies its own capacity to develop further (Fig. 17.1). Later in development the organism begins to change through the process of learning as well as growth. The two processes are combined in Figure 17.2 as if one preceded the other, but they actually overlap. In this diagram we have indicated that heredity and the environment continually act upon an organism, which is also at any given time the result of their past action. There is a new complication in that behavior may change the physical environment, as well as changing the nature of the organism through the process of learning. This change in environment is reflected in future behavior. Just as in the previous diagram, the activity of the organism modifies itself and hence its future activity, but this time the arrows lead through behavior and the physical environment.

Figure 17.3 shows the result of substituting another individual organism for the physical environment. The result is an enormously complicated diagram. The behavior of an individual in a social relationship depends not only upon his own heredity and past environment (experience), but also on those of the other individual. Furthermore, the arrows from his own heredity and past environment eventually lead through his behavior to the other individual and produce a response, which in turn modifies his own behavior. The result is the mutual modification of behavior and the creation of something which is impossible for either individual to achieve by himself: a social relationship.

Does this make it impossible either theoretically or practically to analyze the effect of heredity? The diagrams indicate that heredity is not, as far as we know, modified by the activity of the organism. The heredity of an individual remains a constant, but the fact that it acts within a continuously changing organism makes for difficulties.

As we found in the case of the development of the dominance relationship in puppies, we can predict the action of heredity, provided we have reasonably pure breeding strains developing within a uniform environment. We can predict that fox terriers reared in large litters with their mothers will develop serious fighting about the age of 7 weeks and will eventually work out dominance orders in which the males are completely dominant over the females. We can also predict some of the other possibilities based on experimental evidence.

However, where we do not have pure breeding stocks and where heredity is unknown except through its expression in an individual, there is a far more difficult problem, and this is the one which we face in a human society. In order to predict the behavior of a human individual we need to know his basic capacities for the development of social behavior, irrespective of how they have been modified by environment and use. For example, we need tests of aggressiveness which will measure a person's capacity to develop aggressive behavior without reference to the amount that he has actually developed. Since such capacities are basically physiological, our hope of success lies in the development of physiological tests. If we could measure the degree of emotional arousal in the central nervous system, we would have a measurement of one trait involved in the capacity to develop aggressiveness. Our studies with dogs have not proceeded far in this direction, but we do have one such physiological measurement connected with the arousal of emotion, namely, the heart rate reaction, in which the terriers were quite different from all the rest.

The other practical need is for tests of different sorts which can be given soon enough in development so that environmental opportunities could, if needful, be adjusted to basic capacities. In this respect our studies of the dog are only moderately encouraging. Differences in aggressive behavior are not readily recognizable until about 7 weeks of age. Other differences in behavior occur earlier, but they are not directly correlated with adult traits. In fact, it looks as if the earliest expression of social behavior is quite variable, indicating that the capacities for behavioral development are at first quite flexible.

If we could get basic information on early development, we would at best have only a list of capacities. Actual behavior would depend on whether or not these capacities were in fact developed. Consequently if we want accurate prediction of the behavior of an adult individual, we must work in terms of the potential development of capacities as expressed in important early social relationships. Fortunately, the number of such relationships is limited, or the task of description would be impossibly difficult, complicated as it is by the phenomenon of cultural change.

Just as biological change is determined by the biological mechanism for transmitting biological heredity from one generation to another, so cultural change depends on the basic mechanism for transmitting learned information from one generation to the next. Thus the mechanism of learning has the same fundamental relationship to theories of cultural change that the mechanism of Mendelian heredity bears to the theory of organic evolution. As we have seen in past chapters, the two basic mechanisms are not entirely independent of each other, and the problem of how heredity affects the capacity to learn consequently has great theoretical importance.

Heredity and the capacity to learn.—Although these experiments were guided by the general concept of social behavior, the problem of individual differences in learning and problem solving seemed so important to us that we devoted a large number of our tests to an attempt to measure the differential learning capacities of dog breeds and individuals.

We began with the idea that the study of development would be important because this would enable us to observe the behavior of young animals before they had been molded by the environment and while they were more completely controlled by heredity. We therefore expected to see the effects of genetics pure and undefiled in the young puppy. This idea turned out to be somewhat naïve, for we found that many of the characteristic breed differences in behavior did not appear in early life, and this eventually led us to the concepts of the development and *differentiation* of behavior, affected by both the processes of heredity and those of learning (Scott, 1957).

Instead of showing stereotyped behavior, a young puppy confronted by a new situation usually tries out one pattern of behavior after another. If one of these seems to work, he is likely to repeat it on subsequent occasions, with the result that his behavior eventually becomes less variable and more consistent. Thus there are two opposing processes in any behavioral adaptation which involves learning. One of these is variation, which permits improvement of performance, and the other is habit formation, which fixes performance. Heredity may alter the balance between these two processes in either direction. For example, Shetland sheep dogs form fixed habits very readily, particularly those involving inhibitory training, whereas beagles form such habits less readily than the average dog. Although heredity thus determines differences between the breeds, the final limits on behavior are set by the process of habit formation.

Effect of learning on hereditary differences.—Once he reaches the age of 5 or 6 weeks, a puppy has a wide choice of behavior patterns which can be used in any given situation. As we have pointed out in earlier chapters, there were several kinds of learning situations included in those we gave the dogs. One of these was a problem with only one correct answer, a solution which we imposed upon the dog. This is the method of "forced training," well-known to professional dog trainers. In a situation like this, heredity can modify the speed with which the dog accepts the training, as in the leash-control test, and also the degree of perfection with which he performs the specified movements, but it has relatively little effect on the end result. When forced training is imposed on two breeds of dogs, the final outcome is that they become more alike than they were in the beginning.

In another type of training the puppy is subjected to less rigid control. When we were training puppies to be weighed, we did little more than hold our hands near the puppy to keep him from falling off the scales. From the puppy's point of view, the goal of his behavior was to get off the scales and back on the floor, and this was achieved at the end of one minute, no matter how he acted. In this situation the cocker spaniels usually developed the behavior pattern of sitting quietly, while terriers and basenjis developed that of constant struggling. Genetic differences influenced the choice of the behavior pattern, and the end result was that the three breeds got further apart as they grew older. However, genetics did not control the animal's behavior so exactly that every cocker spaniel always sat quietly and every terrier always struggled. What it did do was influence the probability that these reactions would appear, rather than control them in any ironclad way.

Most of our performance tests involved finding a goal where food was available. There was a correct solution, but the puppy had to find it himself instead of having it forced upon him. Such "problem-solving" tests are natural ones for hunting animals who must find their food under a variety of situations. Although we tried to design the tests so that only one solution was possible, we usually found that the puppies attempted a great variety of solutions and often perfected ones which we did not consider the most highly adaptive. As the theory of the adaptive function of behavioral variability would predict, they showed much more variation of behavior in the early parts of the test than later on. However, in some of these tests (such as the spatial-orientation problem, in which the dogs were given a series of increasingly difficult problems on the same apparatus), the differences between the breeds remained proportionally the same from first to last. Behavior became a great deal more consistent under the process of learning, but the relative amount of variation produced by heredity remained the same.

Thus the effect of learning can be to either reduce hereditary differences, magnify them, or keep them the same, depending upon the kind of training or problem-solving situation involved.

Complexity of interaction between heredity and the learning process.—Learning is only one mechanism by which the effects of heredity are expressed in highly complex ways. As our experience with the heredity of seasonal breeding shows, it is possible to get a highly complex expression of heredity in the physiological factors which underlie behavior, as well as in the interaction of heredity with learning.

Furthermore, we found that hereditary effects upon behavior were highly specific and that we got clear-cut indications of hereditary differences only when we worked with very simple behavior patterns. In most problem-solving situations, behavior is highly complex and many behavior patterns are used, combined, and recombined with each other. The result is that while complex performance and intelligence tests show indications of the effect of heredity in both pure breeds and hybrids, neither the kinds of solutions nor the degrees of success are inherited in any simple Mendelian fashion.

At this point we might emphasize the fact that our work conflicts with several popular notions of the effect of heredity on behavior. One is that there are a few basic hereditary traits which appear in all the behavior of an individual and make his behavior predictable and understandable. Perhaps this is what is ordinarily meant by the terms "character" and "personality." Although it may be possible for human beings to develop such consistent behavior traits through learning, there is no indication in our results that heredity acts in any such simple fashion.

Another mistaken notion is that there is such a thing as "general intelligence." As we showed in an earlier chapter, a dog breed may rate very high in one test situation and quite low in another, with little uniformity. Our general impression is that an individual from any dog breed will perform well in a situation in which he can be highly motivated and for which he has the necessary physical capacities. Heredity may limit both motivation and capacity, but in highly specific ways. If we compare these results with the human situation, we likewise find no general confirmation of the existence of a factor of general intelligence (Anastasi, 1958). Human beings, however, have still another capacity (or capacities) which is probably affected by heredity, and that is the ability to learn and use language. This may change the situation from that which we found in dogs, for in human beings the lack of only one specific group of capacities; i.e., the ability to use language, would be a general handicap to almost all aspects of learning.

Still a third popular notion is that individuals fall into behavioral "types" and that these are somehow related to physique or special physical characteristics. Our results indicate that even in dogs we are dealing with populations involving continuous variation and that "types" have relatively little importance.

We must conclude that any attempt to improve the capacities for learning and problem solving by genetic methods is likely to be most successful when done in a highly specific way, as it has been done for the dog breeds. Even in these animals, the breed differences usually are not great, provided all breeds are given an equal amount of training and opportunity to learn. There is still an enormous amount of general adaptive capacity in the specialized dog breeds, and a purebred dog can learn many things beyond its own specialty.

The chief differences between the breeds seem to lie in the amount of limitation produced by these special capacities. Certain breeds are quite limited with regard to the situations into which they can adapt well, and others are much more generalized. For example, with only a minimum amount of training, the beagle does quite well as a rabbit hunter and almost equally well as a house pet. Most of the shepherd dogs can become independent hunters of big game (as when collies run deer), but they require a great deal of special training in order to become either successful shepherd dogs or house pets.

There has never been any deliberate selection of human beings corresponding to that in dogs, although modern human social organization can conceivably have this effect indirectly by selecting capable individuals of both sexes for college training and then bringing the selected individuals together at the mating age. However, the capacities useful in higher education are so broad and complex compared to the relatively simple capacities which can be easily modified by a genetic selection, that it is unlikely that even such assortive mating will ever produce a highly specialized and uniform population hereditarily different from the rest. Most individuals are likely to have certain special capacities highly developed and others relatively poorly developed. Because of the wide spectrum of adaptability present in most individuals, heredity should not be the most important limiting factor in determining their lives and behavior. In fact, as we have seen above, the most important *limiting* factor on behavior, and particularly that limiting the differentiation of behavior between individuals, is learning and habit formation, and our chief practical problem is to design a social environment which will provide an opportunity to express in useful and pleasurable ways the enormous amount of variation in behavior of which human beings are capable.

How does heredity affect the capacity to learn?—One of the obvious ways in which heredity affects the capacity to learn is by limiting physical capacities. A dachshund, for example, will never be able to learn to catch a rabbit by chasing it. Such limitations, in turn, have an important effect on learned motivation, which is so largely dependent upon success. In addition, the dog breeds show more direct hereditary differences in motivation, hunting dogs in our strains showing more interest in food than representatives of the working breeds. Differences in emotional responsiveness are important and may make the learning of certain tasks easier or harder.

One of the most interesting modifications of learning is that of hereditary shifting of the balance between variability of behavior and fixation by habit formation. These differences, however, appear to be modifications of the peripheral processes of learning (i.e., motivation to learn) rather than of the central phenomenon of association. All dogs can make associations, and it is possible that this kind of ability is so fundamental that the animal cannot exist without it, that learning is as basically important to existence as is metabolism, and as little subject to change.

CULTURAL EVOLUTION AND BIOLOGICAL CHANGE

This brings us to a third basic problem of social genetics: how does social organization affect the process of biological inheritance and thus affect biological evolutionary change? The existence of social relationships in an animal society usually produces a mating system of some sort, so that genes are not passed from one generation to the next in a random fashion, and this fact in turn may either speed up or slow down evolutionary change.

Stabilizing effect of social organization.—From a theoretical point of view, the general effect of a stable social organization on an animal population should be to slow down genetic changes. We have seen that there are two basic kinds of processes of functional differentiation going on: physical and anatomical differentiation based on the process of growth, and behavioral differentiation based on the process of learning as well as on growth. For a change in heredity to survive, it must produce a favorable effect on at least one of the above processes without producing an overriding unfavorable effect on the other. The addition of still another process, the differentiation of behavior in a social relationship, produces a third form of organization in which the hereditary change must produce a favorable or at least neutral effect. This, of course, still further reduces the limits within which permanent hereditary change can take place. These developmental processes are all partially dependent upon one another, and so we find that the social behavior patterns and the social development of a species are directly related to the normal social organization of the species, and that a disruption of development produces a disruption in social organization.

In addition, any population in a constant environment tends to approach a state of genetic equilibrium, and the more constant the social environment, the more stable the genetic system. Even if the state of equilibrium is temporarily disturbed, as by the change in environment produced by civilized living, a new state of genetic balance is eventually reached.

Cultural evolution and biological evolution compared.—In human populations these processes operating to produce genetic stability are opposed by other forces making for social change. A great many animal societies show the beginnings of cultural inheritance. Sometimes animals inherit structures built by their ancestors, as in the case of prairie dog burrows. Or they may learn behavior from older animals, as young deer learn to be fearful of human beings from their elders. To these abilities, human beings have added the capacity for speech, which enormously magnifies the possibility of transmitting learning from one generation to the next. Furthermore, what is learned through speech becomes almost completely independent of biological heredity, with the result that cultural change ordinarily proceeds much faster than biological change.

Anyone who compares biological and cultural evolution becomes fascinated with the resemblances between the two. Languages seem to evolve in much the same way as species, and one can draw a phylogenetic tree of languages which is no different in form from the phylogenetic tree of the species of the animal kingdom. Yet the two kinds of change rest on entirely different processes, one being Mendelian inheritance and the other the process of learning.

In Mendelian heredity a parent passes along to each child only one of each pair of genes, selected purely by chance. With the usual non-inbred human parent, the result is an arrangement of genetic factors in new combinations in the next generation with the virtual certainty that none of his children will have exactly the same combination as his own. Similarly, the basic characteristic of the process of learning is that each child must start all over again from the beginning. It is impossible for him to acquire exactly what his parents have learned, because part of his environment, as distinct from theirs, is what they attempt to teach him. The process of learning is both selective and organizational; consequently the information learned by the previous generation is never passed along completely and never organized in exactly the same way. Biological and cultural heredity are similar in that they are both fundamentally unstable systems which make change inevitable.

There is one big difference between the two systems. Cultural heredity is cumulative and has become much more so with the invention of written symbols and records. Biological heredity, on the other hand, is non-cumulative. Each individual has only the same number of chromosomes and genes as his immediate ancestors. This is another way of stating that learned behavior can be inherited culturally but not biologically.

The result is that human societies inevitably change, and change with great rapidity—so fast that biological change is inevitably left far behind. We have every indication that the human race is physically quite similar to what it was ten thousand years ago, yet this is the time during which we have progressed from primitive hunting tribes to modern industrial civilization. If anything, the process of cultural change seems to be speeding up rather than slowing down.

Heredity, learning, and cultural evolution.—Evolution toward wider capacities for adaptation is a general tendency in the animal kingdom, although there are many exceptions in the case of species which have become specifically and narrowly adapted to a particular environment. Specialized species are likely to become extinct when the environment changes, with the long term result that those which survive usually have a wide range of behavioral and physiological adaptability.

This general tendency involves a de-emphasis of behavior organized on a purely hereditary basis, for such behavior can only be adaptive in a completely stable environment. Along with this we find an increasing emphasis on variable behavior organized by learning. Moreover, the processes of cultural change and cultural evolution automatically produce an unstable and changing environment. In human societies such changes have always taken place between one generation and the next, and in the current speed-up of cultural change they are beginning to take place even within a single lifetime. Consequently there are and will be increasing demands for persons who are flexible and adaptive or, in short, who have increased learning capacities.

The kinds of effects of heredity upon learning capacities which we have described in dogs chiefly contribute to breed and individual differences within the species, increasing the range of adaptability of the species rather than that of the individual, whose own behavior may, in fact, be limited by them.

There are no breeds in the human species, but there is an enormous amount of individual genetic variability. We have every indication that our species as a whole already has a wide range of adaptability, inhabiting as it does every climate on the globe and taking part in innumerable productive and destructive occupations. Individual hereditary variation has the effect of making this range of adaptability for the species even wider, and this in turn has the effect of facilitating cultural change. Thus we can conclude that there is a synergistic relationship between hereditary variation in behavioral capacities and the speed of cultural evolution. Hereditary variation widens the range of adaptability, permitting cultural change; and cultural change, in turn, permits survival of larger populations, with the consequent increased possibility of hereditary variation.

Now, what effect do these rapid changes in human social environment have upon the dog? Dogs reflect the culture in which they live. In the Middle East there are essentially only three kinds of dogs: salukis used for gazelle hunting in the deserts, large herding dogs used by Kurdish shepherds to protect their flocks from wolves, and mongrels which chiefly act as scavengers in the cities. By contrast, in modern France, where dogs are chiefly used on farms, there are some seventeen breeds of shepherd and stock dogs. In England, where hunting has been a popular sport in all classes from as far back as the Middle Ages, there were in 1959 twenty-six recognized breeds of sporting dogs.

We can think of human social change as providing the opportunity for the phenomenon of canine adaptive radiation. Each social change opens up a whole new environment for dogs, and the expansion of their numbers from small original groups provides an opportunity for rapid biological change. In modern Europe and the United States increased wealth and urban living has opened up a whole new environment for dogs as social companions, and we may expect to see the breeds changed and modified in that direction. Thus we see the great popularity of the beagle as a children's pet, beagles being extremely non-aggressive dogs which adapt well to family life with a minimum of training. Another breed which is becoming highly popular is the Chihuahua, whose small size and sturdy physical constitution make it an ideal animal for apartment dwellers. It will bark as vigorously as any watchdog, but is too small to bite the postman.

Modern dogs also fill more serious social functions. Companion dogs provide a solution for many of the psychological strains imposed by cultural change. The elderly couple whose children have moved away can adopt a puppy as a child substitute, and the lonely bachelor or spinster may keep a dog as a dependable companion. We may expect that in the future there will be deliberate as well as unconscious attempts to develop breeds which will satisfy these social needs.

In the long run, the effect of selecting dogs for first one cultural environment and then another will be to produce random selection, eventually leading to permanent genetic change. Although wolflike characteristics are still preserved in various dog breeds, it would be extremely difficult at this point to re-create anything like a wolf from a single breed like the beagle. In general, we can predict that as long as breeds are maintained as separate populations there will be a continuous process of change in directions determined by the as yet unpredictable course of human cultural change. Biological evolution of the dog will come to a standstill only if breeds and local varieties are no longer maintained separately and all animals allowed to mate at random.

This last is essentially what is now happening to the human population as a result of cultural change. The kind of cultural change presently characteristic of human populations seems to have had the net effect of slowing down human biological change while speeding up that of the dog. It is true that human beings, like dogs, continually have new cultural environments opened up to them in which one kind of individual may be more successful than another, but human generations are so long and cultural changes have become so rapid that there is no time for differential selection to have any considerable biological effect. Even this small possibility is nullified by the enormous numbers in human populations.

To conclude, the effect of cultural change upon human societies has been to reduce differentiation between small human subpopulations and to increase the differentiation between individuals within a population. Indeed, there is a synergistic relationship between cultural evolution and individual biological variability. The principal effect of cultural change has been to widen the limits within which various kinds of human beings can exist and create a wider variety of social environments within these limits. One of our efforts toward conscious social change should be in the direction of recognizing this variety and providing social flexibility which would permit the useful and happy development of inevitably different individuals. One of our basic problems in this task will be to recognize individual capacities and assist people to find fitting and congenial social environments. Such assessments are crudely done at present and perhaps can never be done perfectly. In fact, a perfect job is theoretically unnecessary, since one of the basic capacities shared by the majority of human beings is that of adjustment and adaptation to many different sorts of environments.

Another fundamental contribution of a science of human nature will be the discovery of the limits of general adaptation. This will enable us to recognize and avoid harmful and disappointing social change in advance: to choose instead more beneficial social changes and to bring them about in peaceful and productive ways.

BIBLIOGRAPHY

NOTE: The following list contains all papers published by the authors on the subject of dogs, as well as those specifically mentioned in the text. For those interested in a wider bibliography, Baege's (1934) publication gives an excellent coverage of older work, while Mason's (1959) list gives a more recent but less complete source of references.

AHMED, I. A. 1941. Cytological analysis of chromosome behaviour in three breeds of dogs, Proc. Roy. Soc. Edinburgh B., **61**:107–18.

ALLEN, D. L., and MECH, L. D. 1963. Wolves versus moose on Isle Royale, National Geographic Magazine, pp. 200–19.

ALLEN, G. M. 1920. Dogs of the American aborigines, Bull. Museum Comp. Zool. Harvard Coll., **63**:431–517.

American Kennel Club, 1956. The complete dog book. New York.

ANASTASI, A. 1958. Differential psychology. 3d ed. New York: Macmillan.

———, FULLER, J. L., SCOTT, J. P., and SCHMITT, J. R. 1955. A factor analysis of the performance of dogs on certain learning tests, Zoologica, **40**:33–46.

ANDERSON, W. S. 1939. Fertile mare mules, J. Heredity, **30**:549–51.

ASH, E. C. 1924. Dogs, their history and development. London: Ernest Benn.

BAEGE, B. 1933. Zur Entwicklung der Verhaltungsweisen junger Hunde in den ersten drei Lebensmonaten, Z. Hundeforschung, **3**:3–64.

———. 1934. Kynologische Bibliographie. Berlin: Zoological Garden.

BAHRS, A. M. 1927. Notes on the reflexes of puppies in the first six weeks after birth, Am. J. Physiol., **82**:51–55.

BAKER, J. A., ROBSON, D. S., GILLESPIE, J. H., BURGHER, J. A., and DOUGHTY, M. F., 1959. A nomograph that predicts the age to vaccinate puppies against distemper, Cornell Vet., **49**:158–67.

BAKER, P. T. 1958. Racial differences in heat tolerance, Am. J. Phys. Anthropol., **16**:287–306.

BAYER, L. M., and BAYLEY, N. 1959. Growth diagnosis. Chicago: University of Chicago Press.

BEACH, F. A., and GILMORE, R. W. 1949. Response of male dogs to urine from females in heat, J. Mamm., 30:391–92.

BEADLE, G. W. 1945. Biochemical genetics, Chem. Rev., 37:15–96.

BERG, I. A. 1944. Development of behavior: the micturition pattern in the dog, J. Exp. Psychol., 34:343–68.

BLISS, J. Q., STEWART, P. B. and FULLER, J. L. 1958. The selective cutaneous response to autologous and non-autologous plasma in dogs: observations on the effects of inbreeding and of immaturity, Brit. J. Exp. Pathol., 39:30–36.

BLOME, R. 1686. The gentleman's recreation. London: Roycroft.

BLUME, D. 1956. A study of the eye of the dog. Private communication.

BOWLBY, J. 1951. Maternal care and mental health. Geneva: World Health Organization.

BOYDEN, A. 1942. Systematic serology: a critical appreciation, Physiol. Zool., 15:109–45.

BRACE, C. L. 1961. Physique, physiology, and behavior: an attempt to analyze a part of their roles in the canine biogram. Harvard University Ph.D. thesis.

———. 1962. Refocusing on the Neanderthal problem, Am. Anthropol., 64:729–41.

BRAIDWOOD, R. J., and REED, C. A. 1957. The achievement and early consequences of food production: a consideration of the archeological and natural-historical evidence, Cold Spring Harbor Symposia on Quantitative Biology, 22:19–31.

BRELAND, K., and BRELAND, M. 1961. The misbehavior of organisms, Am. Psychol., 16:681–84.

BROADHURST, P. L. 1960. Applications of biometrical genetics to the inheritance of behaviour. In: EYSENCK, H. J. (ed.), Experiments in personality, 1:1–102. New York: Humanities Press.

———. 1962. Note on some possibilities for primate psychogenetics, Lab. Primate Newsletter, 1(4):1–7.

BROADHURST, P. L., and JINKS, J. L. 1961. Biometrical genetics and behavior: reanalysis of published data, Psychol. Bull., 58:337–62.

BRODBECK, A. J. 1954. An exploratory study on the acquisition of dependency behavior in puppies, Bull. Ecol. Soc. Am., 35:73.

BRUELL, J. H. 1962. Dominance and segregation in the inheritance of quantitative behavior in mice. In: BLISS, E. L. (ed.), Roots of behavior, pp. 48–67. New York: Hoeber-Harper.

BURNS, M. 1952. The genetics of the dog. Farnham Royal, Commonwealth Agricultural Bureaux.

CAIUS, J. 1576. Of Englishe dogges. Copied and reprinted in modern type by A. BRADLEY, London, 1880.

CARPENTER, C. R. 1934. A field study of the behavior and social relations of howling monkeys, Comp. Psychol. Monograph, 10(2):1–168.

CASTLE, W. E. 1921. An improved method of estimating the number of genetic factors concerned in cases of blending inheritance, Science, 54:223.

CHAMBERS, R. M., and FULLER, J. L. 1958. Conditioning of skin temperature changes in dogs, J. Comp. Physiol. Psychol., 51:223–26.

CHAPIN, J. P. 1958. Private communication.

CHARLES, M. S., and FULLER, J. L. 1956. Developmental study of the electroencephalogram of the dog, Electroencephalog. Clin. Neurophysiol., 8:645–52.

CLOUDMAN, A. M., and BUNKER, L. E., JR. 1945. The varitint-waddler mouse, J. Heredity, 36:259–63.

COHEN, C., and FULLER, J. L. 1953. The inheritance of blood types in the dog, J. Heredity, 44:225–28.

COLEMAN, C. E. 1943. The Shetland sheepdog. New York: Gallagher, Levine, and Miller.

COLLIAS, N. E. 1952. The development of social behavior in birds, Auk, 69:127–59.

———. 1956. The analysis of socialization in sheep and goats, Ecology, 37:228–38.

CORBIN, J. E., MOHRMAN, R. K., and WILCKE, H. L. 1962. Purebred dogs in nutrition research, Proc. Animal Care Panel, 12:163–68.

CORNWELL, A. C., and FULLER, J. L. 1960. Conditioned responses in young puppies, J. Comp. Physiol. Psychol., 54:13–15.

CRISLER, L. 1956. Observations of wolves hunting caribou, J. Mamm. 37:337–46.

———. 1958. Arctic wild. New York: Harper.

CROW, J. F. 1961. Mechanisms and trends in human evolution, Daedalus, 90:416–31.

DAHR, E. 1937. Studien über Hunde aus primitiven Steinzeitkulturen in Nordeuropa, Lunds Universitets Årsskrift, 32(4):1–63.

DARWIN, C. 1859. The origin of species. Reprinted. New York: Modern Library.

DAWSON, W. M. 1937. Heredity in the dog, U. S. Dep. Agr. Yearbook, 1315–49.

DEGERBØL, M. 1927. Über prähistorische, dänische Hunde, Vidensk, Meddel. Dansk. Naturhist. For. København, 84:17–72.

DENENBERG, V. H. 1962. The effects of early experience. In: HAFEZ, E. S. E. (ed.), The behaviour of domestic animals. London: Baillière, Tindall & Cox.

DICE, L. R. 1952. A panel discussion: genetic counseling, Am. J. Human Genet., 4:332–46.

DIMASCIO, A., FULLER, J. L., AZRIN, N., and JETTER, W. 1956. The effect of total body x-irradiation on delayed response performance of dogs, J. Comp. Physiol. Psychol., 49:600–604.

DOBZHANSKY, T. 1927. Studies on the manifold effect of certain genes in Drosophila melanogaster, Z. Ind. Abst. Vererbungslehre, 43:330–88.

————. 1962. Mankind evolving. New Haven: Yale University Press.

DUERST, U. 1942. Zur Frage der Herkunft des Haushundes, Anthropos, 37–40:318–19.

ELLERMAN, J. R. and MORRISON-SCOTT, T. C. S. 1951. Check list of Palaearctic and Indian mammals, 1758–1946. London: British Museum.

ELLIOT, O., and KING, J. A. 1960. Effect of early food deprivation upon later consummatory behavior in puppies, Psychol. Rept., 6:391–400.

ELLIOT, O., and SCOTT, J. P. 1961. The development of emotional distress reactions to separation in puppies, J. Genet. Psychol., 99:3–22.

————. 1965. The analysis of breed differences in maze performance of dogs, Animal Behaviour, 13:5–18.

ETKIN, W. 1954. Social behavior and the evolution of man's faculties, Am. Naturalist, 88:129–42.

FABRICIUS, E. 1951. Zur Ethologie junger Anatiden, Acta Zool. Fennica, 68:1–178.

FALCONER, D. S. 1960. Introduction to quantitative genetics. New York: Ronald Press.

FIELDE, A. M. 1904. Power of recognition among ants, Biol. Bull., 7:227–50.

FISHER, A. E. 1955. The effects of differential early treatment on the social and exploratory behavior of puppies. Pennsylvania State University Ph.D. thesis.

FREDERICSON, E. 1952. Perceptual homeostasis and distress vocalization in puppies, J. Personality, 20:472–77.

FREEDMAN, D. G., KING, J. A., and ELLIOT, O. 1961. Critical period in the social development of dogs, Science, 133:1016–17.

FULLER, J. L. 1948. Individual differences in the reactivity of dogs, J. Comp. Physiol. Psychol., 41:339–47.

————. 1950. Instruments for the measurement of physiological reactions of unrestrained animals, Ann. N. Y. Acad. Sci., 51:1051–56.

————. 1951. Genetic variability in physiological constants of dogs, Am. J. Physiol., 166:20–24.

————. 1953. Cross-sectional and longitudinal studies of adjustive behavior in dogs, Ann. N. Y. Acad. Sci., 56:214–24.

————. 1954. Heredity and structural defects in dogs, Gaines Veterinary Symposium, October 20, 1954, pp. 2–5.

————. 1955. Hereditary differences in trainability of purebred dogs, J. Genet. Psychol., 87:229–38.

————. 1956a. Inheritance of structural defects in dogs, Mich. State Univ. Vet., 16(2).

————. 1956b. Photoperiodic control of estrus in the basenji, J. Heredity, 47:179–80.

————. 1956c. The path between genes and behavioral characters, Eugenics Quart., 3:209–12.

————. 1957. The genetic base: pathways between genes and behavioral characteristics. The nature and transmission of the genetic and cultural characteristics of human populations, Proc. Milbank Mem. Fund. Conf., 101–11.

———. 1960. A decade of progress in dog genetics. Gaines Veterinary Symposium: A Decade of Progress in Canine Medicine.

———. 1962. Effect of drugs on psychological development, Ann. N. Y. Acad. Sci., 96:199–204.

———. 1963. Effects of experimental deprivation upon behavior in animals. Proc. World Congr. Psychiat. (Montreal) 3:223–27.

———. 1964. The measurement of alcohol preference in genetic experiments, J. Comp. Physiol. Psychol., 57:85–88.

FULLER, J. L., and CHRISTAKE, A. 1959. Conditioning of leg flexion and cardio-acceleration in the puppy, Fed. Proc., 18:49.

FULLER, J. L., CLARK, L. D., and WALLER, M. B. 1960. Effects of chlorpromazine upon psychological development in the puppy, Psychopharmacologia, 1:393–407.

FULLER, J. L., and DUBUIS, E. M. 1962. The behaviour of dogs. In: HAFEZ, E. S. E. (ed.), The behaviour of domestic animals. London: Ballière, Tindall & Cox.

FULLER, J. L., EASLER, C., and BANKS, E. 1950. Formation of conditioned avoidance responses in young puppies, Am. J. Physiol., 160:462–66.

FULLER, J. L., EASLER, C., and SMITH, M. E. 1950. Inheritance of audiogenic seizure susceptibility in the mouse, Genetics, 35:622–32.

FULLER, J. L., and GILLUM, E. 1950. A study of factors influencing performance of dogs on a delayed response test, J. Genet. Psychol., 76:241–51.

FULLER, J. L., and GORDON, T. M., JR. 1948. The radio inductograph, a device for recording physiological activity in unrestrained animals, Science, 108:287–88.

FULLER, J. L., ROSVOLD, H. E., and PRIBRAM, K. H. 1957. The effect on affective and cognitive behavior in the dog of lesions of the pyriform-amygdala-hippocampal complex, J. Comp. Physiol. Psychol., 50:89–96.

FULLER, J. L., and SCOTT, J. P. 1954. Heredity and learning ability in infrahuman mammals, Eugenics Quart., 1:28–43.

FULLER, J. L., and THOMPSON, W. R. 1960. Behavior genetics. New York: Wiley.

GEORGE, W. C. 1962. The biology of the race problem. By Commission of the Governor of Alabama.

GINSBURG, B. E. 1963. Genetic variables in sexual behavior. In: BEACH, F. A. (ed.), 1965. Sex and behavior. New York: Wiley.

GOWER, J. C. 1962. Variance component estimation for unbalanced hierarchical classifications, Biometrics, 18:537–42.

GRANT, D. A. 1947. Additional tables of the probability of "runs" of correct responses in learning and problem-solving, Psychol. Bull., 44:276–79.

GRAY, A. P. 1954. Mammalian hybrids. Slough: Commonwealth Agricultural Bureaux.

GRAY, P. H. 1958. Theory and evidence of imprinting in human infants, J. Psychol., 46:155–66.

———. 1960. Evidence that retinal flicker is not a necessary condition of imprinting, Science, 132:1834–35.

GUILFORD, J. P. 1950. Fundamental statistics in psychology and education. 2d ed. New York: McGraw-Hill.

GUTTMAN, L. 1955. A generalized simplex for factor analysis, Psychometrika, 20:173–92.

HAAG, W. G. 1948. An osteometric analysis of some aboriginal dogs, Univ. of Kentucky Rept. in Anthropol., 8:107–264.

HALL, C. S. 1941. Temperament: a survey of animal studies, Psychol. Bull., 38:909–43.

HAMMOND, W. H. 1957. The status of physical types. Human Biol., 29:223–41.

HAMMONS, H. G. (ed.), 1959. Heredity counseling. New York: Hoeber-Harper.

HARLOW, H. F. 1958. The nature of love. Am. Psychol., 13:673–85.

HARLOW, H. F., HARLOW, M. K., and HANSEN, E. W. 1963. The maternal affectional system of rhesus monkeys. In: RHEINGOLD, H. (ed.), Maternal behavior in mammals. New York: Wiley.

HARMAN, P. J. 1958. Private communication.

HATT, R. T. 1959. The mammals of Iraq, Misc. Publ. Museum Zool. (University of Michigan), 106:1–113.

HAYES, C. 1951. The ape in our house. New York: Harper.

HAYES, K. J. 1962. Genes, drives, and intellect, Psychol. Rept. 10:299–342.

HEBB, D. O. 1947. The effects of early experience on problem solving at maturity, Am. Psychol., 2:306–7.

———. 1949. The organization of behavior. New York: Wiley.

HESS, E. H. 1957. Effects of meprobamate on imprinting in waterfowl, Ann. N. Y. Acad. Sci., 67:724–32.

———. 1959. The relationship between imprinting and motivation. In: Nebraska Symposium on Motivation, 7:44–77. Lincoln: University of Nebraska Press.

HILDEBRAND, M. 1954. Comparative morphology of the body skeleton in recent Canidae. Berkeley: University of California Press.

HILZHEIMER, M. 1908. Beitrag zur Kenntniss der nordafrikanischen Schakale; nebst Bemerkungen ueber deren Verhältnis zu den Haushunden insbesondere nordafrikanischen und altägyptischen Hunderassen, Zool. Heft, 20:1–111.

HIRSCH, J., and ERLENMEYER-KIMLING, L. 1962. Studies in experimental behavior genetics IV. Chromosome analysis for geotaxis, J. Comp. Physiol. Psychol., 55:732–39.

HOWELLS, W. W. 1952. A factorial study of constitutional type, Am. J. Phys. Anthropol., 10:91–118.

HUMPHREY, E., and WARNER, L. 1934. Working dogs: an attempt to produce a strain of German shepherds which combines working abilities with beauty of conformation. Baltimore: Johns Hopkins Press.

HUNTER, W. S. 1913. The delayed reaction in animals and children, Animal Behavior Monograph, 2(6):1–86.

IGEL, G. J., and CALVIN, A. D. 1960. The development of affectional responses in infant dogs, J. Comp. Physiol. Psychol., 53:302–5.

ILJIN, N. A. 1941. Wolf-dog genetics, J. Genet., 42:359–414.

JAMES, H. 1959. Flicker: an unconditioned stimulus for imprinting, Can. J. Psychol., 13:59–67.

JAMES, W. T. 1952. Observations on the behavior of newborn puppies. I. Method of measurement and types of behavior involved, J. Genet. Psychol., 80:65–73. II. Summary of movements involved in group orientation, J. Comp. Physiol. Psychol., 45:329–35.

JOLICOEUR, P. 1959. Multivariate geographical variation in the wolf, *Canis lupus* L., Evolution, 13:283–99.

KAISER, H. F. 1958. The varimax criterion for analytic rotation in factor analysis, Psychometrika, 23:187–200.

KEELER, C. E. 1942. The association of the black (non-agouti) gene with behavior in the Norway rat, J. Heredity, 33:371–84.

KELLOGG, W. N., and KELLOGG, L. A. 1933. The ape and the child. New York: McGraw-Hill.

KING, J. A. 1954. Closed social groups among domestic dogs, Proc. Am. Phil. Soc., 98:327–36.

KLYAVINA, M. P., KOBAKOVA, E. M., STELMAKH, L. N., and TROSHIKHIN, V. A. 1958. On the speed of formation of conditioned reflexes in dogs in ontogenesis, J. Higher Nervous Activity, 8:929–36.

KOEHLER, W. 1927. The mentality of apes. 2d ed. New York: Harcourt, Brace.

KRUSHINSKII, L. V. 1962. Animal behavior, its normal and abnormal development. New York: Consultant's Bureau.

KULP, J. L. 1961. Geologic time scale, Science, 133:1105–14.

LACEY, J. I., and LACEY, B. C. 1958. Verification and extension of the principle of autonomic response-stereotypy, Am. J. Psychol., 71:50–73.

LAWRENCE, B. 1956. Cave fauna, Peabody Museum Papers, 48(2): 80–81.

LERNER, I. M. 1958. The genetic basis of selection. New York: Wiley.

LEVINE, S. 1962. The psychophysiological effects of early stimulation. In: BLISS, E. (ed.), Roots of behavior. New York, Hoeber-Harper.

LITTLE, C. C. 1957. Inheritance of coat color in dogs. Ithaca: Cornell University Press.

LORENZ, K. 1935. Der Kumpan in der Umwelt des Vogels, J. Ornithol., 83:137–213, 289–413.

LUMER, H. 1940. Evolutionary allometry in the skeleton of the domestic dog, Am. Naturalist, 74:439–67.

LUSH, J. L. 1943. Animal breeding plans. Ames: Iowa State University Press.

McCay, C. M. 1946. Nutrition of the dog. Ithaca: Comstock.

McKusick, V. A. 1960. Heritable disorders of connective tissue. 2d ed. St. Louis: Mosby.

Martins, T. 1947. The body posture of the dog during micturition, and sex hormones, Mem. Inst. Oswaldo Cruz, 44:343–61.

———. 1949. Disgorging of food to the puppies by the lactating dog, Physiol. Zool., 22:169–78.

Mason, M. W. 1959. Bibliography of the dog. Ames: Iowa State University Press.

Mason, W. A., and Harlow, H. F. 1958a. Formation of conditioned responses in infant monkeys, J. Comp. Physiol. Psychol., 51:68–70.

———. 1958b. Performance of infant rhesus monkeys on a spatial discrimination problem, J. Comp. Physiol. Psychol., 51:71–74.

Mather, K. 1949. Biometrical genetics. London: Methuen.

Matthew, W. D. 1930. The phylogeny of dogs, J. Mammalogy, 11:117-38.

Matthey, R. 1954. Chromosomes et systématique des Canidés, Mammalia, 18:225–30.

Mech, L. D. 1962. The ecology of the timber wolf (Canis lupus L.) in Isle Royale National Park. Purdue University Ph.D. thesis.

———. 1963. Timber wolf and moose of Isle Royale, Naturalist, 14(2):12–15.

Menzel, R., and Menzel, R. 1937. Welpe und Umwelt, Kleintier u. Pelztier, 13:1–65.

Miller, N. E., and Dollard, J. 1941. Social learning and motivation. New Haven: Yale University Press.

Morton, N. E. 1960. The mutational load due to detrimental genes in man, Am. J. Human Genet., 12:348–64.

Murie, A. 1940. Ecology of the coyote in the Yellowstone. U.S.D.I. Fauna series no. 4, Washington: Government Printing Office.

———. 1944. The wolves of Mt. McKinley. U.S.D.I. Fauna series no. 5, Washington: Government Printing Office.

Nice, M. M. 1943. Studies in the life history of the song sparrow, Trans. Linnaean Soc. New York, 1–328.

Ogburn, W. F., and Bose, W. K. 1959. On the trail of the wolf-children, Genet. Psychol. Monographs, 60:117–93.

Packard, A. S. 1885. Origin of the American varieties of the dog, Am. Naturalist, 19:896–901.

Parry, H. B. 1953. Degenerations of the dog retina: structure and development of retina of normal dog, Brit. J. Ophthamol., 37:385–404.

Pauly, L. K., and Wolfe, H. R. 1957. Serological relationship among members of the order Carnivora, Zoologica, 42:159–66.

Pavlov, I. P. 1927. Conditioned reflexes: an investigation of the physiological activity of the cerebral cortex. London: Oxford University Press; Humphrey Milford.

———. 1928-41. Lectures on conditioned reflexes. 2 vols. New York: International Publishers.

PAWLOWSKI, A. A., and SCOTT, J. P. 1956. Hereditary differences in the development of dominance in litters of puppies, J. Comp. Physiol. Psychol., 49:353–58.

PFAFFENBERGER, C. J. 1963. The new knowledge of dog behavior. New York: Howell.

PFAFFENBERGER, C. J., and SCOTT, J. P. 1959. The relationship between delayed socialization and trainability in guide dogs, J. Genet. Psychol., 95:145–55.

REED, C. A. 1959. Animal domestication in the prehistoric Near East, Science, 130:1629–39.

———. 1960. A review of archeological evidence on animal domestication in the prehistoric Near East. In: BRAIDWOOD, R. J. (ed.), Prehistoric investigations in Iraqi Kurdistan. Oriental Institute, University of Chicago. Studies in ancient oriental civilization, 31:119–45.

REED, S. C. 1963. Counseling in medical genetics. 2d ed. Philadelphia: Saunders.

RHEINGOLD, H. 1963. Maternal behavior in the dog. In: Maternal behavior in mammals. New York: Wiley.

RIBBANDS, C. R. 1953. The behaviour and social life of honeybees. New York: Dover.

ROSS, S. 1950. Some observations on the lair dwelling behavior of dogs, Behaviour, 2:144–62.

ROSS, S., SCOTT, J. P., CHERNER, M., and DENENBERG, V. H. 1960. Effects of restraint and isolation on yelping in puppies, Animal Behaviour, 8:1–5.

ROYCE, J. R. 1950. The factorial analysis of animal behavior, Psych. Bull., 47:235–59.

———. 1955. A factorial study of emotionality in the dog. Psychol. Monograph, 69(22):1–27.

———. 1957. Factor theory and genetics, Educational Psychol. Measurement, 17:361–76.

———. 1963. Factorial studies in comparative physiological psychology. In: CATTELL, R. B. (ed.), 1966. Handbook of multivariate experimental psychology. Chicago: Rand McNally.

SCHENKEL, R. 1947. Ausdrucks-studien an Wölfen, Behaviour, 1:81–129.

SCHWAB, J. J. 1940. A study of the effects of a random group of genes on shape of spermatheca in Drosophila melanogaster, Genetics, 25:157–77.

SCOTT, J. P. 1937. The embryology of the guinea pig, Am. J. Anat., 60:397–432; J. Morphol. (1938), 62:299–321; J. Expl. Zool. (1937), 77:123–57.

———. 1944. An experimental test of the theory that social behavior determines social organization, Science, 99:42–43.

———. 1945. Social behavior, organization, and leadership in a small flock of domestic sheep, Comp. Psychol. Monograph, 18(4):1–29.

———. 1950. The social behavior of dogs and wolves; an illustration of sociobiological systematics, Ann. N. Y. Acad. Sci., 51:1009–21.

————. 1952. (ed.), Minutes of the conference on the effects of early experience on mental health. Bar Harbor: Roscoe B. Jackson memorial laboratory.

————. 1953*a*. The process of socialization in higher animals. In: Milbank conference report: Interrelations between the social environment and psychiatric disorders. New York: Milbank Memorial Fund.

————. 1953*b*. Implications of infra-human social behavior for problems of human relations. In: SHERIF, M., and WILSON, M. O. (eds.), Group relations at the crossroads. New York: Harper.

————. 1954. The effects of selection and domestication upon the behavior of the dog, J. Natl. Cancer Inst., **15**:739–58.

————. 1957. The genetic and environmental differentiation of behavior. In: HARRIS, D. (ed.), The concept of development. Minneapolis: University of Minn. Press.

————. 1958. Critical periods in the development of social behavior in puppies, Psychosomat. Med., **20**:42–54.

————. 1960. Comparative social psychology. In: WATERS, R. H. *et al.* (eds.), Principles of comparative psychology. New York: McGraw-Hill.

————. 1962*a*. Genetics and the development of social behavior in mammals, Am. J. Orthopsychiat., **32**:878–93.

————. 1962*b*. Critical periods in behavioral development, Science, **138**:949–58.

————. 1963*a*. The process of primary socialization in canine and human infants, Monograph Soc. Res. Child Develop., **28**(1):1–47.

————. 1963*b*. Dog. Encyclopaedia Britannica.

SCOTT, J. P. and BEACH, F. A. 1947. (eds.), Minutes of the conference on genetics and social behavior. Bar Harbor: Roscoe B. Jackson memorial laboratory.

SCOTT, J. P. and CHARLES, M. S. 1953. Some problems of heredity and social behavior, J. Gen. Psychol., **48**:209–30.

————. 1954. Genetic differences in the behavior of dogs: a case of magnification by thresholds and by habit formation, J. Genet. Psychol., **84**:175–88.

SCOTT, J. P., DESHAIES, D., and MORRIS, D. D. 1962. The effect of emotional arousal on primary socialization in the dog. In preparation.

SCOTT, J. P., and FULLER, J. L. 1950. Manual of dog testing techniques. Mimeographed. Bar Harbor: Roscoe B. Jackson memorial laboratory.

————. 1951. Research on genetics and social behavior at the Jackson laboratory, 1946–51, a progress report, J. Heredity, **42**:191–97.

————. 1954. Experimental investigation of hereditary differences in learning ability in mammalian populations, Proc. World Population Congr. (Rome), **6**:641–52.

SCOTT, J. P., FULLER, J. L., and FREDERICSON, E. 1951. Experimental exploration of the critical period hypothesis, Personality, **1**:162–83.

SCOTT, J. P., FULLER, J. L., and KING, J. A. 1959. Inheritance of annual breeding cycles in hybrid basenji-cocker spaniel dogs, J. Heredity, 50:254–61.

SCOTT, J. P., GINSBURG, B. *et al.* 1962. The use of purebred dogs in research, Proc. Animal Care Panel, 12:149–86.

SCOTT, J. P., and MARSTON, M. V. 1950. Critical periods affecting normal and maladjustive social behavior in puppies, J. Genet. Psychol., 77:25–60.

SCOTT, J. P., ROSS, S., FISHER, A. E., and KING, D. J. 1959. The effects of early enforced weaning on sucking behavior of puppies, J. Genet. Psychol., 95:261–81.

SCOTT, J. P., VOGEL, H. H., and MARSTON, M. V. 1950. Social facilitation and allelomimetic behavior in dogs: I. Social facilitation in a non-competitive situation. II. The effects of unfamiliarity, Behaviour, 2:121–43.

SEARLE, L. V. 1949. The organization of hereditary maze-brightness and maze-dullness, Genet. Psychol. Monograph, 39:279–325.

SEITZ, A. 1959. Ein Bastard Nordafrikanischer Goldschakal- ♂ × Coyote- ♀ , unsw., Zool. Garten, 25:79–95.

SETON, E. T. 1925. The lives of game animals. New York: Doubleday.

SHELDON, W. H., and STEVENS, S. S. 1942. The varieties of temperament, a psychology of constitutional differences. New York: Harper.

SINGH, J. A. L., and ZINGG, R. M. 1939. Wolf-children and feral man. New York: Harper.

SKINNER, B. F. 1938. The behavior of organisms. New York: Appleton-Century-Crofts.

SMITH, A. C. 1945. British dogs. London: Collins.

STANLEY, W. C. 1962. Private communication.

————. 1963. Private communication.

STANLEY, W. C., CORNWELL, A. C., POGGIANI, C., and TRATTNER, A. 1963. Conditioning in the neonatal puppy, J. Comp. Physiol. Psychol., 56:211–14.

STANLEY, W. C., and ELLIOT, O. 1962. Differential human handling as reinforcing events and as treatments influencing later social behavior in basenji puppies, Psychol. Rept., 10:775–88.

STOCKARD, C. R. 1941. The genetic and endocrine basis for differences in form and behavior. Philadelphia: Wistar Institute.

STRANDSKOV, H. H. 1953. A twin study pertaining to the genetics of intelligence, Proc. 9th Int. Congr. Genet., 811–13.

STUDER, T. 1901. Die praehistorische Hunde in ihrer Beziehung zu den gegenwärtigen lebenden Hunderassen, Abhandl. Schweiz. paläontol. Ges., 28.

TEMBROCK, G. 1957. Zur Ethologie des Rotfuchses (*Vulpes vulpes* L.), unter besonderer Berücksichtigung der Fortpflanzung, Zool. Garten, 23:289–532.

THURSTONE, L. L. 1947. Multiple-factor analysis. Chicago: University of Chicago Press.

TINBERGEN, N. 1958. Curious naturalists. London: Country Life.

TREVELYAN, G. M. 1943. English social history. New York: Longmans, Green.

TROSHIKHIN, V. A. 1956. Die Entwicklung einiger Funktionen der höheren Nerventätigkeit bei Tieren in der frühen postnatalen Periode. In: Erste Arbeitstagung über zentrale Regulation der Funktionen des Organismus. Berlin: Verl. Volk Gesundheit.

TRYON, R. C. 1930–41. Studies in individual differences in maze ability, J. Comp. Psychol. 11:145–170; 12:1–22, 95–114, 303–45, 401–20; 28:361–415; 30:283–335, 535–82; 32:407–35, 447–73.

———. 1940. Genetic differences in maze learning ability in rats, Yearbook Nat. Soc. Stud. Education, 39:111–19.

———. 1963. Results of the complete bipolar selective breeding design, and a polygenic theory. Unpublished MS.

TUDOR-WILLIAMS, V. 1946. Basenjis, the barkless dogs. London: James Heap.

VAN DER MERWE, N. J. 1953. The jackal. Fauna and Flora, Transvaal Prov. Admin. Publication No. 4.

VOLOKHOV, A. A. 1959. Comparative physiological study of unconditioned and conditioned reflexes in development, J. Higher Nervous System, 9:53–62.

VON UEXKÜLL, J., and SARRIS, E. G. 1931. Das Duftfeld des Hundes, Z. Hundeforschung, 1:55–68.

WAGNER, K. 1930. Rezente Hunderassen; eine osteologische Untersuchung, Oslo Videnskaps-Akademi, 3(9):1–157.

WALK, R. D., and GIBSON, E. J. 1961. A comparative and analytical study of visual depth perception, Psychol. Monographs, 75(15):1–44.

WALLER, M. B., and FULLER, J. L. 1961. Preliminary observations on early experience as related to social behavior, Am. J. Orthopsychiat., 31:254–66.

WELKER, W. I. 1959. Factors affecting aggregation of neonatal puppies, J. Comp. Physiol. Psychol., 52:376–80.

WERTH, E. 1944. Die primitiven Hunde und die Abstammungsfrage des Haushundes, Z. Tierzüchtung Züchtungsbiologie, 56:213–60.

WHITE, G. 1842. The natural history of Selborne. New York: Harper.

WRIGHT, S. 1934. The results of crosses between inbred strains of guinea pigs, differing in number of digits, Genetics, 19:537–51.

———. 1935. Evolution in populations in approximate equilibrium, J. Genet., 30:257–66.

———. 1950. The genetical structure of populations, Ann. Eugenics, 15:323–54.

———. 1952. The genetics of quantitative variability. In: WADDINGTON, C. H. (ed.), Quantitative inheritance. London: H. M. Stationery Off., pp. 5–41.

————. 1953. Gene and organism, Am. Naturalist 87:5–18.

————. 1955. Classification of the factors of evolution, Cold Spring Harbor Symp. Quant. Biol., 20:16–24.

YOUNG, S. P., and GOLDMAN, E. A. 1944. The wolves of North America. Washington: American Wildlife Institute.

YOUNG, S. P., and JACKSON, H. H. T. 1951. The clever coyote. Washington: Wildlife Management Institute.

AUTHOR INDEX

Ahmed, I. A., 53, 435
Allen, D. L., 31, 74–75, 435
Allen, G. M., 33, 35, 399, 435
American Kennel Club, 50, 435
Anastasi, A., 354, 370, 427, 435
Anderson, W. S., 52, 435
Arndt, B., 124
Ash, E. C., 399, 435
Azrin, N., 437

Baege, B., 435
Bahrs, A. M., 435
Baker, J. A., 15, 435
Baker, P. T., 82, 354, 435
Banks, E., 97, 226, 439
Bayer, L. M., 418, 435
Bayley, N., 418, 435
Beach, F. A., 5, 69, 436, 439, 444
Beadle, G. W., 333, 436
Berg, I. A., 436
Bliss, E. L., 436, 441
Bliss, J. Q., 436
Blome, R., 47–48, 436
Blume, D., 93, 436
Bose, W. K., 148, 442
Bowlby, J., 149, 436
Boyden, A., 44, 436
Brace, C. L., 323, 336, 342–43, 347, 350–
    51, 371–74, 436
Braidwood, R. J., 34, 436, 443
Breland, K., 216, 436
Breland, M., 216, 436
Broadhurst, P. L., 324, 436
Brodbeck, A. J., 144, 436
Bruell, J. H., 299, 324, 436
Bunker, L. E., Jr., 338, 437

Burgher, J. A., 435
Burns, M., 4, 389, 436

Caius, J., 45–47, 51, 436
Calvin, A. D., 146, 441
Carpenter, C. R., 152, 436
Castle, W. E., 263, 266, 436–37
Cattell, R. B., 443
Chambers, R. M., 437
Chapin, J. P., 48–49, 437
Charles, M. S., 93, 121, 437, 444
Chaucer, G., 45
Cherner, M., 443
Christake, A., 100, 439
Clark, F., 93
Clark, L. D., 105, 439
Cloudman, A. M., 338, 437
Cohen, C., 295, 437
Coleman, C. E., 50, 437
Collias, N. E., 129, 143, 437
Corbin, J. E., 327, 437
Cornwell, A. C., 98, 226, 437, 445
Crisler, L., 62–66, 71, 75, 79, 140, 437
Crow, J. F., 409, 437

Dahr, E., 35, 37–38, 437
Darwin, C., 29, 51, 274, 404, 437
Dawson, W. M., 4, 334, 437
Degerbøl, M., 34–35, 437
Denenberg, V. H., 385, 437, 443
Deshaies, D., 444
Dice, L. R., 437
DiMascio, A., 437
Dobzhansky, T., 333, 410–11, 437
Dollard, J., 144, 442
Doughty, M. F., 435

Dubuis, E., 439
Duerst, U., 45, 438

Easler, C., 97, 226, 314, 439
Ellerman, J. R., 438
Elliot, O., 73, 102, 105, 109, 124, 128, 144–45, 177, 193, 236, 438, 445
Erlenmeyer-Kimling, L., 324, 440
Etkin, W., 81, 438
Eysenck, H. J., 436

Fabricius, E., 142–43, 438
Falconer, D. S., 297, 298, 391–92, 438
Feider, A., 144
Fielde, A. M., 438
Fisher, A. E., 105, 132, 145, 438, 445
Fredericson, E., 122, 305, 438, 444
Freedman, D. G., 105, 124, 128, 438
Fuller, J. L., 4, 21–22, 67, 93, 97–98, 100, 105, 106, 121–23, 132, 197, 226, 282, 295, 314, 324, 389, 391, 435–39, 444, 446
Freud, S., 3, 144

George, W. C., 323, 439
Gessner, C., 45
Gibson, E. J., 93, 446
Gillespie, J. H., 435
Gillum, 0., 439
Gilmore, R. W., 69, 436
Ginsburg, B. E., 62, 67, 74, 76–77, 131, 141, 148, 394, 402, 439, 445
Goldman, E. A., 31, 37, 51, 65, 447
Gordon, T. M., Jr., 439
Gower, J. C., 359, 439
Grant, D. A., 241, 439
Gray, A. P., 52, 439
Gray, P. H., 143, 146–47, 440
Guilford, J. P., 186, 440
Guttman, L., 210, 440

Haag, W. G., 35, 440
Hafez, E. S. E., 437, 439
Hall, C. S., 6, 194, 440
Hammond, W. H., 339, 440
Hammons, H. G., 400, 440
Hansen, E. W., 440
Harlow, H. F., 143, 145, 226, 440, 442
Harlow, M. K., 440
Harman, P. J., 86, 88, 440
Harris, D., 444
Hatt, R. T., 34, 440
Hayes, C., 143, 440
Hayes, K. J., 387, 440
Hebb, D. O., 26, 353, 440
Heinroth, O., 142

Hess, E. H., 143, 440
Hildebrand, M., 31, 37, 51, 440
Hilzheimer, M., 29, 440
Hirsch, J., 324, 440
Howells, W. W., 339, 440
Humphrey, E., 350, 440
Hunter, W. S., 242, 441

Igel, G. J., 146, 441
Iljin, N. A., 52, 441

Jackson, H. H. T., 31, 37, 447
James, H., 146, 441
James, W. T., 86, 441
Jetter, W., 437
Jinks, J. L., 324, 436
Jolicoeur, P., 43, 51, 83, 441

Kaiser, H. F., 348, 441
Keeler, C. E., 334, 441
Kellogg, L. A., 143, 441
Kellogg, W. N., 143, 441
King, J. A., 62, 73, 105, 124–28, 144, 167, 177, 282, 438, 441, 444–45
Klyavina, M. P., 98–99, 441
Kobakova, E. M., 441
Koehler, W., 226, 441
Krushinskii, L. V., 59, 385, 441
Kulp, J. L., 33, 441

Lacey, B. C., 204, 441
Lacey, J. I., 204, 441
Lawrence, B., 38, 441
Lerner, I. M., 391, 441
Levine, S., 385, 441
Linnaeus, C., 29–30, 56
Little, C. C., 6, 338, 389, 441
Lorenz, K., 142–44, 441
Lumer, H., 441
Lush, J. L., 391–92, 441

McCay, C. M., 72, 442
McKusick, V. A., 354, 442
Marston, M. V., 445
Martins, T., 109, 442
Mason, M. W., 435, 442
Mason, W. A., 226, 442
Mather, K., 187, 264, 324, 442
Matthew, W. D., 33, 442
Matthey, R., 52–53, 442
Mech, L. D., 74–75, 435, 442
Menzel, R., 442
Miller, N. E., 144, 442
Mohrman, R. K., 437
Morris, D. D., 444
Morrison-Scott, T. C. S., 438

Morton, N. E., 409–10, 442
Murie, A., 51, 61–65, 75–76, 140, 398, 416, 442

Nice, M. M., 143, 442

Ogburn, W. F., 148, 442

Packard, A. S., 29, 442
Parry, H. B., 93, 442
Pauly, L. K., 45, 442
Pavlov, I. P., 97, 144, 195, 442
Pawlowski, A. A., 159, 443
Pfaffenberger, C. J., 17, 109, 218, 443
Poggiani, C., 445
Pribram, K. H., 439

Reed, C. A., 34–35, 436, 443
Reed, S. C., 400, 410, 443
Rheingold, H., 171, 440, 443
Ribbands, C. R., 142, 443
Robson, J. A., 435
Ross, S., 90–91, 102, 443, 445
Rosvold, H. E., 439
Royce, J. R., 323, 368–70, 443

Sarris, E. G., 68, 446
Schenkel, R., 62–65, 79, 443
Schlager, G., 359
Schmidt, J. R., 370, 435
Schwab, J. J., 333, 443
Scott, J. P., 5, 51, 57, 59, 90, 102, 109, 113, 116, 119–20, 122, 129, 143, 153, 159, 169, 197, 236, 265–66, 282, 324, 333, 394, 414, 416–17, 425, 435, 438, 439, 443–45
Searle, L. V., 6, 194, 256, 388, 445
Seitz, A., 53, 445
Seton, E. T., 69, 445
Shade, C., 341
Sheldon, W. H., 339, 340–43, 445
Sherif, M., 444
Singh, J. A. L., 147–48, 445
Skinner, B. F., 100, 216, 445
Smith, A. C., 50, 314, 445
Smith, M. E., 439

Stanley, W. C., 87, 100, 144, 445
Stelmakh, L. N., 441
Stevens, S. S., 341, 445
Stewart, P. B., 436
Stockard, C. R., 6, 42, 323, 445
Strandskov, H. H., 368, 445
Studer, T., 35, 445

Tembrock, G., 62–65, 414, 445
Terhune, A. P., 163
Thompson, W. R., 324, 439
Thurstone, L. L., 348, 368–69, 445
Tinbergen, N., 109, 446
Trattner, A., 445
Trevelyan, G. M., 45, 446
Troshikhin, V. A., 86, 441, 446
Tryon, R. C., 5, 6, 446
Tudor-Williams, V., 48, 446

Van der Merwe, N. J., 50, 446
Vogel, H. H., 445
Volokhov, A. A., 446
Von Uexküll, J., 68, 446

Waddington, C. H., 446
Wagner, K., 38–41, 446
Walk, R. D., 93, 446
Waller, M. B., 105, 439, 446
Warner, L., 350, 440
Waters, R. H., 444
Watson, J. B., 3
Welker, W. I., 87, 446
Werth, E., 32, 38, 54–55, 446
White, G., 50, 446
Wilcke, H. L., 437
Wilson, M. O., 444
Wolfe, H. R., 45, 442
Wright, S., 5, 263, 266, 297, 314, 392, 398, 408, 446

Yerkes, R. M., 4
Young, S. P., 31, 36–37, 51, 63–65, 447

Zahn, T., 167
Zingg, R. M., 147–48, 445

SUBJECT INDEX

"Acquired drive," does not explain socialization, 144, 146
Activity-success factor, in performance tests, 372–73
Adaptation, limited by specialization, 427–28
Adaptive radiation
  dog, 36, 55–56, 398, 432
  human, 400
Adoption, in relation to socialization process, 149
Africa
  disappearance of aboriginal breeds, 399
  distribution of basenji in, 49
  distribution of jackals in, 31
African hunting dog, 30
  fossils, 33
Age
  changes in heritability with, 201
  effect on emotional reactivity, 198–200, 201, 312–13, 315
Aggression; see Agonistic behavior
Aggressive behavior, breed comparisons, 136–37
Aggressiveness
  independent of timidity, 369–70
  measured by dominance test, 137
  problem of measurement in early development, 423–24
Agonistic behavior
  basenji, 261
  cocker spaniel, 261
  dog, 60, 64, 75–78
  in emotional reactivity test, 202
  inheritance of patterns, 267–68
  in juvenile period, 109

limits socialization in dog, 131–32
  in socialization period, 104–6
  in transition period, 91–92
  wolf, 64, 75–78
Agricultural revolution, and dog, 34
Airedale, history, 51
Allelomimetic behavior
  dog, 63, 74–75
  effect on socialization in dog, 131
  in juvenile period, 110
  in socialization period, 106
  wolf, 63, 74–75
America, aboriginal dog breeds in, 399
Analysis of variance
  applied to matings, 298
  methods, 189–93, 298
  simplified schema, 192
  see also Variance
Anatomy, comparative, evidence of dog origin, 36–44
Ants, socialization in, 141–42
Arrhythmia; see Heart rate
Assortive mating, effect on human behavior, 428
Attraction score, obedience test, 212, 214
Australia, dingo in, 32

Barking and barklessness, inheritance of, 273–78
Barrier tests; see Detour test; Maze test
Basenji
  agonistic behavior, 261
  behavior
    in leash test, 306
    in motivation test, 307
    in spatial orientation test, 319

452

body proportions, 341
breeding cycle, 49, 50, 67, 279
climbing ability of, 219
fear of apparatus in, 247
growth curves, 328, 330–31
history, 48–50
inguinal hernia in, 389–90
inheritance of barking and barkless-
    ness, 273–78
manipulative ability of, 229
origin, 296
postural responses, 288–91
problem-solving ability of, 258, 322
reaction to passive handler test, 145
skull measurements, 40–41, 43–44
variation in body length, 371
see also Breed comparisons
Basset hound, skull shape, 42
Beagle
  adaptive capacities, 428
  compared to wolf, 402
  history, 47
  otocephaly in, 390
  as research subject, 395
  skull measurements, 40–41, 43–44
  slow habit formation in, 237–38, 425
  use as children's pet, 432
  use of senses, 79, 80
  used in critical period experiment, 122
  vocalization in strange room, 122
  weight gain, 327
  see also Breed comparisons
Bees, socialization in, 142
Behavior, 3–7
Behavior genetics
  biometrical goals for, 323–25
  field of, 413
Behavior patterns
  in Canidae, 63–65, 414–15
  definition, 59
  of dogs and wolves, 383–84
  inheritance, 261–94
  methods of observation, 57–58
  in relation to instinct, 60
  single factor and polygenic inheritance,
      292–93
  as units of study, 58–61
Behavioral systems
  definition, 59–60
  of dog and wolf, 61–80
  in relation to instinct, 60
  relationships between, 61
Biometrical genetics, goals of, 323–25
Birds, process of socialization in, 115,
    142–43
Biting, in reactivity test, 313–14

Blindness, inheritance in dog, 389–90
Blood hound, history, 46–47
Body position, in emotional reactivity
    test, 313
Body proportions, breed comparisons,
    341–42
Body size, inheritance, 286–87
Bolognese, skull measurements, 40–41
Borzoi
  head shape, 29
  skull measurements, 40–42
  use of senses, 79
Boston terrier, skull shape, 42
Boxer, skull measurements, 40–42
Brain, size in dwarf breeds, 42
Breed, proportion of variance associated
    with, 360–67
Breed associations, objectives for, 393–94
Breed comparisons
  aggressive behavior, 136–37
  barking in dominance test, 274–75
  body proportions, 341–42
  coefficients of inbreeding, 48–49
  congenital defects, 406–7
  cue-response test, 241–42
  delayed-response test, 243
  detour test, 228–29
  development of allelomimetic behavior,
      106
  development of heart rate, 121–22
  development of startle response, 94
  distress vocalization, 92
  dominance-subordination relationship,
      156–63, 419
  emotional reactivity test, 198–203
  eruption of teeth, 96
  fear responses, 134–36
  following test, 174–75, 177
  forced training, 215–16
  goal-orientation test, 217–18
  growth curves, 329–31
  handling test, 134–38
  investigatory behavior, 137–38
  leash-training test, 208–9
  manipulation test, 230–31
  maternal nursing behavior, 171
  maternal retrieving test, 172–73
  maze test, 234–35, 237–38
  in mode of motivation, 387
  mortality rates, 405–6
  motivation test, 240
  motor development, 96
  motor-skill test, 220–21
  obedience test, 213–14
  physique and behavior, 345–47
  problem-solving behavior, 257–58

quieting test, 206–7
retrieving test, 218–19
size, 329
skull shape, 37–44
spatial-orientation test, 250–52
statistical methods for, 185–93
summary table, 360–61
tail-wagging, 137–40
time of opening of eye, 89–90
trailing test, 246–47
variance components, 360–67
see also Inheritance
Breed differences
in endocrine glands, 6
and rearing practices, 385–86
see also Breed comparisons
Breed symbols, key, 10
Breeding, scientific, in dogs, 399
Breeding cycle
basenji, 49–50, 67
dingo, 67
dog, 67
inheritance, 278–83
in wolf, dog, and coyote, 278–79
Breeding plan, 7–11
Breeds
aboriginal, in America and Eurasia, 399
effect of panmixis on, 399
factor analysis applied to, 371–75
genetic heterogeneity in, 29, 185, 295, 378, 386–87
history, 45–51
not equivalent to races, 82
as populations, 393–94
use in research, 394–95
see also under name of breed
Brussels griffon, skull shape, 42
Bulldog
bull-baiting, 77
head shape, 29
jaw shape, 38
short-leggedness in, 42
skull shape, 40–43, 400
Bulldog, French, skull measurements, 40–42
Bulldog head, probably a mutation, 42
Bush-dog, South American, 30
fossils, 33

Canidae
chromosome numbers, 52–53
classification, 30–31
fossils, 33
geographical origin and distribution, 31–32

serological relationships, 44–45
Canis
ecology of, 31
genetics of, 51–53
interspecific hybrids, 52–53
Care-dependency relationship
development, 176–78
dog, 153, 174
Care-soliciting behavior; see Et-epimeletic behavior
Carnivora, classification, 30
Castle-Wright formula, for analysis of quantitative inheritance, 263–64
Catching time, test results, 168
Chick, critical period for socialization in, 143
Chihuahua
popularity in cities, 432
size, 56
skulls, 43
weight, 29
China, home of chow, 50
Chow, history, 50
Chromosomes, in family Canidae, 52–53
Classification, biological, see Taxonomy or name of animal
Classification, of social relationships, 152–53
Cleft palate, inheritance, 389
Club foot, in fox terriers, 390
Cocker spaniel
agonistic behavior, 261
behavior
in leash test, 306
in motivation test, 307
body proportions, 341
breeding cycle, 278–79
early learning in, 97
growth curves, 330–31
history, 48
hydrocephaly in, 390
origin, 296
postural responses, 288–91
problem-solving ability of, 322
skull measurements, 40–41, 43–44
variation in body length, 371
see also Breed comparisons
Collie, adaptive capacities, 428
Comfort-seeking behavior
effect on socialization process, 146
see also Shelter-seeking behavior
Conditioned response, experiments in transition period, 97–100
Conditioning
relationship to socialization process, 144, 146

see also Learning

Confidence, effect on maze performance, 237

Confidence score, obedience test, 212, 214

Confinement test
affected by hair length, 336–37
variance components, 360

Co-ordinated attack relationship, dog, 154

Correlational methods, applied to genetic variance, 376–78

Covariance, method of analysis, 236

Coyote
appearance and proportions, 37
behavior patterns, 63–65
classification, 31
distribution in North America, 31
fossils, 33
interspecific hybrids, 52–53
skull measurements, 40–41, 44
social behavior patterns, 62–65
social groups, 415

Critical period concept, 117–18

Critical period hypothesis, tests of, 122–29

Critical period for learning
in retrieving test, 22

Critical period for primary socialization
in birds, 142–43
boundaries of, 118–29
dog, 108, 111, 117–50
general phenomenon, 417
in guinea pig, 143
in human infants, 115, 147–50
in rhesus monkey, 143
in sheep, 143

Critical periods, in development, 110–12

Cross-fostering, results of, 186, 191

Cue-response test
breed comparisons, 241–42
hybrid comparisons, 308–9
methods, 238–42

Cultural change, based on process of learning, 424

Cultural evolution; *see* Evolution, cultural

Dachshund
compared to wolf, 402
leg length, 29, 42
skull measurements, 40–41

Defects, structural
in dog breeding management, 389–90

Delayed-response test
breed comparisons, 243–44

hybrid comparisons, 309–10
methods, 242

Denmark
domestication of dogs in, 54
possible origin of dog in, 35
stone age dog in, 34–35

Dependency; *see* Care-dependency relationship

Detour test
behavior during, 227–28
breed comparisons, 228–29
correlation with trailing test, 374
effect of maternal environment, 284–85
factor loadings
for activity-success and heart-rate, 372
for reactivity, 374
methods, 19–20, 226–27
in test schedule, 24
variance components, 361–62, 366

Development of behavior
care-dependency relationship, 176–78
changes in handling test, 176–77
concept of, 16–17
differentiation of behavior in, 287–92
dog, 84–116
dog-human relationship, 175–80
normal variation in, 119–20
observational methods, 15–17
problem-solving behavior, 225–26
processes of, 111–12

Developmental genetics, 413

Dhole, 30
fossils, 33

Differentiation, growth curves, 332

Differentiation of behavior
during development, 287–92, 425
functional, 421–22
and genetics, 165–69, 181
in social relationships, 152

Differentiation of social relationships, law, 168–69, 181–82

Dingo
breeding cycle, 67
classification, 31
origin, 32
result of adaptive radiation, 398
skull measurements, 40–41

Disease control, methods of, 13–15

Distemper, canine, control of, 13–15

Doberman pinscher
history, 51
skull measurements, 40–41

Dog
adaptive radiation of, 36, 55–56, 398, 432

agonistic behavior, 60, 64, 75–78
and agricultural revolution, 34
allelomimetic behavior, 63, 74–75
behavior patterns, 63–65, 383–84
behavioral systems, 61–80
comparison with human, 80–83
breeding cycle, 67, 279
care dependency relationship, 174
chromosome number, 53
classification, 29–31
critical period for primary socialization,
    117–50
development of behavior, 84–116
development of social relationships in,
    151–82
domestication from wolf, 54–56
dominance-subordination   relationship
    in, 77
effect of human cultural change upon,
    432–33
effect of isolation upon genetic change,
    398–99
in Egypt, 34
eliminative behavior, 65, 68–70
epimeletic behavior, 63, 70–71, 169–
    75
et-epimeletic behavior, 63, 72
evolution, 81–82, 397–411, 433
genetic variation, 4, 400–401
    in development, 418
    magnitude, 358–66
geographical distribution, 31–32
growth and growth curves, 326–32
human social needs for, 432
hunting behavior, 75–76, 78–80
hybrids with other canids, 52–53
ingestive behavior, 60, 65, 72–73
investigative behavior, 63, 78–80
in Iraq, 34
in Jarmo, 34
in Jericho, 34–35
leader-follower relationship, 175
maternal behavior, 169–75
in Mesopotamia, 34–35
mother-offspring relationship, 169–75
mutations in, 51–52
origin, 29–56
    evidence from pre-history, 33–36
    evidence from taxonomy, 30–33
    fossil evidence, 33
    in Palestine, 34
performance tests compared with hu-
    man IQ tests, 257
periods of development, 84–112
as pilot experiment for human race,
    397

polymorphism in, 83
serological relationship with other ca-
    nids, 44–45
sex-linked inheritance not found, 283
sexual behavior, 62–68
shelter-seeking behavior, 65, 73–74
skull shape, 38–44
social behavior, 57–83
social groups, 61–62, 414–15
social organization in, 414–29
social relationships, 153
socialization in, 118–41
somatotyping, 339–43
specialization in, 402–3
stone age, 34–35
territory, 62, 69, 110
use in behavior genetics, 325
variation in tooth size, 39, 41
"wild," 30
Dog breeding, 383–96
environmental control in, 384
genetic management, 390–91
history, 399
hybridization in, 392–93
methods for associations, 393–94
methods for kennel owners, 393
for physical type, 389
for research, 394–96
selection methods, 391–92
strain improvement, 391–96
and structural defects, 389–90
see also Rearing practices
Dog-human comparison, of heritabilities,
    324
Dog-human social relationship
development, 175–80
novelty in, 415–16
optimum time for initiating, 385
Domestication
estimated date and location, 54–56
effects on dog, 400–3
Dominance, genetic, effects on behavior,
    356–58
Dominance organization of group, genetic
effects upon, 163–64
Dominance-subordination relationship
breed comparisons, 156–63
development, 155–69
dog, 77, 154, 415
effect of genetic segregation upon,
    165–66
effect of heredity on, 418–19
effect of sex on, 164–65
effect of size on, 164–65
effect on expression of "jealousy," 167
effect on fighting, 156–59

in juvenile period, 109–10
between mothers and offspring, 175
wolf, 77, 415
Dominance test
breed comparisons of barking, 274–75
breed and hybrid comparisons, 360
measure of aggressiveness, 137
methods, 155–56
in test schedule, 18–19, 24
Drinking; *see* Ingestive behavior
Duck, socialization in, 143
Dwarf breeds, brain size, 42
Dwarf pinscher, skull measurements, 40–41

EEG, development, 93, 107
Ear, first function of, 94
Early experience
dog development suitable for experiments, 112–13
East Indies, dogs in, 54
Eating; *see* Ingestive behavior
Ecological niche, wolf and human, 397
Ecology, *Canis,* 31
Egypt, dogs in, 34, 54
Eliminative behavior
in emotional reactivity test, 202
in juvenile period, 109
in neonatal period, 85
patterns in dog, 65, 68–70
patterns in wolf, 65, 68–70
in socialization period, 101–2
in transition period, 91
Emotional behavior
effect of genetics on, 6
effect on maze performance in rats, 6
effect on performance, 246–47, 256, 387
effects on trainability, 222
importance in selection for performance, 387, 394
Emotional reactivity
definition, 194
heritability of, 204
Pavlovian types of, 195
sex differences in, 202
unitary nature of, 203, 315–16
Emotional reactivity factor in F₁ hybrids, 374
Emotional reactivity test
age changes in, 198–201, 312–13, 315
biting in, 313–14
body position in, 313
breed comparisons, 198–203
correlation with trailing test, 374

effect of heredity at different ages, 201
effect of maternal environment, 285
heart rate, 199, 313–14
hybrid comparisons, inheritance, 312–17
loadings in activity-success and heat-rate factors, 372
loading in reactivity factor, 374
mating differences, 316–17
methods, 194–97
observer reliability, 196
response categories, 196–97
results in home-reared dogs, 180
results in "wild dog experiment," 128
tail-wagging in, 313–15
in test schedule, 21–22, 24
variance components, 360, 366
Emotional response, effect of conditioning upon, 100
Emotional tests
factor analysis of, 369–70
variance components, 360
Endocrine glands, differences among breeds, 6
England
development of dog breeds in, 45–48
dogs of, 432
first dog show, 399
English setter, hemophilia in, 407
Environment
effect of variation on behavior, 236, 256
enriched effects on dogs and rats, 26–27
physical, 25–28
variance components, 190
Environmental improvement, use in dog breeding, 384
Epimeletic behavior
dog, 63, 70–71
self grooming, 107
wolf, 63, 71
Eskimo dog
compared to wolf, 403
result of adaptive radiation, 398
Et-epimeletic behavior
dog, 63, 72
in neonatal period, 85
in socialization period, 102–4
in transition period, 91–92
wolf, 63, 72
Eurasia
aboriginal dog breeds in, 399
distribution of jackal in, 31
wolf in, 31
Europe, dogs of, 432

Evolution
  behavior a conservative trait in, 415
  dog, 397–411
  human, 399–400, 409–12
Evolution, cultural
  compared to biological evolution, 430–31
  effect on biological change, 429–33
  synergistic relationship with biological variation, 431–33
Evolution of behavior, dog and human, 81–82
Experimental design; *see* Methods
Eye
  development in socialization period, 107
  development in transition period, 93–94
  inheritance of defects, 389–90
  time of opening in breeds and hybrids, 89–90

Factor analysis
  breed and hybrid populations, 371–75
  emotional and physiological tests, 369–70
  evidence of general traits, 367–68
  and genetics, 343–44, 368–99, 376
  performance tests, 370–71
  physique, 340, 343–44, 347–51
Failure, effect on motivation, 230–31, 256
Fear of apparatus
  in spatial orientation test, 320
  in trailing test, 246–47, 256
Fear responses
  breed comparisons, 134–36
  development, 104–5
  see also Agonistic behavior
Feeblemindedness, correlated with low fecundity, 410
Fennec, chromosome number, 53
Fertility
  decline with age, 110
  increase under domestication, 401–2
  problem in human populations, 410
Fighting
  effect of dominance relationship upon, 156–59
  see also Agonistic behavior; Dominance-subordination relationship
Fighting, playful
  development of, 105–6
  effect on socialization in dog, 130
  inheritance, 269–73
Following test

breed comparisons, 174–75, 177
  methods, 173–74
Food and feeding, 11–12, 14
Food rewards, effect on dog-human relationship, 177–78
Forced training, 205–16
  breed comparisons, 215–16
  non-unitary nature of reactions, 215–16
Fossils, from *Canidae*, 33
Fox
  chromosome numbers, 53
  compared to wolf, 403
  fossils, 33
  serological relationship with other canids, 44–45
  social behavior patterns, 62–65
  social groups, 414–15
Fox terrier
  club foot in, 390
  early learning in, 97–98
  group attacks by, 106
  history, 50
  rate of development, 418
  reaction to punishment, 145
  size factor in, 349
  skull measurements, 40–41, 43–44
  use of senses, 80
  weight gain, 327
  see also Breed comparisons
France, dogs of, 432

Gene action, theory of specific, 333–34
Genetic effects on behavior, 3–7
Genetic equilibrium, theory of, 407–8
Genetic variation, in development of dog and human, 418
Genetics
  and differentiation of behavior, 165–69, 181
  and factor analysis, 367–69, 376
  of genus *Canis*, 51–53
  sub-divisions, 413
Geographical distribution, dog and other Canidae, 31–32
German shepherd
  few correlations of physique and behavior, 350
  hip dysplasia in, 407
  jaw shape, 39
  skull measurements, 40–42
Goal-orientation test
  breed comparisons, 217–18
  correlation with trailing test, 374
  methods, 217
  variance components, 361
Golden retriever, history, 50

Goose, primary socialization in, 142
Great dane
  agonistic behavior, 78
  jaw shape, 39
  size, 56, 400–401
  skull measurements, 40–42
Greyhound
  compared to wolf, 402
  history, 45–47, 51
  jaw shape, 38
  leg length, 29
  origin, 399
  skull, 400
  skull measurements, 40–43
Growth
  dog, 326–32
  in juvenile period, 108
  model for learning process, 332, 352–53
  relation to periods of development, 326–29
Growth curves, dog, 328–31
Guard dog, Kurdish, 34
Guinea pig
  critical period in, 143
  polydactylous monster, 333

Habit formation
  effect of heredity, 425
  limits variability of behavior, 151
  weak tendency in beagle, 237–38
Habit-formation test; see Goal-orientation test
Hair color, inheritance and relation to behavior, 338–39
Hair length
  inheritance, 262, 334–36
  in relation to size and behavior, 336–37
Handling test
  breed comparisons, 134–38
  correlation with trailing test, 374
  developmental changes, 176–77
  effects of hand feeding upon, 177–78
  effect of maternal environment, 284–85
  of home-reared dogs, 179–80
  inheritance
    of fear reactions, 267–68
    of playful aggressiveness, 269–73
  loading in
    activity-success and heart-rate factors, 372
    reactivity factor, 374
  methods
  results in "wild dog experiment," 125

  in test schedule, 18, 24
  variance components, 360, 366
Handler, effect on obedience test, 212, 214
Hardy-Weinberg law, 404–5
Heart rate
  changes with age, 199
  effect of maternal environment, 285
  inheritance, 287
  in reactivity test, 199, 313–14
  relation to arrhythmia, 203
  response to quieting, 199
  variance components, 360, 366
Heart-rate development, breed comparisons, 121–22
Heart-rate factor, of physiological tests, 372–73
Hepatitis, infectious canine, control of, 15
Hemophilia
  in English setter, 407
  inheritance, 389
Heredity
  action through developmental processes, 113
  cultural and biological compared, 430–31
  effects
    of cultural evolution upon, 431–33
    on development of social relationships, 421–24
    on differentiation of behavior, 152
    on dominance organization of group, 163–64
    on learning capacities, 424–25, 428–29
    on social relationships, 113, 418–19
    on socialization process, 133–41
    on trainability, 222–23
    on variation and habit formation, 425
  magnitude of effect on behavior and physique, 358–66
  synergistic relationship with cultural evolution, 431–33
  variance components, 190, 321
Heritability
  of behavior, and selection, 391–92
  effect of practice on, 254–56
  emotional reactivity test, 201, 316
  leash training test, 299, 301, 305
  limitations and usefulness, 324–25
  measured by intraclass correlation, 192
  methods of estimation, 185–86
  motivation test, 308
  obedience test, 311

problem solving behavior, 321, 388
regression analysis, 297–98
specificity of estimates, 323–25
summary table, 360–61
Hernia, inguinal, in basenji, 389, 406
Heterogeneity, in pure breeds, 185, 295, 317, 386–87
Heterosis; *see* Hybrid superiority
Hip dysplasia
in German shepherds, 407
inheritance, 389, 391
History, dog breeds, 45–51
Home-reared dogs, social relationships, 178–82
Homozygosity, rare in pure breeds, 378
Hostility, toward similar and unlike dogs, 167–68
Hounds, agonistic behavior, 77
House breaking, behavior patterns useful in, 102
Housing
design of nursery room and runs, 25–26
plan for all-weather kennel, 27
Human
behavioral systems, 80–83
critical period in infant, 147–50
danger from mutations, 408–9
development of social organization, 417
ecological niche, 397
effect of
assortive mating on, 428
isolation upon development of, 147–48
evolution, 81–82, 399–400, 409–12
fertility and sterility problems, 410
genetic change in, 399–400
genetic variation in development, 418
IQ tests compared with dog tests, 257
periods of development in, 113–16
polymorphism, usefulness of, 83, 411–12
prenatal mortality rate, 410
social needs for dogs, 432
somatotyping, 339–40
variation, 401
Hunting, behavior patterns in dog and wolf, 75–76, 78–80
Hunting dogs, historical records, 45–48
Hybrid comparisons
cue-response test, 308–9
delayed-response test, 309–10
development of startle response, 94
emotional reactivity, 312–17
eruption of teeth, 96

leash training, 299–306
mortality rates, 405–6
motivation test, 307–8
motor development, 96
motor-skill test, 221–22
obedience test, 311
physique and behavior, 345–46
spatial-orientation test, 318–20
statistical methods, 297–98
summary, 360–61
time of opening of eye, 89–90
variance components, 360–66
see also Inheritance
Hybridization, in dog breeding, 392–93, 396
Hybrid superiority
in basenji-cocker cross, 323
in motivation test, 307
in spatial orientation, 318–19
Hybrids
behavior of, 295–325
factor analysis of, 371–75
reactivity factor in F_1, 374
for research, 395
between species in genus *Canis*, 52–53
Hydrocephaly, in cocker spaniel, 390

Icelandic dog, skull measurements, 40–41
Imprinting
equivalent to primary socialization, 142
see also Socialization, primary; Socialization, period; Socialization, process
Inbreeding, in foundation stock, 48–49
Individual variation
in delayed-response test, 243–44, 256, 310
in emotional reactivity, 202
in maze test, 237
in spatial-orientation test, 251
variance components within litters, 190–91, 321, 364–67
Ingestive behavior
dog, 60, 65, 72–73
effect on socialization process, 131, 144–46
in socialization period, 101
in transition period, 89–90
wolf, 65, 72–73
Inheritance, mode of
avoidance and vocalization patterns, 267–68
barking and barklessness, 273–78
behavior patterns, 261–94
body size, 286–87
breeding cycle, 278–83

emotional reactivity test scores, 312–17
hair color, 338
hair length, 262, 334–36
heart rate, 287
leash fighting, 268–69
multiple- and single-factor theories, 367–68
no clear patterns in performance tests, 285–86
patterns of agonistic behavior, 267–78
playful fighting, 269–73
postural responses, 288–92
quieting test, 287–88
selection and, 392
structural defects, 389–90
wildness and tameness, 266–69
see also, Single-factor inheritance; Polygenic inheritance
Insight learning, development in detour test, 228–29, 256
Instinct, in relation to behavior patterns and systems, 60
"Intelligence" concept
no evidence of general factor, 256, 388, 427
see also Problem-solving behavior
Intelligence tests; see Test; Performance test
Interaction, genotype-life history, 325
Intraclass correlation, measure of heritability, definition, 192
Investigative behavior
breed comparisons, 137–38
in dog and wolf, 63, 78–80
effect on socialization in dog, 130
in emotional reactivity test, 200
in neonatal period, 85
in socialization period, 104
in transition period, 91
Iraq, dog in, 34
Irish wolfhound, skull measurements, 40–41
Isolation
effects on socialization process and human development, 145–48
and genetic change in dogs, 398–99

Jackal
appearance and size, 37
chromosome number, 53
classification, 31
distribution in Africa, Eurasia, 31
fossils, 33
hybrids, 52–53
serological relationships with other canids, 45
skull measurements 40–41, 44
social groups, 415
vocalization, 50
Jackdaw, socialization in, 142
Jarmo, dog in, 34
Jaw shape, of dogs and wolves, 38–41
"Jealousy," in relation to dominance-subordination relationships, 167
Jericho, dog in, 34–35, 54
Juvenile period, dog, 108–10

Kasper Hauser, case of partial isolation, 147–48
King Charles spaniel, history, 46

Labrador retriever, history, 50
Lapland dog, skull measurements, 40–41
Leader-follower relationship, dog, 154
in dog and wolf, 175, 415
Learning
analogy with growth, 332, 352–53
basic process of cultural change, 424, 430
complex interaction with heredity, 426–28
effect of heredity upon, 424–25, 428–29
effects on expression of genetic variation, 425–26
effects on socialization process, 144
limiting factor on behavioral variation, 428
see also; Training; Performance
Learning capacities
in juvenile period, 109
in neonatal period, 87–88, 98
in socialization period, 107–8
in transition period, 97–100
Learning curves, maze test, 235
Leash-control test; see Leash-training test
Leash fighting
inheritance of, 268–69
in leash-training test, 301–5
Leash-training test
breed comparisons, 208–9
demerit scores in hybrids, 299–300
effect of maternal environment, 285
heritability, 299, 301, 305
hybrid comparisons, 298–306
inheritance of leash fighting, 268–69
intercorrelations of demerits, 210
loadings on reactivity factor, 374
methods, 207

results in "wild dog experiment," 127–28

in test schedule, 22, 24

variance components, 361–62, 366

Legs, mutation for shortness, 42

Levels of organization, in relation to behavior, 5

Lip-licking, in emotional reactivity test, 199

Litter differences
in cue-response test, 242, 309
in emotional reactivity, 203
in leash-control test, 301–2
proportion of variance associated with, 190–91, 364–67
in spatial-orientation test, 319

Localization, process of, 102–4, 112

Mammals, socialization in, 143–44

Man; see Human

Manchester terrier, chromosome number, 53

Manipulation test
breed comparisons, 230–31, 361
correlation with trailing test, 374
hybrid comparisons, 361
loading in activity-success factor, 372
method, 229–30

Mastiff
agonistic behavior, 78
history, 46

Master list of experimental animals, 10

Maternal behavior
dog, 169–75, 181
limits period of socialization in sheep, 130
weaning test, 173
see also Epimeletic behavior

Maternal environment
effects on tests, 283–85, 363
on leash training, 300–301
on reactivity test, 317

Matings, differences between
analysis of variance, 298
in cue-response test, 309
in leash-control test, 302–5
in reactivity test, 316–17
variance associated with, 190–91, 364–66

Maturation, interacts with genetic differences, 225

Maze performance, rats, effect of emotional behavior, 6

Maze test
behavior during, 234
breed comparison, 234–35, 237–38

correlation with trailing test, 374
loading in activity-success factor, 372
methods, 232–34
and "positional habit" factor, 370
variance components, 236–37, 361

Mendelian genetics, 413

Merle gene, genetic management, 390

Mesopotamia, dog in, 34, 35
domestication of dogs in, 54

Methods
analysis
of qualitative inheritance, 264–66
of single-factor inheritance, 262–66
of variance, 189–93, 236, 298
breeding plans, 7–11
correlational, applied to genetic variance, 376–78
cue-response test, 238–42
delayed-response test, 242
detour test, 19–20, 226–27
disease control, 13–15
dominance test, 155–56
emotional reactivity test, 194–97
following test, 173–74
genetic, 7–11, 296–97
goal-orientation test, 216–17
handling test, 133–34
leash-training test, 207
manipulation test, 229–30
maternal retrieving test, 171
maze test, 232–34
motivation test, 240
motor-skill test, 219–20
nutrition, 11–12
obedience test, 211–12
observation of behavior patterns, 57–58
observation of development, 15–17
performance tests, 19–25
physique measures, 340–41
quieting test, 206
retrieving test, 218
sanitation, 13–14
of scaling, 324
social behavior tests, 17–19
spatial orientation, 248–50
statistical
breed comparisons, 185–93
hybrid comparisons, 297–98
trailing test, 244–45
variance component computation, 359
weaning test, 173

Mexican hairless, 29, 400

Middle East, dogs of, 432

Mixed litters, experiment on breed differences, 191

Molecular genetics, 413

Molossian hound, 56
Monorchidism, in Shetland sheepdog, 390
Mortality rates
    breed and hybrid comparisons, 405–6
    human prenatal, 410
Mother-offspring relationship, dog, 169–75
Motivation
    breed comparisons, 387
    effect
        of failure on, 230–31, 256
        on performance, 387–88, 428–29
        on problem-solving, 230–31
        method used in performance tests, 20–21
Motivation test
    breed comparisons, 240
    effect of maternal environment, 285
    hybrid comparisons, 307–8
    method, 240
    stanine scaling, 188–89
Motor capacities
    in juvenile period, 108–9
    in neonatal period, 87
    in socialization period, 107
Motor development
    breed and hybrid comparisons, 96
    In transition period, 94–96
Motor-skill test
    breed comparisons, 220–21
    correlation with trailing test, 374
    hybrid performance on, 221–22
    loading in activity-success factor, 372
    methods, 219–20
    in test schedule, 22, 24
    variance components, 361
Multiple-factor inheritance; see Polygenic inheritance
Mutations
    accumulation in dogs, 405–8
    basis of variation in dog, 51–52
    danger in human populations, 408–9
Mutual care relationship, in dog, 154
Mutual defence relationship, dog, 154
Myelination, in neonatal period, 88

Natural selection
    and degeneration, 404, 408
    role in human populations, 403, 410
Neonatal period
    characterized by neonatal nutrition, 416
    decline of nursing in, 170–71
    dog, 84–89
    rearing practices for, 384

Newfoundland breed, skull measurements, 40–41
"Nicking," use in dog breeding, 392–93
North America
    distribution of coyote in, 31
    dogs in, 55
    stone age dogs in, 35
    wolves in, 31
Norwegian hare hound, skull measurements, 40–41
Nursing behavior, 170–71
Nutrition, methods of, 11–12

Obedience test
    breed comparisons, 213–14
    hybrid comparison, 311
    loadings in activity-success and heart-rate factors, 372
    methods, 211–12
    in test schedule, 22, 24
    variance components, 361
Obesity, in Shetland sheepdog, 390, 406
Observer reliability, in emotional reactivity test, 196
Olfaction, in neonatal period, 86
"One-man dog," lack of evidence for, 213–14, 223
Organization of basic capacities, in training, 222
Otocephaly, in beagle, 390
Outbreeding, protects recessive genes, 405

Palestine, dog in, 34
Panmixis
    effect on dog breeds, 399
    human, 400
Parent-offspring correlation; see Regression analysis
Passive-handler test
    reaction of basenjis, 145
    results in "wild dog experiment," 125–26
Pekingese, skull measurements, 40–42
Performance, effects of emotional and motivational differences, 387–88
Performance tests
    correlation with trailing test, 374
    factor analysis of, 370–71
    of home-reared dogs, 180
    loadings in activity-success factor, 372–73
    methods of, 19–25
    motivation in, 20–21
    no clear patterns of inheritance, 285–86

relationships between, 21
schedule of testing, 22–24
Periods of development
dog, 84–116
and growth, 326–29
human, 113–16
and problem-solving, 224
processes of, 416
Persistence test; see Spatial-orientation test
"Personality"
and behavioral organization, 80
expressed as social relationships, 166–67, 419–20
Physiology, comparative, evidence on dog origin, 44–45
Physiological genetics, 413
Physiological tests
factor analysis of, 369–70
heart-rate factor, 372–73
Physique
and behavior, breed and hybrid comparisons, 344–47
factor analysis, 346–51
general size factor, 348–49, 371
magnitude of genetic effects on, 358–66
methods of measurement, 340–41
relation to behavior, 326, 333–55, 389
variance components, 360, 362, 366
Pleiotropy
in behavior, 375
hypotheses concerning, 333, 371
limited occurrence of, 352
and type concept, 353–54
Pointers
agonistic behavior, 78
compared to wolf, 403
history, 48
skull measurements, 40–41
Polydactyly, guinea pig, 333
Polygenic inheritance
of behavior patterns, 293
expression of, 262–63
theory, 367–68
Polymorphism
in dogs, wolves, humans, 83
useful in human societies, 411–12
wolf, 398
Poodle
hair length, 29
hair structure, 400
skull measurements, 40–41
Population concept
applied to dog breeds, 393
applied to dog skeletons, 37–44

applied to human populations, 354–55
of species, 30
use in taxonomy and genetics, 351–52
Population genetics, 413–14
Populations, wolf, variation in, 43
Postural responses, inheritance, 288–92
Practice, effect on heritability, 254–56
Predation; see Hunting
Prehistory, evidence of dog origin, 33–36
Problem-solving behavior, 224–58
as adaptation, 388
breed comparisons, 257–58
dog-human comparison, 257
dual process theory of, 322–23
heritability of, 321, 361, 388
motivation and, 255, 320–21
unitary nature of, 256–57, 388–89
Problem-solving tests
schedule of testing, 22–24
variance of components, 361
Processes, of development, 111–12
Psychological processes, in social relationships, 151–52
Pug
head shape, 29
skull measurements, 40–42
Punishment, effect on socialization process, 145–46

Qualitative inheritance, method of analysis, 264–66
Quantitative inheritance, method of analysis, 263–64
Quieting test
breed comparisons, 206–7, 360
differentiation of behavior in, 425–26
hybrid comparisons, 360
inheritance, 287–88
loading in activity-success factor, 372
loading in reactivity factor, 374
method, 206

Raccoon dog, chromosome number, 53
Race concept, in relation to type and population concepts, 354–55
Races, not equivalent to dog breeds, 82
Radiation, effects on mutation, 409
Rat
effect of enriched environment, 26–27
effects of genetics on behavior of, 5–6
Rattle pinscher, skull measurements, 40–41
Reactivity test; see Emotional reactivity test
Rearing practices, 384–87
Regression analysis

of heritability, 297–98
of leash-control scores, 305
of motivation test, 308
of obedience test, 311
Reliability
  leash-training test, 207
  obedience test, 213
  spatial-orientation test, 254–55
Research, use of pure breeds and hybrids in, 394–96
Retrievers
  agonistic behavior, 78
  use of senses, 79
Retrieving test
  breed comparisons, 219
  critical period, 22
  methods, 218
  in test schedule, 22, 24
Retrieving test (maternal)
  breed comparisons, 172–73
  method, 171
Reward training, 216–23
  as method of social control, 155
Rhesus monkey
  critical period in, 143
  effect of rearing in isolation, 145
Russia, stone age dog in, 35

Saint Bernard
  jaw shape, 38, 39
  size, 400
  skull measurements, 40–42
  weight, 29
Saluki
  gazelle hunting, 34
  history, 51
  leg length, 29
  skull shape, 42
  use in Middle East, 432
  use of senses, 79
Sanitation, methods of, 13–14
Scaling
  methods of, 324
  stanine system, 186–89
Schedule of testing, 24
Schnauzer, skull measurements, 40–41
Scottish terriers, use of senses, 80
Sealyham terriers, chromosome number, 53
Segregation, genetic
  effect on variance, 357–58
  evidence for, 355–58
Selection, effect limited by heritability, 391
Selection procedures, in dog breeding, 391–92, 396

Sensory capacities
  in neonatal period, 86–87
  use in hunting, 79
Separation, effects upon socialization process, 149–50
Serological relationships, between *canidae*, 44–45
Setters
  agonistic behavior, 78
  compared to wolf, 403
  history, 46
  skull measurements, 40–41
Sex differences
  effect on dominance-subordination relationship, 164–65, 419
  growth curves, 329–32
  none in emotional reactivity, 202–3
Sex-linked inheritance, 283
  not found in data, 362
Sexual behavior
  dog, 62–68
  effect on socialization in dog, 130–31
  effect on socialization in sheep, 129
  home-reared dogs, 179
  inheritance of breeding cycle, 278–83
  in juvenile period, 110
  in socialization period, 106–7
  wolf, 62–68
Sexual relationship, dog, 154
Sheep
  critical period in, 143
  development of flock organization, 417
  effect of human socialization upon, 129–30
Sheep dogs
  agonistic behavior, 78
  compared to wolf, 403
Sheep dogs; *see* Shetland sheepdog
Shelter-seeking behavior
  dog, 65, 73–74
  in transition period, 92
  wolf, 65, 73–74
Sheltie; *see* Shetland sheepdog
Shetland sheepdog
  early learning in, 97
  history, 50
  management of merle gene in, 390
  obesity and monorchidism in, 390
  ready habit formation in, 425
  size factor in, 349
  skull measurements, 40–41, 43–44
  stereotyped behavior in, 237
  see also Breed comparison
Single-factor inheritance, 367–68
  behavior patterns, 292–93
  methods of analysis, 262–66

Size
  breed and sex differences, 329–32
  correlation with behavior, 344–51
  effect on dominance-subordination re-
      lationship, 164–65
  in relation to hair length, 337
  variation in dog, 401
Skeleton
  dog, 37–44
  wolf, 51
Skull measurements, wolf, 40–41, 51
Skull shape, of dogs, wolves, and other
      canids, 38–44
Smell, sense of; see Olfaction
Social behavior patterns
  changes in transition period, 89–93
  coyote, 62–65
  dog, 57–83
  fox, 62–65
  methods of measuring, 17–19
  in neonatal period, 84–86
  in socialization period, 101–7
  wolf, 57–83
Social control, in dog-human relationship,
      154–55
Social development, correlated with so-
      cial organization, 416–17
Social environment, experimental design,
      17–18
Social genetics, subject matter, 413–33
Social groups, 61–62
Social organization
  in canine societies, 414–29
  correlated with social development,
      416–17
  stabilizing effect on genetic change,
      429–30
Social relationships
  classification, 152–53
  definition, 151
  development in dog, 151–82, 416
  differentiation of behavior in, 152, 416
  dog and human, 152–53
  effect of heredity upon development,
      113, 421–24
  expression of "personality" in, 166–67,
      419–20
  psychological processes in, 151–52
Social relationship, tests
  schedule of, 18–19
  variance components, 360
Socialization, dog-human, optimum time
      for, 385
Socialization, primary
  in ants, 141–42
  in bees, 142

  in birds, 142–43
  in dogs, 118–41
  in mammals, 143–44
Socialization period
  characterized by development of social
      relationships, 416
  dog, 101–8
  rearing practices in, 385
  variation in boundaries, 120–22
Socialization process
  analysis of, 144–47
  behavioral mechanisms, 129–33
  effects of adoption upon, 149
  effects of genetic differences, 133–41
  effects of separation upon, 149–50
  largely independent of outside stimuli,
      146–47
Somatotype
  in dog and human, 339–43
  evaluation of concept, 351
Song sparrow
  rapid development in, 115
  socialization in, 143
Spain, origin of spaniel in, 46
Spaniels
  chromosome number, 53
  compared to wolf, 403
  history, 45–48
  see also Cocker spaniel
Spatial-orientation test
  breed comparisons, 250–52
  correlation with trailing test, 374
  effect of maternal environment, 285
  fear of apparatus in, 320
  genetic variance constant in, 426
  hybrid comparisons, 318–20
  loading in activity-success factor,
      372
  methods, 248–50
  persistence in, 250
  reliability of, 254–55
  in test schedule, 23–24
  threshold effect, 320
  time and errors in, 251, 253–254
  variance components, 361
Specialization
  effect of domestication upon, 402–3
  imposes limits on adaptation, 427–28
Species, population and type concepts
      of, 30
Sport, historical records, 45–48
Springer spaniel
  early learning in, 97
  history, 48
Stanine scaling system, description and
      procedure, 186–89

Startle response, breed and hybrid comparisons of development, 94
Stereotyped behavior, in maze test, 237
Sterility, in human populations, 410
Stone age dogs
    geographical distribution, 34–35
    skull shape, 37–38
Strain selection, effect of accidental, 283–84
Suckling, effect of sudden weaning on, 90
Suckling behavior
    in neonatal period, 85
    in transition period, 89–90
"Superman" myth, 403, 411

T-maze test
    apparatus, 239
    common variance with trailing test, 377
    correlation with trailing test, 374
    loading in activity-success and heart-rate factors, 372
    loading in reactivity factor, 374
    and "positional habit factor," 370
    in test schedule, 23–24
    variance components, 361, 366
    See also Cue-response, motivation, and delayed-response tests
Tail-wagging
    breed comparisons, 136–40
    in emotional reactivity test, 202, 313–15
    equivalent to smile, 104
Tameness, inheritance of, 266–69
Taxonomy, evidence of dog origin from, 30–33
Teeth
    eruption of first canines and incisors, 95–96
    eruption of permanent teeth, 108
    size variation in dogs, 39, 41
Temperature, control and recording of, 26
Terriers
    agonistic behavior, 77
    compared to wolf, 402–3
    history, 46–47
    see also Fox terrier
Territory
    dog, 62, 69, 110
    wolf, 61, 69, 76
Test
    effects of maternal environment on, 283–85
    see also Schedule of testing, 24, and under name of specific test; see also; Emotional tests; Performance

tests; Physiological tests; Physique, Problem-solving tests; Social relationship; Training tests
Threshold effects
    in emotional reactivity test, 314
    in spatial orientation, 320
Timidity, independent of aggressiveness, 369–70
Timidity factor, 375–77
Trailing test
    breed comparisons, 246–47
    common variance with T-maze test, 377
    correlation with performance tests, 374
    effect of maternal environment, 285
    fear of apparatus in, 246–47
    loading in activity-success factor, 372
    method, 244–45
    in test schedule, 22, 24
    variance components, 361, 366
Trainability, 205–23
Trainability tests, variance components, 360–61
Training
    effect on genetic variance, 210–11, 213, 222–23, 254–55
    effect on obedience test, 213
    as form of problem-solving, 225
Training, forced, reduces expression of genetic variation, 425
Trait concept, inadequacy of, 427
Traits, little generality of, 323–24, 375, 378
Transition period
    characterized by development of adult behavior patterns, 416
    dog, 89–101
    rearing practices for, 385
    variation in boundaries, 120
"Tryon distribution," 264, 266, 270–73, 280–82
"Type," result of selection for, 393–94
"Type" concept
    applied to behavior, 371
    and emotional reactivity, 195
    inadequacy of, 351–52
    little evidence for behavioral "types," 427
    related to pleiotropy, 353–54
    in relation to race concept, 354–55
    of species, 30
    use and abuse in dog breeding, 389

Undershot jaw
    in basenji, 406
    inheritance, 389
United States, dogs of, 432

Variance
between breeds, 185, 190–91, 295
effect of training on genetic, 210–11, 213, 254–55
proportion attributed to matings, litters, and individuals, 364–67
in segregating and non-segregating populations, 357–58
Variance components
breed and hybrid comparisons, 360–67
in maze test, 236–37
method of computing, 359
theoretical analysis, 190–91
Variation
extent in form and behavior of dog, 4, 400–401
use in dog breeding, 396
Visual cliff test, 93–94
Vocalization
development of reaction to strange pen, 102–4
in emotional reactivity test, 202
in leash training test, 305–6
on scale, breed comparisons, 92
on scale, loading in heart-rate factor, 372
on scale, loading in reactivity factor, 374
on scale, variance components, 360
in strange room by beagles, 122

Walking, time of development, 95
Weaning
effect on non-nutritive sucking, 90
process of, 101
Weaning test, method, 173
Weight, gain per week in puppies, 327
Whippet, skull measurements, 40–41
"Wild dog experiment," on critical period hypothesis, 124–29
Wildness
development, 105
effect of domestication upon, 402

inheritance of, 266–69
Wire-haired fox terrier; see Fox terrier
Wolf
agonistic behavior, 64, 75–78
allelomimetic behavior, 63, 74–75
appearance and proportions, 36–37
behavior patterns, 63–65, 383–84
behavioral systems, 61–80
breeding cycle, 278
classification, 29, 31
development of pack organization, 416
domestication of dog from, 54–56
dominance-subordination relationship in, 77
ecological niche, 397
eliminative behavior, 65, 68–70
epimeletic behavior, 63, 70–71
et-epimeletic behavior, 63, 72
fossils, 33
hunting behavior, 75–76, 78–80
hybrids, 52–53
Indian, as ancestor of dogs, 38
ingestive behavior, 65, 72–73
investigative behavior, 63, 78–80
leader-follower relationship, 175
in North America, Eurasia, 31
polymorphism, 83, 398
relationship to modern dog breeds, 433
results of human socialization, 140–41
serological relationship with other canids, 45
sexual behavior, 62–68
shelter-seeking behavior, 65, 73–74
skeleton, 51
skull shape and measurements, 38–41, 43, 51
social behavior, 57–83
social groups, 61–62, 414–15
territory, 61, 69, 76
unspecialized compared to dog, 402–3
variation in populations, 43, 51
"Wolf children," 148–49
Working dogs, breeding of, 394
Worms, intestinal, control of, 13